Harmonic Analysis and Fractal Analysis over Local Fields and Applications

Harmonic Analysis and Fractal Analysis over Local Fields and Applications

Weiyi Su
Nanjing University, China

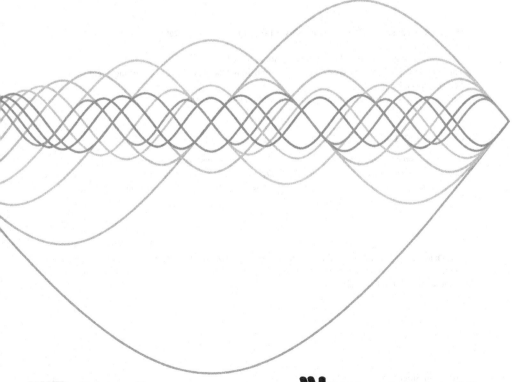

Science Press
Beijing

World Scientific

Published by

World Scientific Publishing Co. Pte. Ltd.
5 Toh Tuck Link, Singapore 596224
USA office: 27 Warren Street, Suite 401-402, Hackensack, NJ 07601
UK office: 57 Shelton Street, Covent Garden, London WC2H 9HE

Library of Congress Cataloging-in-Publication Data
Names: Su, Weiyi.
Title: Harmonic analysis and fractal analysis over local fields and applications /
    by Weiyi Su (Nanjing University, China).
Description: New Jersey : World Scientific, 2016. | Includes bibliographical references and index.
Identifiers: LCCN 2016052212 | ISBN 9789813200494 (hardcover : alk. paper)
Subjects: LCSH: Local fields (Algebra) | Algebraic fields. | Harmonic analysis. | Fractal analysis.
Classification: LCC QA247 .S87 2016 | DDC 512.7--dc23
LC record available at https://lccn.loc.gov/2016052212

British Library Cataloguing-in-Publication Data
A catalogue record for this book is available from the British Library.

Harmonic Analysis and Fractal Analysis over Local Fields and Applications
@ Weiyi Su
The Work is originally published by Science Press in 2011.
This edition is published by World Scientific Publishing Company Pte Ltd by arrangement with Science Press, Beijing, China.
All rights reserved. No reproduction and distribution without permission.

Copyright © 2018 by World Scientific Publishing Co. Pte. Ltd.

All rights reserved. This book, or parts thereof, may not be reproduced in any form or by any means, electronic or mechanical, including photocopying, recording or any information storage and retrieval system now known or to be invented, without written permission from the publisher.

For photocopying of material in this volume, please pay a copying fee through the Copyright Clearance Center, Inc., 222 Rosewood Drive, Danvers, MA 01923, USA. In this case permission to photocopy is not required from the publisher.

Printed in Singapore

Preface

1975, M. H. Taibleson[100] published a consummate book *Fourier Analysis on Local Fields* in which he has given a nice summary of research on local fields, and it formed the foundation of harmonic analysis on local fields. "Harmonic analysis on local fields" is one of important branches of abstract harmonic analysis, one may take local field as underlying space to study lots of interesting and important topics in the area of abstract harmonic analysis, including many hotspots in theory and applications.

Almost at the same period, B. B. Mandelbrot[36],[37] introduced a quite new concept "fractal", in fact, he has explored a new area, and promoted formation of a new bright spot of the geometric measure theory. The subjects in fractal geometry are those, called "fractal" by Mandelbrot, which often appear in modern science areas, or industry application areas, but could not be dealt by classical mathematical methods. We may mention some of them: functions are not differentiable, or even not continuous, such as the Weierstrass function, Cantor function, ...; sets are not regular, such as Koch curve, Cantor set, ... These functions and sets appear in many scientific areas, such as, physics, chemistry, astronomy, computer science, life science, clinical medicine science, geology, meteorology, as well as signal analysis and transfer in some industry applications. Then scientists accept the idea "fractal" very quickly, and they have recognized that "fractal" is one of the essential properties of non-linear phenomena. So that the "non-linear science" has appeared in 90s of last century.

A local field is a locally compact, totally disconnected, non-Archimedean valued, complete topological field, it has extensive application prospects, for instance, the dyadic numbers and p-adic numbers are used widely, and indeed they are special cases of local fields. More interesting fact is that lots of fractal sets can be described as sets in local fields. Also, some fractal functions can be described as functions defined on local fields.

Now a new cross-scientific area between pure mathematics and applied mathematics emerges combining "harmonic analysis on local fields" with "fractal analysis". This new cross-scientific branch reveals one of essential properties of our universe — non-linearity, and becomes one of the main topics of non-linear science.

Meanwhile, a lager number new problems have appeared in life science and clinical medicine science area, for instance, what kinds of human genes control hearts of persons, and what kinds of genes control their lungs, control their livers, ...? And what role does each gene play? More interesting and significant problem is: can we determine the malignancy of liver cancer exactly by the boundary shape of lesion of cancer, or by fractal dimension of boundary? Many research results show that lots of new problems appeared in the various science areas can be described by virtue of establishing mathematical models on local fields, and use the tools of local fields to find the corresponding solutions.

The propose for writing this book is to exhibit a new research direction — "harmonic analysis and fractal analysis over local fields and applications", including fundamentals of local fields and fractals, and new research results in more than 30 years since 1980s. From the mathematics foundation theory to the applications of clinical medicine science. It forms a self-contained body, and shows that in this new cross research area what has been done, what researchers are doing now, and what will be done in the future.

The book is divided as 3 parts and 7 chapters. The first part (chapter 1 and 2) consists fundamentals of local fields. The second part (chapter 3 and 4) is the foundation theory of harmonic analysis over local fields. The third part (chapter 5, 6, and 7) is the theory and applications of fractal analysis over local fields.

In Chapter 1, the basic knowledge of Galois field $GF(p)$ and the structures of local fields are introduced. In Chapter 2, the character groups of local fields are exhibited, and some important formulas about local fields and their character groups are evaluated. Chapter 3 and Chapter 4 contain the foundation theory of harmonic analysis: Fourier analysis and function spaces on local fields. Specially, the new calculus based on local fields as underlying spaces is established, the p-type derivatives and p-type integrals are defined by virtue of so called pseudo-differential operators, and important properties of p-type calculus are proved. Then, a comparison between Euclidean spaces and local fields is given. Chapter 5 discusses the fractal analysis over local fields, including basic knowledge of fractals. Some fractal sets and fractal functions are constructed on local fields, for example, Cantor type sets, Weierstrass type functions, and so on. Then the features of analysis on local fields and that on Euclidean spaces are pointed out. Chapter 6 devotes to the topic on fractal partial differential equations — a challenging research topic in fractal analysis. Finally, in Chapter 7, problems and applications in clinical medicine science are shown.

Comparing with classical harmonic analysis on Euclidean spaces, the harmonic analysis and fractal analysis over local fields are quite young, only more than 30 years, it is just a moment in history, a drop in ocean of knowledge. However, it plays an important role to reveal the essences of universe, life regularities of living beings, and becomes one of the cores of non-linear science. We may expect that the harmonic analysis and fractal analysis on Euclidean spaces and over local fields, both will be powerful nice tools for humankinds to recognize and study universe and nature.

I summarize and cite the research results in the period of more than 30 years of the group of harmonic analysis and fractal analysis in the Department of Mathematics, Nanjing University, obtained by my colleagues Professors Weixing Zheng, Zelin He, Huikun Jiang, Zhaoxi Wang, Zhaojin Wu and Ph.D students Huojun Ruan, Kui Yao, Hua Qiu, Yueping Zhu, Shijun Zheng, Guangcai Zhou, Ning Xu, Bo Wu, Baoyi Wu, Yin Li, Lintao Ma; as well as undergraduate students Yun Peng, Kaiming Shen, Qingsong Gu, et al. It is my pleasure to work with them. I would like to express my heartfelt thanks to all of them for pleasant co-working.

I am greatly indebted to Professor Silei Wang in Zhejiang University and Professor Shenzhen Lu in Beijing Normal University for their abundant help. Moreover, I would like to thank Professor Yudong Qiu of Nanjing Gulou Hospital and his graduate students for fruitful cooperation.

Special thanks due to the editors of Science Press in Beijing, especially, the chief editor Yanchao Zhao for the effort during the publication of this book.

I try to show readers a development of a new science area, and its splendid structure, nice theory and applications based on strict mathematics theory. Meanwhile to make smooth reading for readers, I have tried to organize the contents in order of difficulty, to write in a pithy style, and to make the key point stand out. Hope these efforts can benefit newer in the area.

Definitely there are possible mistakes and short-comings in such a book about so rapidly developing research area. I hope experts and readers can kindly feedback with valued instructions. All suggestions and criticisms are welcome.

Weiyi Su
Nanjing University, Nanjing, China
December 2016

Contents

Preface   v

1. Preliminary   1
   - 1.1 Galois field $GF(p)$ 1
     - 1.1.1 Galois field $GF(p)$, characteristic number p 1
     - 1.1.2 Algebraic extension fields of Galois field $GF(p)$ .. 3
   - 1.2 Structures of local fields 5
     - 1.2.1 Definitions of local fields 5
     - 1.2.2 Valued structure of a local field K_q 6
     - 1.2.3 Haar measure and Haar integral on a local field K_q 7
     - 1.2.4 Important subsets in a local field K_q 9
     - 1.2.5 Base for neighborhood system of a local field K_q . 10
     - 1.2.6 Expressions of elements in K_q and operations ... 11
     - 1.2.7 Important properties of balls in a local field K_p . 14
     - 1.2.8 Order structure in a local field K_p 15
     - 1.2.9 Relationship between local field K_p and Euclidean space $\mathbb{R}$ 17
     - Exercises 19

2. Character Group Γ_p of Local Field K_p 21
   - 2.1 Character groups of locally compact groups 21
     - 2.1.1 Characters of groups 21
     - 2.1.2 Characters and character groups of locally compact groups 22
     - 2.1.3 Pontryagin dual theorem 23

		2.1.4	Examples	23

- 2.2 Character group Γ_p of K_p 25
  - 2.2.1 Properties of $\chi \in \Gamma_p$ and Γ_p 26
  - 2.2.2 Character group of p-series field S_p 29
  - 2.2.3 Character groups of p-adic field A_p 32
- 2.3 Some formulas in local fields 35
  - 2.3.1 Haar measures of certain important sets in K_p . . 35
  - 2.3.2 Integrals for characters in K_p 36
  - 2.3.3 Integrals for some functions in K_p 38
  - Exercises 39

3. Harmonic Analysis on Local Fields 41

- 3.1 Fourier analysis on a local field K_p 41
  - 3.1.1 L^1-theory 42
  - 3.1.2 L^2-theory 60
  - 3.1.3 L^r-Theory, $1 < r < 2$ 67
  - 3.1.4 Distribution theory on K_p 69
  - Exercises 81
- 3.2 Pseudo-differential operators on local fields 82
  - 3.2.1 Symbol class $S^\alpha_{\rho\delta}(K_p) \equiv S^\alpha_{\rho\delta}(K_p \times \Gamma_p)$ 82
  - 3.2.2 Pseudo-differential operator T_σ on local fields . . . 85
- 3.3 p-type derivatives and p-type integrals on local fields . . . 88
  - 3.3.1 p-type calculus on local fields 88
  - 3.3.2 Properties of p-type derivatives and p-type integrals of $\varphi \in \mathbb{S}(K_p)$ 89
  - 3.3.3 p-type derivatives and p-type integrals of $T \in \mathbb{S}^*(K_p)$ 92
  - 3.3.4 Background of establishing for p-type calculus . . 94
- 3.4 Operator and construction theory of function on Local fields 102
  - 3.4.1 Operators on a local field K_p 102
  - 3.4.2 Construction theory of function on a local field K_p 105
  - Exercises 128

4. Function Spaces on Local Fields 129

- 4.1 B-type spaces and F-type spaces on local fields 129
  - 4.1.1 B-type spaces, F-type spaces 129
  - 4.1.2 Special cases of B-type spaces and F-type spaces . 135

	4.1.3	Hölder type spaces on local fields	136
	4.1.4	Lebesgue type spaces and Sobolev type spaces	141
	Exercises		147
4.2	Lipschitz class on local fields		148
	4.2.1	Lipschitz classes on local fields	148
	4.2.2	Chains of function spaces on Euclidean spaces	153
	4.2.3	The cases on a local field K_p	157
	4.2.4	Comparison of Euclidean space analysis and local field analysis	159
	Exercises		162
4.3	Fractal spaces on local feilds		162
	4.3.1	Fractal spaces on K_p	163
	4.3.2	Completeness of $(\mathbb{K}(K_p), h)$ on K_p	164
	4.3.3	Some useful transformations on K_p	171
	Exercises		181

5. **Fractal Analysis on Local Fields** — 183

5.1	Fractal dimensions on local fields		183
	5.1.1	Hausdorff measure and dimension	183
	5.1.2	Box dimension	190
	5.1.3	Packing measure and dimension	196
	Exercises		200
5.2	Analytic expressions of dimensions of sets in local fields		200
	5.2.1	Borel measure and Borel measurable sets	200
	5.2.2	Distribution dimension	201
	5.2.3	Fourier dimension	210
	Exercises		213
5.3	p-type calculus and fractal dimensions on local fields		213
	5.3.1	Structures of K_p, 3-adic Cantor type set, 3-adic Cantor type function	213
	5.3.2	p-type derivative and p-type integral of $\vartheta(x)$ on K_3	218
	5.3.3	p-type derivative and integral of Weierstrass type function on K_p	226
	5.3.4	p-type derivative and integral of second Weierstrass type function on K_p	233
	Exercises		242

6. Fractal PDE on Local Fields — 243

- 6.1 Special examples — 244
  - 6.1.1 Classical 2-dimension wave equation with fractal boundary — 244
  - 6.1.2 p-type 2-dimension wave equation with fractal boundary — 255
- 6.2 Further study on fractal analysis over local fields — 266
  - 6.2.1 Pseudo-differential operator T_α — 266
  - 6.2.2 Further problems on fractal analysis over local fields — 281
- Exercises — 282

7. Applications to Medicine Science — 283

- 7.1 Determine the malignancy of liver cancers — 284
  - 7.1.1 Terrible havocs of liver cancer, solving idea — 284
  - 7.1.2 The main methods in studying of liver cancers — 287
- 7.2 Examples in clinical medicine — 291
  - 7.2.1 Take data from the materials of liver cancers of patients — 291
  - 7.2.2 Mathematical treatment for data — 291
  - 7.2.3 Compute fractal dimensions — 300
  - 7.2.4 Induce to obtain mathematical models — 303
  - 7.2.5 Other problems in the research of liver cancers — 304

Bibliography — 305

Index — 315

Chapter 1

Preliminary

1.1 Galois field $GF(p)$

Galois field is a foundation of study of local fields, we introduce some basic theory and knowledge of Galois fields firstly.

1.1.1 Galois field $GF(p)$, characteristic number p

We recall the idea of "Abelian group" and "field".

Definition 1.1.1 (Abelian group)   Let G be a set with an operation for its elements, denoted by $\times$, satisfy:
  (i) *Operation is closed*: $x, y \in G \Rightarrow x \times y \in G$;
  (ii) *Combination law*: $x, y, z \in G \Rightarrow (x \times y) \times z = x \times (y \times z)$;
  (iii) *There exists "unit"*: $\exists e \in G$, such that $\forall x \in G \Rightarrow x \times e = x$;
  (iv) *There exists "inverse"*: $\forall x \in G, \exists x^{-1} \in G$, such that $x \times x^{-1} = e$;
  (v) *Commutative law*: $\forall x, y \in G \Rightarrow x \times y = y \times x$.
Then the set G is said to be an **Abelian group**, denoted by $(G, \times)$, or simply, G. The set G with operation $\times$ satisfying (i)$\sim$(iv) is said to be a **group**.

Usually, the product $x \times y$ is denoted by $x \cdot y$, or xy, for simply.

The operation of a group may be multiplication, addition, or any operation satisfying (i)$\sim$(v). For instance, for real number set $\mathbb{R} = (-\infty, +\infty)$, the set $(\mathbb{R}, +)$ is an Abelian group under the addition of real numbers, the unit of $+$ is the zero 0. For positive real number set $\mathbb{R}^+ = (0, +\infty)$, the set $(\mathbb{R}^+, \times)$ is an Abelian group under the multiplication of real numbers, the unit of $\times$ is natural number 1. The complex number set $\mathbb{C}$ is an Abelian group $(\mathbb{C}, +)$ with addition of complex numbers, the unit is the $(0, 0)$.

In the sequel of this book, we denote $\mathbb{N}$, $\mathbb{Z}$, $\mathbb{Q}$, $\mathbb{R}$, $\mathbb{C}$ are the positive integer set (the natural number set), the integer set, the rational number set, the real number set and the complex number set, respectively.

Definition 1.1.2 (Order of non-zero elements in a group) *In an Abelian group $(G, \times)$, for a non zero element $a \in G \backslash \{0\}$, the smallest positive integer $m \in \mathbb{N}$ satisfying*

$$a^m = \underbrace{a \times a \times \cdots \times a}_{m} = e \tag{1.1.1}$$

*is said to be the **order** of a. If there is no this m, then a is said to have **infinite order**.*

Definition 1.1.3 (Field) *Let a set F have two operations for its elements, addition $+$ and multiplication $\times$, satisfy the following:*

(i) *The set F is an Abelian group for $+$;*
(ii) *The set $F \backslash \{0\}$ is an Abelian group for $\times$;*
(iii) *The distribution law for $+$ and $\times$ holds:*

$$x, y, z \in F \Rightarrow x \times (y + z) = x \times y + x \times z.$$

*Then the set F is said to be a **field**, denoted by $(F, +, \times)$, or simply, F.*

$(\mathbb{R}, +, \times)$, $(\mathbb{C}, +, \times)$ are the real number field and complex number field, respectively.

Definition 1.1.4 (Characteristic number of a field) *Let F be a field with operations addition and multiplication, $(F, +, \times)$. If for the addition $+$, all non zero elements of F have a same order, denoted by p, then this p is said to be the **characteristic number** of F.*

The characteristic number of a field is important in studying of constructions and properties of a field. We only list some properties of characteristic numbers without proofs[6],[103].

Theorem 1.1.1 *Let $(F, +, \times)$ be a field with characteristic number p. We have*

(i) *p is a finite number or infinite. If p is finite, then it is a positive integer. If p is infinite, then we say that the characteristic number of F is $p = 0$.*

(ii) *If $p = 0$, then the field F is isomorphic with the rational number field (that is, the cardinal number of F is $\aleph_0$), and F has countable infinite members.*

(iii) *If p is finite, $0 < p < \infty$, then p is a prime, and F is a finite field.*

Definition 1.1.5 (Galois field) *If F is a finite field, then its characteristic number p is defined by the smallest prime integer satisfying*

$$p \times x \equiv p \cdot x \equiv \underbrace{x + x + \cdots + x}_{p} = 0, \quad \forall x \in F, \tag{1.1.2}$$

*where $0 \in F$ is a unit of addition $+$ of F, and there are p elements in F. Then F is said to be a **Galois field**, denoted by $GF(p)$.*

For the structure of Galois field $GF(p)$, we have

Theorem 1.1.2 *Galois field $GF(p)$ is isomorphic with the **congruence class** of p, denoted by $\mathbb{Z}/p$, that is*

$$GF(p) \sim \mathbb{Z}/p = \{0, 1, \cdots, p-1\},$$

where $\mathbb{Z}$ is the set of all integers, $p \geqslant 2$ is a prime.

1.1.2 Algebraic extension fields of Galois field $GF(p)$

Definition 1.1.6 (Extension field) *An extension field F of a Galois field $GF(p)$ is defined as follows:*

*(i) If a field F contains a Galois field, $F \supset GF(p)$, then F is said to be an **extension field** of $GF(p)$.*

*(ii) If each element of extension field F is an **algebraic element** of $GF(p)$, i.e., $\forall \gamma \in F, \exists c_k \in GF(p), k = 0, 1, \cdots, n$, satisfies*

$$\sum_{k=0}^{n} c_k \gamma^k = 0, \tag{1.1.3}$$

*then F is said to be an **algebraic extension field** of $GF(p)$, denoted by*

$$GF(p)[\alpha], \quad \alpha \in F \backslash GF(p). \tag{1.1.4}$$

About an algebraic extension field F of $GF(p)$, we have

Theorem 1.1.3 *Let $GF(p)$ be a Galois field. Then*

(i) Any finite extension field F of $GF(p)$ is an algebraic extension field, and F can be structured by virtue of adding finite algebraic elements; Inversely, any extension field F generated by adding finite algebraic elements to $GF(p)$ is a finite algebraic extension field.

*(ii) An algebraic extension field F of $GF(p)$ is a linear space on $GF(p)$. If the dimension of F is n, then this n is said to be the **degree** of F on $GF(p)$, denoted by*

$$n = (F : GF(p)).$$

When the degree of an extension field F is finite, then F is said to be a **finite extension field**; otherwise, it is said to be an **infinite extension field**.

(iii) If F is a finite algebraic extension field of $GF(p)$, then
$$F = GF(q),$$
where $q = p^c, c \in \mathbb{N}$, and
$$c = (GF(q) : GF(p)) \tag{1.1.5}$$
is the **degree of extension field** $F = GF(q)$.

Hence, the finite algebraic extension field $F = GF(q)$ is a c-dimension linear space on Galois field $GF(p)$. There exists a base in $GF(q)$ associated with $GF(p)$
$$\rho_0, \rho_1, \cdots, \rho_{c-1}, \tag{1.1.6}$$
where
$$\rho_0 \in GF(p), \quad \rho_1, \cdots, \rho_{c-1} \in GF(q) \backslash GF(p),$$
such that $\forall \alpha \in GF(q), \exists \gamma_0, \gamma_1, \cdots, \gamma_{c-1} \in GF(p)$, holds
$$\alpha = \gamma_0 \rho_0 + \gamma_1 \rho_1 + \cdots + \gamma_{c-1} \rho_{c-1} \tag{1.1.7}$$
for $\alpha \neq 0$, and $\gamma_0, \gamma_1, \cdots, \gamma_{c-1}$ are not zero simultaneously.

(iv) *Galois field $GF(p)$ is a **prime field**, it does not contain any true subfields.*

Since a prime field is either isomorphic to the rational number field $\mathbb{Q}$ (if it is an infinite field), or isomorphic to the congruence class $\mathbb{Z}/p$ of p (if it is a finite field), so for the Galois field, we have $GF(p) \sim \mathbb{Z}/p$.

(v) *A finite algebraic extension field $GF(q)$ of Galois field $GF(p)$ contains a prime field which isomorphic to the congruence class $\mathbb{Z}/p$ of p, and has p elements. Also contains $GF(p)$.*

(vi) *$GF(q)$ has $q = p^n$ elements, and each $\alpha \in GF(q)$ is the zero of function $x^q - x$.*

For given p and c, all fields which have $q = p^c$ elements are isomorphic. Moreover, all subfields in $GF(q)$ are Galois fields $GF(p^m)$, where m is a factor of $c \in \mathbb{N}$.

The element of $GF(q)$ has the following expansions.

Theorem 1.1.4 *Let $GF(q) = GF(p^c)$, $p \geqslant 2$ be a prime, $c \in \mathbb{N}$.*

(i) If $c = 1$, then $GF(q) = GF(p) \xleftrightarrow{\text{isomor.}} \{0, 1, \cdots, p-1\}$; $\forall \alpha \in GF(p)$ is a zero point of the p-power function $x^p - x$.

(ii) *If $c = 2$, choose an algebraic element ρ of $GF(q)$, and $\rho \notin GF(p)$, for it, $\exists a_0, a_1, a_2 \in GF(p)$, not zero simultaneously, such that the following equation*

$$a_0 + a_1\rho + a_2\rho^2 = 0 \qquad (1.1.8)$$

holds. Then $GF(p^2)$ has a base (ρ_0, ρ_1) ($\rho_0 = 1$ is the unit element of $GF(p)$), such that

$$GF(p^2) = \{\gamma_0 \cdot 1 + \gamma_1 \cdot \rho : \gamma_0, \gamma_1 \in GF(p)\}. \qquad (1.1.9)$$

Moreover, $\forall \alpha \in GF(p^2)$ is a zero point of p^2-power function $x^{p^2} - x$.

(iii) *If $c \in \mathbb{N}$, then $GF(p^c)$ has a base $\{1, \rho_1, \cdots, \rho_{c-1}\}$, such that*

$$GF(p^c) = \{\gamma_0 \cdot 1 + \gamma_1 \cdot \rho_1 + \cdots + \gamma_{c-1}\rho_{c-1} : \gamma_0, \gamma_1, \cdots, \gamma_{c-1} \in GF(p)\}. \qquad (1.1.10)$$

Moreover, $\forall \alpha \in GF(p^c)$ is a zero point of p^c-power function $x^{p^c} - x$.

Galois theory is very deep and complex, we just need some foundation parts of it, and for others we refer to [6], [30], [103].

1.2 Structures of local fields

Let $q = p^c$, $p \geqslant 2$ be a prime, $c \in \mathbb{N}$. The basic knowledge of local fields[100] is introduced in this section.

1.2.1 Definitions of local fields

Definition 1.2.1 (Locally compact field) *Let K be a field with operations $+, \times$, and a T_2-type topological space with topology τ. If the addition group $(K^+, +) \equiv (K, +)$ and multiplication group $(K^*, \times) = (K \setminus \{0\}, \times)$ both are locally compact Abelian group; moreover, the operations addition $+$ and multiplication $\times$ are **compatible** with topology τ, i.e., the mappings*

$$(x, y) \in K \times K \to x + y \in K, \qquad (1.2.1)$$

$$(x, y) \in K^* \times K^* \to x \times y \in K^* \qquad (1.2.2)$$

*are continuous in τ, then K is said to be **a locally compact topological field**, or for simply, **locally compact field**.*

By the field theory, we have

Theorem 1.2.1 *Let K be a non-trivial locally compact complete topological field.*

(i) *If K is connected, then it is either the real number field $\mathbb{R}$, or the complex number field $\mathbb{C}$.*

(ii) *If K is not connected, then it is totally disconnected. So that K is* **a non-trivial, locally compact, totally disconnected, and complete topological field,** *called **local field**.*

There are 4 cases in local fields.

① *If K has finite characteristic number $p \neq 0$, then K is a local field of formal power series over the finite field $GF(p^c)$, and it contains a prime field which is isomorphic to Galois field $GF(p)$.*

(i) *When $c = 1$, then K is said to be a **p-series field**.*

(ii) *When $c > 1$, then K is a **finite algebraic extension of c-degree of the p-series field**.*

② *If K has characteristic number zero, $p = 0$, then K is a local field of formal power series over the finite field $GF(p^c)$, and it contains a prime field which is isomorphic to the rational number field $\mathbb{Q}$.*

(iii) *When $c = 1$, then K is said to be a **p-adic number field** (simply, p-adic field).*

(iv) *When $c > 1$, then K is a **finite algebraic extension of c-degree of the p-adic field**.*

In the sequel, we denote K_p as a local field K for both p-series and p-adic field. And K_q, $q = p^c$, $c > 1$, for the finite algebraic extension of K_p.

1.2.2 *Valued structure of a local field K_q*

Let $K_q = (K, \oplus, \otimes, \tau)$ be a T_2, non-trivial, locally compact, totally disconnected, complete topological field. In the sequel, we agree for convenience that: the addition $x \oplus y$ and multiplication $x \otimes y$ of a local field K_q are denoted by $x + y$ and $x \times y$, or $x \cdot y$, or xy; and $K_q = (K, +, \times, \tau)$ by K_q, for simply, if there is no confusion.

A non-Archimedean valued norm can be endowed to K_q.

Definition 1.2.2 (Non-Archimedean valued norm) *Let F be a field. If a mapping $x \in F \to |x| \in [0, +\infty)$ from F to $[0, +\infty)$ satisfies*

(i) $|x| \geqslant 0$; *and* $|x| = 0 \Leftrightarrow x = 0$;

(ii) $|x \cdot y| = |x| \, |y|$;

(iii) $|x+y| \leqslant \max\{|x|, |y|\}$, *and if $|x| \neq |y|$, then $|x+y| = \max\{|x|, |y|\}$, then $|x|$ is said to be a **non-Archimedean valued norm** of $x \in F$, and F*

is said to be a **non-Archimedean valued field**.

The inequality in (iii) is said to be an **ultra-metric inequality**.

We may prove that, the local field K_q is an non-Archimedean valued field.

Theorem 1.2.2 *Let K_q be a local field (p-series field, p-adic field, the finite algebraic extensions of them). Then it is a non-Archimedean valued field. The valued norm denoted by $|x|$ for $x \in K_q$, and*

$$\forall x \in K_q \Rightarrow |x| \in \{q^k : k \in \mathbb{Z}\} \cup \{0\}. \tag{1.2.3}$$

If $c = 1$, the range of the valued norm of K_p is

$$\forall x \in K_p \Rightarrow |x| \in \{p^k : k \in \mathbb{Z}\} \cup \{0\}. \tag{1.2.4}$$

We will give calculating methods for this non-Archimedean valued norm in the sequel.

1.2.3 Haar measure and Haar integral on a local field K_q

1. Haar measure and Haar integral on locally compact groups

There exist the Haar measures and Haar integral with invariance of translation on each locally compact group.

Theorem 1.2.3 *Let G be a locally compact group, then there exist left invariant measure and right invariant measure of translations μ_l and μ_r respectively, such that for each Borel set F in G satisfy*

invariance of left translation $-\forall a \in G \Rightarrow \mu_l(aE) = \mu_l(E)$;

invariance of right translation $-\forall a \in G \Rightarrow \mu_r(Ea) = \mu_r(E)$.

If G is a locally compact Abelian group, then the invariant measures in left translation and right translation are equal each other, $\mu \equiv \mu_l \equiv \mu_r$, and μ is said to be a **Haar measure**:

$$\forall a \in G \Rightarrow \mu(aE) = \mu(Ea) = \mu(E).$$

Theorem 1.2.4 *There exists invariant integral of translations corresponding to the Haar measure μ of invariance of translations on a locally compact Abelian group G for a Haar measurable function $f : G \to \mathbb{C}$ denoted by*

$$\int_G f(x)dx \equiv \int_G f(x)d(\mu x), \tag{1.2.5}$$

where $dx \equiv d(\mu x)$ is the Haar infinitesimal of μ. The invariance of translations of the Haar integral is

$$\int_G f(ax)dx = \int_G f(xa)\,dx = \int_G f(x)dx, \quad a \in G.$$

And the Haar integral on G is unique except for a strictly positive proportion factor.

2. Haar measure and Haar integral on a local field K_q

There are two locally compact Abelian groups, addition group K_q^+ and multiplication group K_q^* in K_q, thus there are Haar integrals on both groups K_q^+ and K_q^* which play important roles in abstract harmonic analysis. We only study topics in harmonic analysis and fractal analysis on K_q^+, but show some motivations and open problems for K_q^* in which there are more special, more interesting, and more valuable topics.

For $K_q^+ \equiv K_q$, the Haar measure of a Borel set $E \subset K_q$ is denoted by $\mu(E) \equiv |E|$ satisfying the invariance of translation

$$\forall a \in K_q \Rightarrow |aE| = |Ea| = |E|.$$

The Haar integral for a Haar measurable function $f : K_q \to \mathbb{C}$ is denoted by $\int_{K_q} f(x)dx$ satisfying the invariance of translation

$$\int_{K_q} f(x+a)\,dx = \int_{K_q} f(x)dx, \quad \forall a \in K_q.$$

3. Modular function for an element $a \in K_q$

We define a so called modular function for $a \in K_q$.

Definition 1.2.3 (Modular function) *Let K_q be a local field. For $a \in K_q$, a function $a \to |a|$ is defined by*

$$|a| = 0 \Leftrightarrow a = 0;$$

$$|a| \neq 0, \quad |a| = \frac{\int_{K_q} f(a^{-1}x)\,dx}{\int_{K_q} f(x)dx} = \frac{\int_{K_q} f(x)d(ax)}{\int_{K_q} f(x)dx}, \quad (1.2.6)$$

where $f \in C_C^+(K_q)$, that is, f is a positive function on K_q with compact support, and $d(ax)$ satisfies

$$d(ax) = |a|dx, \quad (1.2.7)$$

then $|a|$ is said to be a **modular function** of a.

We can prove that: the modular function is, in fact, a non-Archimedean valued norm, it is depending only on $a \in K_q$, but independent of $f \in C_C^+(K_q)$.

Haar measure and non-Archimedean valued norm both are unique except for a strictly positive proportion factor.

If we define
$$d(x,y) = |x - y|, \qquad (1.2.8)$$
then $d(x,y) = |x - y|$ is an ultra-metric on K_q such that K_q is an **ultra-metric space**.

Hence, a local field K_q is a non-trivial, T_2-type, locally compact, totally disconnected, non-Archimedean valued norm, completed and ultra-metric topological field.

Remark. A local field K_p can be regarded as a completion of the rational number field $\mathbb{Q} = \left\{\dfrac{q}{p} : q \in \mathbb{Z}, p \in \mathbb{N}\right\}$ with respect to non-Archimedean valued norm $|x|, x \in K_p$.[60]

1.2.4 Important subsets in a local field K_q

It is suitable to define some important sets in K_q by non-Archimedean valued norm.

The following notions and sets, such as, prime group, ring, sub-ring, integer-ring, idea, prime idea, maximal idea, unit prime group, fractional idea, etc., are standard in courses of modern algebra. We refer to [6], [30], [103].

In the analysis over local fields, it is very important to determine Haar measures of some sets, and to show topological properties of those sets in K_q.

1. Some sets in K_p

(i) D, **the ring of integers in** K_p
$$D = \{x \in K_p : |x| \leqslant 1\}, \qquad (1.2.9)$$
it is the unique maximal compact sub-ring, and it is open, closed, as well as compact. The Haar measure is $|D| = 1$.

(ii) B, **the prime ideal in** K_p

$$B = \{x \in K_p : |x| < 1\}, \tag{1.2.10}$$

it is the unique maximal ideal in D, and it is also principal ideal and prime ideal. Moreover, it is an open, closed and compact set, and its Haar measure is $|B| = p^{-1}$.

(iii) D^*, **the unit prime group in** K_p^*

$$D^* = \{x \in D : |x| = 1\} = D \backslash B, \tag{1.2.11}$$

it is also open, closed and compact, the Haar measure is $|D^*| = 1 - p^{-1}$.

(iv) B^k, **the fractional ideal in** K_p

$$B^k = \{x \in K_p : |x| \leqslant p^{-k}\} = \beta^k D, \quad k \in \mathbb{Z}, \tag{1.2.12}$$

they are open, closed and compact sets, the Haar measures are $|B^k| = p^{-k}$, $k \in \mathbb{Z}$.

On the other hand, B^k is a sub-ring of K_p for $k \geqslant 0$. And we have

$$B = \{x \in K_p : |x| < 1\} = \{x \in K_p : |x| \leqslant p^{-1}\} = B^1.$$

Remark. We will calculus the Haar measures of the above sets in the Section 2.3.

2. Sets in K_q

For the extension field K_q, the above sets can be defined similarly, and denoted by

$$D(q) = \{x \in K_q : |x| \leqslant 1\},$$
$$B(q) = \{x \in K_q : |x| < 1\},$$
$$D^*(q) = \{x \in K_q : |x| = 1\},$$
$$B^k(q) = \{x \in K_q : |x| \leqslant q^{-k}\}, \quad k \in \mathbb{Z}.$$

1.2.5 Base for neighborhood system of a local field K_q

1. Base for neighborhood system of K_p

By using non-Archimedean valued norm we define a topological base of zero in K_p, i.e., a base for neighborhood system of K_p.

The set $\{B^k \subset K_p : k \in \mathbb{Z}\}$ and B^k satisfy:

(i) $\{B^k \subset K_p : k \in \mathbb{Z}\}$ is a base for neighborhood system of zero K_p, and $B^{k+1} \subset B^k, k \in \mathbb{Z}$;

(ii) $B^k, k \in \mathbb{Z}$, is open, closed and compact in K_p;

(iii) $K_p = \bigcup_{k=-\infty}^{+\infty} B^k$ and $\{0\} = \bigcap_{k=-\infty}^{+\infty} B^k$;

(iv) The quotient group D/B is

$$D/B = \{0 \cdot \beta^0 + B, 1 \cdot \beta^0 + B, \cdots, (p-1) \cdot \beta^0 + B\},$$

with $j \cdot \beta^0 + B, j = 0, 1, \cdots, p-1$. The set D/B is the p **cosets** of B in D, and

$$\{j \cdot \beta^0 + B : j = 0, 1, \cdots, p-1\} \tag{1.2.13}$$

is isomorphic with the Galois field $GF(p)$.

Thus, $\{B^k\}_{k \in \mathbb{Z}}$ is a base for neighborhood system of zero in K_p, that is, a base for neighborhood system of K_p.

2. Base for a neighborhood system of K_q

Use $B^k(q)$ instead of B^k.

1.2.6 Expressions of elements in K_q and operations

1. Expressions of elements in K_p

There exists an element $\beta \in K_p$ in K_p, with non-Archimedean valued norm $|\beta| = p^{-1}$, called a **generator** of K_p (also, called a **prime element**), such that $\forall x \in K_p$,

$$x = x_s \beta^s + x_{s+1} \beta^{s+1} + \cdots, \tag{1.2.14}$$

where $x_s, x_{s+1}, \cdots \in \{0, 1, \cdots, p-1\}, s \in \mathbb{Z}$; and $\forall x \in K_p$, $x_s \neq 0$, has $|x| = p^{-s}$ for x in (1.2.14) as its unique non-Archimedean valued norm.

We emphasize that: p-series field, denoted by S_p with characteristic number $p \geqslant 2$; p-aidc field, denoted by A_p with characteristic number $p = 0$, both as local fields, have quite different properties under different operations, whereas their elements have similar expresses.

(i) **p-series field S_p**: $\forall x \in S_p$ has express as in (1.2.14)

$$x = x_s \beta^s + x_{s+1} \beta^{s+1} + \cdots, \quad x_s, x_{s+1}, \cdots \in \{0, 1, \cdots, p-1\}, \quad s \in \mathbb{Z},$$

the addition of S_p is defined as **addition of sequence coordinate-wise**, mod$p$, **no carrying**, that is, for $x, y \in S_p$,

$$x = x_s\beta^s + x_{s+1}\beta^{s+1} + \cdots, \quad y = y_s\beta^s + y_{s+1}\beta^{s+1} + \cdots,$$
the addition is defined as
$$x + y = (x_s + y_s)\beta^s + (x_{s+1} + y_{s+1})\beta^{s+1} + \cdots,$$
with
$$x_j + y_j = x_j + y_j \,(\text{mod}\,p), \quad \forall j \geqslant s, \; s \in \mathbb{Z}.$$

The multiplication of S_p can be defined by the Cauchy product, $\text{mod}\,p$, no carrying.

(ii) **p-adic field A_p**: $\forall x \in A_p$ has express as in (1.2.14):
$$x = x_s\beta^s + x_{s+1}\beta^{s+1} + \cdots, \quad x_s, x_{s+1}, \cdots \in \{0, 1, \cdots, p-1\}, \quad s \in \mathbb{Z},$$
the addition of A_p is defined as **addition of sequences coordinate-wise, $\text{mod}\,p$, carrying from left to right**, that is, for $x, y \in A_p$,
$$x = x_s\beta^s + x_{s+1}\beta^{s+1} + \cdots, \quad y = y_s\beta^s + y_{s+1}\beta^{s+1} + \cdots,$$
the addition is defined as
$$x + y = (\bar{x}_s + \bar{y}_s)\beta^s + (\bar{x}_{s+1} + \bar{y}_{s+1})\beta^{s+1} + \cdots,$$
with
$$\bar{x}_j + \bar{y}_j = \begin{cases} x_j + y_j, & 0 \leqslant x_j + y_j < p, \\ p - (x_j + y_j), & x_j + y_j \geqslant p, \text{ and } \bar{x}_{j+1} + \bar{y}_{j+1} = x_{j+1} + y_{j+1} + 1, \end{cases}$$
for $j \geqslant s, s \in \mathbb{Z}$.

The multiplication of A_p can be defined by the Cauchy product, $\text{mod}\,p$, with carrying from left to right.

Remark. (1.2.14) is interpreted as a formal power series on Galois field $GF(p)$. In fact, rewrite (1.2.14) as
$$x = \sum_{j=s}^{+\infty} x_j \beta^j, \quad s \in \mathbb{Z},$$
with
$$x_j \in \{0, 1, \cdots, p-1\} \leftrightarrow GF(p), \quad j = s, s+1, \cdots,$$
$\beta \in K_p$, $|\beta| = p^{-1}$, that is $\forall x \in K_p$ can be expressed as a formal power series on Galois field.

On the other hand, the quotient set B^s/B^{s+1} of cosets of B^{s+1} on B^s is
$$B^s/B^{s+1} = \{0 \cdot \beta^s + B^{s+1}, 1 \cdot \beta^s + B^{s+1}, \cdots, (p-1) \cdot \beta^s + B^{s+1}\},$$
then,
$$B^s/B^{s+1} \leftrightarrow GF(p),$$
the corresponding relationship is
$$j \cdot \beta^s + B^{s+1} \leftrightarrow j, \quad j = 0, 1, \cdots, p-1, \quad s \in \mathbb{Z}.$$
Thus, (1.2.14) can be rewritten as
$$x = x_s\beta^s + \bar{x}, \quad \bar{x} \in B^{s+1}, \quad x_s \neq 0, \quad s \in \mathbb{Z}.$$

We use notations $x + y$ and $x \cdot y$ (or xy) instead of $x \oplus y$ and $x \otimes y$, respectively, if there is no confusion.

2. Expressions of elements in $K_q, q = p^c, c \in \mathbb{N}$

Recall the theory of finite algebraic extension of Galois field: for the Galois field, we have $GF(p) \leftrightarrow \{0, 1, \cdots, p-1\}$, and its finite c-degree algebraic extension is

$$GF(q), \quad q = p^c, \quad c \in \mathbb{N},$$

$GF(q)$ is a c-dimensional linear space on $GF(p)$.

Let a base of $GF(q)$ on $GF(p)$ be

$$\{\rho_0, \rho_1, \cdots, \rho_{c-1}\}, \quad \rho_0 \in GF(p), \tag{1.2.15}$$

where $\rho_1, \cdots, \rho_{c-1}$ are algebraic elements of $GF(p)$, and the elements in $\{\rho_0, \rho_1, \cdots, \rho_{c-1}\}$ are linear independent. Thus, any element $z \in GF(q)$ in a finite algebraic extension field $GF(q)$ can be expressed as

$$z = \gamma_0 \rho_0 + \gamma_1 \rho_1 + \cdots + \gamma_{c-1} \rho_{c-1}, \tag{1.2.16}$$

with $\gamma_0, \gamma_1, \cdots, \gamma_{c-1} \in GF(p) = \{0, 1, \cdots, p-1\}$.

Take the quotient set in $K_q, q = p^c$,

$$D^*(q) = D(q)/B(q). \tag{1.2.17}$$

Let a base of $D(q)/B(q)$ in D/B be

$$\{\varepsilon_0, \varepsilon_1, \cdots, \varepsilon_{c-1}\} \subset D^*(q),$$

such that

$$\{\rho(\varepsilon_0), \rho(\varepsilon_1), \cdots, \rho(\varepsilon_{c-1})\}$$

is the base of $GF(q)$ in $GF(p)$

$$\{\rho_0, \rho_1, \cdots, \rho_{c-1}\}.$$

Then $\forall x \in B^k(q) \subset K_q$, $k \in \mathbb{Z}$, there is $\widetilde{\beta} \in B(q)$ with $|\widetilde{\beta}| = q^{-1}$, such that

$$x = \sum_{j=s}^{+\infty} c_j \widetilde{\beta}^j, \quad s \in \mathbb{Z}, \tag{1.2.18}$$

with $c_j \in D(q)/B(q)$, and

$$c_j = \gamma_0^j \varepsilon_0 + \gamma_1^j \varepsilon_1 + \cdots + \gamma_{c-1}^j \varepsilon_{c-1}, \tag{1.2.19}$$

with $\gamma_0^j, \gamma_1^j, \cdots, \gamma_{c-1}^j \in D/B$, $j = s, s+1, \cdots$, such that $|x| = |\widetilde{\beta}|^s = q^{-s}$.

We emphasize that: in the finite algebraic extensions of p-series fields, the addition and multiplication are no carrying, but in the finite algebraic extensions of p-adic fields, the addition and multiplication both are carrying from left to right.

K_q is also a formal power series field on Galois field $GF(q) = GF(p^c)$. If $c = 1$, then $K_q = K_p$ is S_p, or A_p; $c \neq 1$, K_q is the c-degree finite algebraic extensions of S_p or A_p.

1.2.7 Important properties of balls in a local field K_p

A ball in a local field K_p has following important properties.

Theorem 1.2.5 (i) Let S, T be two balls in local field K_p. Then either S and T are disjoint, or one ball contains the other one;

(ii) Any ball in K_p has multi-centers;

(iii) Any ball in K_p is open, closed, and compact.

Proof. (i) Suppose that $S = x_1 + B^{k_1}$, $T = x_2 + B^{k_2}$ are two balls in K_p with $k_1, k_2 \in \mathbb{Z}$. If $S \cap T = \varnothing$, then S and T are disjoint each other. If $S \cap T \neq \varnothing$, then either $S \subset T$, or $S \supset T$.

In fact, if $k_1 \geqslant k_2$, then $\forall x \in S \cap T$, by the ultra-metric inequality

$$|x_1 - x_2| \leqslant \max\{|x - x_1|, |x - x_2|\} \leqslant \max\{p^{-k_1}, p^{-k_2}\} \leqslant p^{-k_2}$$

holds. This implies $x_1 \in T$.

On the other hand, $\forall y \in S$ holds

$$|y - x_2| \leqslant \max\{|y - x_1|, |x_1 - x_2|\} \leqslant \max\{p^{-k_1}, p^{-k_2}\} \leqslant p^{-k_2},$$

thus, $\forall y \in S \Rightarrow y \in T$, i.e., $S \subset T$.

If $k_1 \leqslant k_2$, similarly, $T \subset S$.

(ii) Suppose that $S = x_1 + B^{k_1}$, $k_1 \in \mathbb{Z}$. For any $x \in S$, denotes by S_x for the ball centered x, with radius p^{-k_1}, then $\forall y \in S$ holds

$$|x - y| \leqslant \max\{|x - x_1|, |x_1 - y|\} = p^{-k_1}.$$

Hence $y \in S_x$, and thus $S \subset S_x$; By symmetry, $S_x \subset S$. This implies $S_x = S$.

(iii) For any ball $S = x_1 + B^{k_1}$, $k_1 \in \mathbb{Z}$, then $\forall y \in S = x_1 + B^{k_1}$ is determined by inequality

$$|y - x_1| \leqslant p^{-k_1}.$$

Thus, the ball $S = x_1 + B^{k_1}$ is closed, so it is compact.

On the other hand, since the non-Archimedean valued norm of local field K_p is discrete, thus

$$S = \{y : |y - x_1| < p^{-k_1+1}\}.$$

This implies that $S = x_1 + B^{k_1}$ is open. The proof is complete.

This theorem is quite different from that in Euclidean space $\mathbb{R}^n$, it is a characteristic and important property of the structure of local field K_p, thus it is also an essential reason for difference of two underlying spaces, the local field K_p and the Euclidean space $\mathbb{R}^n$.

1.2.8 Order structure in a local field K_p

We introduce so called order structure in K_p which plays important role in the analysis over local fields.

Suppose that the coset representatives of B in D is $\{e_0, e_1, \cdots, e_{p-1}\}$, it is one to one corresponding to $GF(p) = \{0, 1, \cdots, p-1\}$. The "order" in K_p is defined as follows.

Definition 1.2.4 (Order in K_p) For $\{e_0, e_1, \cdots, e_{p-1}\}$, an order is defined as

$$e_0 < e_1 < \cdots < e_{p-1}. \tag{1.2.20}$$

Thus, the order in K_p can be defied as: let $\alpha, \lambda \in K_p$,

$$\alpha = \alpha_k \beta^k + \alpha_{k+1} \beta^{k+1} + \cdots, \quad \alpha_l \in \{e_0, e_1, \cdots, e_{p-1}\}, \quad l = k, k+1, \cdots,$$

$$\lambda = \lambda_k \beta^k + \lambda_{k+1} \beta^{k+1} + \cdots, \quad \lambda_l \in \{e_0, e_1, \cdots, e_{p-1}\}, \quad l = k, k+1, \cdots,$$

then

(i) If $|\alpha| < |\lambda|$, then define $\alpha < \lambda$ (or $\lambda > \alpha$);

(ii) If $|\alpha| = |\lambda|$, and $\alpha \neq \lambda$, so $\exists r \in \mathbb{N}$, such that

$$\alpha_k = \lambda_k, \alpha_{k+1} = \lambda_{k+1}, \cdots, \alpha_{k+r} = \lambda_{k+r}, \alpha_{k+r+1} < \lambda_{k+r+1},$$

then define $\alpha < \lambda$ (or $\lambda > \alpha$);

(iii) If all coordinates of α, λ are equal each other, then define $\alpha = \lambda$.

Thus, we may define "interval" in K_p.

Definition 1.2.5 (Intervals in K_p) For $\alpha, \lambda \in K_p$, if $\alpha < \lambda$, we define

$$\begin{aligned}
\text{open interval} \quad & (\alpha, \lambda) = \{x \in K_p : \alpha < x < \lambda\}, \\
\text{closed interval} \quad & [\alpha, \lambda] = \{x \in K_p : \alpha \leqslant x \leqslant \lambda\}, \\
\text{half-open interval} \quad & [\alpha, \lambda) = \{x \in K_p : \alpha \leqslant x < \lambda\}, \\
\text{half-open interval} \quad & (\alpha, \lambda] = \{x \in K_p : \alpha < x \leqslant \lambda\}.
\end{aligned}$$

For example,

$$[0, \beta^1) = \{x \in K_p : 0 \leqslant x < \beta^1\},$$

$$[0, \beta^1] = \{x \in K_p : 0 \leqslant x \leqslant \beta^1\}.$$

If $p = 3$, then

$$[\beta^1, 2\beta^1] = \{x \in K_3 : \beta^1 \leqslant x \leqslant 2\beta^1\};$$

$$[2\beta^1, \beta^0] = \{x \in K_3 : 2\beta^1 \leqslant x \leqslant \beta^0\}.$$

We evaluate the Haar measures of some intervals: denote the **characteristic function** of a measurable set $E \subset K_p$ by

$$\Phi_E(x) = \begin{cases} 1, & x \in E, \\ 0, & x \notin E. \end{cases}$$

We have the Haar measure of E,

$$\mu E \equiv |E| = \int_{K_p} \Phi_E(x) dx = \int_E dx.$$

Now, the Haar measure of $[0, \alpha]$ for $\alpha = \alpha_s \beta^s$, $\alpha_s \in \{1, 2, \cdots, p-1\}$, can be evaluated. Since

$$\mu[0, \alpha] = \int_{K_p} \Phi_{[0,\alpha]}(x) dx = \int_{[0,\alpha]} dx$$

and

$$[0, \alpha] = [0, \beta^{s+1}] \cup (\beta^{s+1}, \alpha] = [0, \beta^{s+1}] \cup (\beta^{s+1}, \alpha_s \beta^s],$$

thus

$$\mu[0, \alpha] = \mu[0, \beta^{s+1}] + \mu(\beta^{s+1}, \alpha_s \beta^s],$$

where $\mu[0, \beta^{s+1}] = \mu B^{s+1} = p^{-(s+1)}$.

On the other hand, for $\mu(\beta^{s+1}, \alpha_s \beta^s]$, notice that

$$B^s/B^{s+1} = \{0 \cdot \beta^s + B^{s+1}, 1 \cdot \beta^s + B^{s+1}, \cdots, (p-1) \cdot \beta^s + B^{s+1}\},$$

thus

$$\mu(0 \cdot \beta^s + B^{s+1}) = \mu(1 \cdot \beta^s + B^{s+1}) = \cdots = \mu((p-1) \cdot \beta^s + B^{s+1}) = \frac{1}{p} p^{-s}.$$

Hence

$$\mu(\beta^{s+1}, \alpha_s \beta^s] = \alpha_s p^{-(s+1)},$$

we have

$$\mu[0, \alpha] = (\alpha_s + 1) p^{-(s+1)}$$

with $\alpha = \alpha_s \beta^s$.

By virtue of similar method, Haar measures of other sets in K_p can be evaluated.

The Cantor type set C_3 on K_3:

Take $p = 3$, $\beta \in K_3$ with non-Archimedean valued norm $|\beta| = 3^{-1}$. Let
$$V_0 = D = \{x \in K_3 : |x| \leqslant 1\},$$
$$V_1 = B^1 = \{x \in K_3 : |x| \leqslant 3^{-1}\},$$
$$V_2 = (1 \cdot \beta^0 + B^2) \cup (2 \cdot \beta^0 + B^2),$$
$$\cdots\cdots$$

Denote
$$C_3 = D \setminus \bigcup_{j=1}^{+\infty} V_j.$$

The set $C_3 = D \setminus \bigcup_{j=1}^{+\infty} V_j$ is called **a Cantor type set** on local field K_3.
Thus
$$\mu C_3 = \mu \left(D \setminus \bigcup_{j=1}^{+\infty} V_j \right) = \mu D - \mu \left(\bigcup_{j=1}^{+\infty} V_j \right)$$
$$= 1 - \left(\frac{1}{3} + \frac{2}{3^2} + \frac{2^2}{3^3} + \cdots \right) = 1 - \frac{1}{2} \left(\frac{2}{3} + \left(\frac{2}{3}\right)^2 + \left(\frac{2}{3}\right)^3 + \cdots \right) = 0,$$
so the Haar measure of the Cantor type set C_3 is $\mu\, C_3 = |C_3| = 0$.

More sets in local field K_p can be constructed, as well as functions defined on sets in K_p can be given. We will define function spaces underlying on local fields and study their properties of harmonic analysis and fractal analysis, such as, various measures, various dimensions, p-type calculus, Fourier transforms, fractal partial differential equations, etc. For instance, the Hausdorff dimension of the above $C_3 \subset K_3$ is $s = \dim_H(C_3) = \frac{\ln 2}{\ln 3}$, and the Hausdorff measure is $H^s(C_3) = 1$ (see Section 5.1).

1.2.9 Relationship between local field K_p and Euclidean space $\mathbb{R}$

Is there an isomorphic mapping between a local field K_p and Euclidean space $\mathbb{R}$?

Let the characteristic number of K_p be a prime $p \geqslant 2$. Take a base $\{\beta^k : k \in \mathbb{Z}\}$ of K_p with
$$B^k = \{x \in K_p : |x| \leqslant p^{-k}\}, \quad k \in \mathbb{Z}$$

and $\forall x \in B^k$, $x \neq 0$,
$$x = x_k \beta^k + x_{k+1} \beta^{k+1} + \cdots,$$
with $x_k \in \{1, 2, \cdots, p-1\}$, $x_{k+1}, x_{k+2}, \cdots \in \{0, 1, 2, \cdots, p-1\}$.

Take $p = 3$, and the interval $[0, 1)$ is divided in 3 subintervals: $\left[0, \frac{1}{3}\right)$, $\left[\frac{1}{3}, \frac{2}{3}\right)$, $\left[\frac{2}{3}, 1\right)$, see Fig. 1.2.1. Then
$$B^1 = \{x \in K_3 : |x| \leqslant 3^{-1}\} : \forall x \in B^1 \Rightarrow x = x_1 \beta^1 + x_2 \beta^2 + \cdots, x_j \in \{0, 1, 2\};$$
$$B^2 = \{x \in K_3 : |x| \leqslant 3^{-2}\} : \forall x \in B^2 \Rightarrow x = x_2 \beta^2 + x_3 \beta^3 + \cdots, x_j \in \{0, 1, 2\};$$
......

and
$$\left[0, \frac{1}{3}\right) \leftrightarrow 0 \cdot \frac{1}{3} + B^2, \quad \left[\frac{1}{3}, \frac{2}{3}\right) \leftrightarrow 1 \cdot \frac{1}{3} + B^2, \quad \left[\frac{2}{3}, 1\right) \leftrightarrow 2 \cdot \frac{1}{3} + B^2.$$

We have found that: the tri-adic rational number in Euclidean space $\mathbb{R}$ has two expressions, finite and infinite decimals, for example, $\frac{1}{3} = 1 \cdot \frac{1}{3} + 0 \cdot \frac{1}{3^2} + \cdots$ and $\frac{1}{3} = 2 \cdot \frac{1}{3^2} + 2 \cdot \frac{1}{3^3} + \cdots$. However, $1 \cdot \beta^1 + 0 \cdot \beta^2 + \cdots \in 1 \cdot \beta^1 + B^2$ and $2 \cdot \beta^2 + 2 \cdot \beta^3 + \cdots \in 0 \cdot \beta^1 + B^2$, in local field K_3, they are different elements since $(1 \cdot \beta^1 + B^2) \cap (0 \cdot \beta^1 + B^2) = \varnothing$.

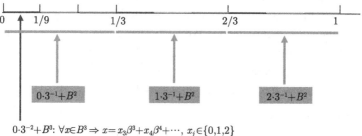

$0 \cdot 3^{-2} + B^3$: $\forall x \in B^3 \Rightarrow x = x_3 \beta^3 + x_4 \beta^4 + \cdots$, $x_j \in \{0, 1, 2\}$

Figure. 1.2.1

A comparison

Local field K_3		Euclidean space $\mathbb{R}$
$1 \cdot \beta^1 + 0 \cdot \beta^2 + \cdots \in 1 \cdot \beta^1 + B^2$	$\longleftrightarrow$	$1 \cdot \frac{1}{3} + 0 \cdot \frac{1}{3^2} + \cdots = \frac{1}{3}$
$2 \cdot \beta^2 + 2 \cdot \beta^3 + \cdots \in 0 \cdot \beta^1 + B^2$	$\longleftrightarrow$	$2 \cdot \frac{1}{3^2} + 2 \cdot \frac{1}{3^3} + \cdots = \frac{1}{3}$

Figure. 1.2.2

We can not expect to deduce some problems over local fields to those over Euclidean spaces by virtue of some isomorphic mappings, since the topological structures of local fields are quite different form those of the Euclidean spaces, they are not topological isomorphic.

However, we may expect that they are different tools for dealing with research topics on the macro-scopical and micro-scopical scientific universe, respectively, both are indispensable[4],[38].

Exercises

1. Construct a Cantor type set on local field K_p, and evaluate its Haar measure.
2. Construct a devil's ladder on local field K_p.
3. Dose it can define Weierstrass function on a local field K_p as on Euclidean space?
   Sierpinski carpet, Sierpinski gasket, etc.?
4. Estimate Haar measures of D, D^*, B^k, $k \in \mathbb{N}$, in K_p (see the Section 2.3).
5. Consider the topological structure and Haar measure of multiplication group K_p^* of K_p.
6. Consider the Haar integral on the finite algebraic extension $K_q, q = p^c, c \in \mathbb{N}$, of K_p.

Chapter 2

Character Group Γ_p of Local Field K_p

The idea of characters and character groups plays very important roles in harmonic analysis.

We introduce concepts of characters of a group, then discuss the structures of the character groups of local fields.

2.1 Character groups of locally compact groups

2.1.1 Characters of groups

Definition 2.1.1 (Character of a group) *Let G be a group with operation '·'. If a complex function $\chi : G \to \mathbb{C}$ on G satisfies*
  (i) $\forall x, y \in G \Rightarrow \chi(x \cdot y) = \chi(x)\chi(y)$;
  (ii) $\exists M > 0$, s.t. $\forall x \in G \Rightarrow |\chi(x)| \leqslant M$, and $\chi(x) \neq 0$,
*then χ is said to be a **character of group** G*[30].

A character χ of a group G has following properties.

Theorem 2.1.1 *If $\chi : G \to \mathbb{C}$ is a character of a group G, then it is a non-zero homomorphism from G to the multiplication group*

$$\mathbb{T} = \left\{z = e^{2\pi i x} \in \mathbb{C} : 0 \leqslant x < 1\right\}. \tag{2.1.1}$$

That is, χ is a complex function $\chi : G \to \mathbb{C}$ on G with properties:
  (i) $\forall x, y \in G \Rightarrow \chi(x \cdot y) = \chi(x)\chi(y)$;
  (ii) $\forall x \in G \Rightarrow |\chi(x)| = 1$.
For the proofs, we leave in Exercises.

2.1.2 Characters and character groups of locally compact groups

Definition 2.1.2 (Character of a locally compact group) *Suppose that G is a locally compact Abelian topological group, for simply, locally compact group, with operation $+$. If a complex function $\chi : G \to \mathbb{C}$ on G satisfies*

(i) $\forall x_1, x_2 \in G \Rightarrow \chi(x_1 + x_2) = \chi(x_1)\chi(x_2)$; (2.1.2)
(ii) $\forall x \in G \Rightarrow |\chi(x)| = 1$. (2.1.3)

*Then χ is said to be a **character** of G.*

We know by Theorem 2.1.1 (ii) that $\chi : G \to \mathbb{T}$, where

$$\mathbb{T} = \{z \in \mathbb{C} : |z| = 1\},$$

and $\mathbb{T}$ is the unit circle on complex plane $\mathbb{C}$, moreover, it becomes a group with multiplication, called a ***circle group***.

Definition 2.1.3 (Character group Γ_G of a locally compact group G) *Suppose that G is a locally compact group with operation $+$. Denote by*

$$\Gamma_G = \{\chi : \chi \text{ is continuous character of } G\}$$

all continuous characters of G, where $\chi : G \to \mathbb{T}$ satisfies $(2.1.2), (2.1.3)$. Then, as a homomorphism from G to $\mathbb{T}$, each $\chi \in \Gamma_G$ satisfies

$$\chi \in \Gamma_G \Rightarrow \chi(x_1 + x_2) = \chi(x_1)\chi(x_2), \quad \forall x_1, x_2 \in G.$$

Endow with multiplication operation "$\times$" to Γ_G as

$$\forall \chi_1, \chi_2 \in \Gamma_G \Rightarrow (\chi_1 \times \chi_2)(x) = \chi_1(x)\chi_2(x), \quad \forall x \in G. \quad (2.1.4)$$

*For simply, denote by $(\chi_1 \times \chi_2)(x) \equiv (\chi_1 \cdot \chi_2)(x) \equiv (\chi_1\chi_2)(x)$. Thus, Γ_G is an Abelian group with multiplication "$\times$", and is said to be the **character group of group** G (or, the **dual group** of G).*

Γ_G becomes a group since it is easy to see that the multiplication on Γ_G defined in (2.1.4) satisfies

(i) Operation is closed: $\forall \chi_1, \chi_2 \in \Gamma_G \Rightarrow \chi_1 \cdot \chi_2 \in \Gamma_G$;
(ii) Combination law: $\forall \chi_1, \chi_2, \chi_3 \in \Gamma_G \Rightarrow (\chi_1 \cdot \chi_2) \cdot \chi_3 = \chi_1 \cdot (\chi_2 \cdot \chi_3)$;
(iii) Unit element: $\exists \chi \equiv I, I(x) = 1$, s.t., $\forall \chi \in \Gamma_G \Rightarrow \chi \cdot I = I \cdot \chi = \chi$;
(iv) Inverse element: $\forall \chi \in \Gamma_G, \exists \chi^{-1} = \overline{\chi}$, s.t., $\chi^{-1} \in \Gamma_G$, and

$$\chi \cdot \chi^{-1}(x) = \chi \cdot \overline{\chi}(x) = \chi(x) \cdot \overline{\chi}(x) = 1;$$

(v) Commutative law: $\forall \chi_1, \chi_2 \in \Gamma_G \Rightarrow \chi_1 \cdot \chi_2 = \chi_2 \cdot \chi_1$.

2.1.3 Pontryagin dual theorem

The famous Pontryagin dual theorem shows the depiction of Γ_G.

Theorem 2.1.2 *Suppose that G is T_2-type Abelian topological group, Γ_G is its character group.*

*(i) If G is a compact group, then Γ_G is a **discrete group**; and if G is discrete, then Γ_G is compact;*

(ii) If G is a locally compact group, then Γ_G is locally compact; moreover, Γ_G and G are topological isomorphic; that is, T_2-type locally compact Abelian group is self dual.

The proof of Pontryagin dual theorem is more complex, and needs more knowledge of algebra and harmonic analysis, we refer to [30]. For a local field K_p, the operation structure and topological structure of Γ_p will show in Section 2.2.

2.1.4 Examples

1. Characters and Character group $\Gamma_\mathbb{R}$ of the addition group of $\mathbb{R}$

Firstly, a character $\chi : \mathbb{R} \to \mathbb{T}$ of the addition group of $\mathbb{R}$ is evaluated.

$\forall \chi : \mathbb{R} \to \mathbb{T}$ implies $\chi : x \in \mathbb{R} \to \chi(x) \in \mathbb{T} = \{e^{2\pi i x}\}$, and χ is continuous. By Definition 2.1.2, there exists a constant $h > 0$, s.t., the Haar integral on $\mathbb{R}$ (it is, as a matter of fact, Lebesgue integral on $\mathbb{R}$) satisfies

$$\int_0^h \chi(x) dx = c \neq 0$$

(other wise, if for all $h > 0$ holds $\int_0^h \chi(x) dx = 0$, then by the continuity of χ, it follows $\chi \equiv 0$, this is contradict with definition of characters). Thus,

$$\forall t \in \mathbb{R} \Rightarrow \int_0^h \chi(x+t)\, dx = \int_0^h \chi(x)\chi(t) dx = \chi(t) \int_0^h \chi(x) dx = c\chi(t),$$

then it holds

$$\chi(t) = \frac{1}{c} \int_0^h \chi(x+t)\, dx = \frac{1}{c} \int_t^{t+h} \chi(u)\, du. \qquad (2.1.5)$$

This implies that $\chi(t)$ in (2.1.5) is continuous and differentiable. Thus evaluate derivatives about t both sides of equation

$$\chi(x+t) = \chi(x)\chi(t),$$

and get

$$\chi'(x+t) = \chi(x)\chi'(t).$$

Set $t = 0$, it follows

$$\chi'(x) = \chi(x)\chi'(0) \equiv a\chi(x)$$

with $a = \chi'(0) \neq 0$. Hence we have the differential equation of $\chi(x)$, i.e., eigen-equation,

$$\chi' = a\chi.$$

The solution of differential equation $\chi' = a\chi$ is $\chi(x) = e^{ax}$, clearly. Moreover, since χ requests $|\chi(x)| = 1$, so $a = 2\pi i$, thus, a character $\chi(x) = e^{2\pi i x}$ is determined.

To determine character group $\Gamma_\mathbb{R}$ of the addition group of $\mathbb{R}$, we look for the unit of $\Gamma_\mathbb{R}$: it is $\chi(x) \equiv 1$, denoted by $I(x) \equiv 1$, since

$$I(x) \equiv \chi(0 \cdot x) = e^{2\pi i x \cdot 0} = 1.$$

Then, $\forall m \in \mathbb{N}$,

$$\chi_m(x) \equiv \chi(m \cdot x) = \chi(x + \cdots + x) = \underbrace{\chi(x) \cdot \cdots \cdot \chi(x)}_{m} = e^{2\pi i x \cdot m},$$

thus $\chi_m(x), m \in \mathbb{N}$, is well-defined.

To define $\chi_{\frac{m}{n}}(x)$ for a rational number $\dfrac{m}{n} \in \mathbb{Q}$, we take

$$\chi_{\frac{m}{n}}(x) = e^{2\pi i \frac{m}{n} x}.$$

Finally, to define $\chi_y(x)$ for a real number $y \in \mathbb{R}$, we take a rational number sequence $r_n \in \mathbb{Q}$ with $\lim\limits_{n \to +\infty} r_n = y \in \mathbb{R}$, by the continuity of characters, we may define for $\forall y \in \mathbb{R}$

$$\chi_y(x) = \lim_{r_n \to y} e^{2\pi i r_n x} = e^{2\pi i y x}. \qquad (2.1.6)$$

Till now, we have all elements in $\Gamma_\mathbb{R}$, that is $\chi_y(\cdot) = e^{2\pi i y \cdot} \in \Gamma_\mathbb{R}$, and it is clearly that $\Gamma_\mathbb{R}$ is isomorphic with $\mathbb{R}$

$$\chi_y(\cdot) \in \Gamma_\mathbb{R} \xleftrightarrow{\text{iso.}} y \in \mathbb{R} \qquad (2.1.7)$$

and

$$\Gamma_\mathbb{R} = \{\chi_y(\cdot) : y \in \mathbb{R}\} = \{e^{2\pi i y \cdot} : y \in \mathbb{R}\}.$$

2. Characters and character group $\Gamma_{\mathbb{Z}/p}$ of the congruence class $\mathbb{Z}/p$

As well known, $\forall p \in \mathbb{N}$, $p \geqslant 2$, the congruence class $\mathbb{Z}/p = \{0, 1, \cdots, p-1\}$ with addition mod p is a finite discrete group, thus its character group $\Gamma_{\mathbb{Z}/p}$ is a finite compact group by Pontryagin dual theorem.

Let a character of $\mathbb{Z}/p$ be $\chi : \mathbb{Z}/p \to \mathbb{T}$. Thus, $I : \mathbb{Z}/p \to \mathbb{T}$

$$I(t) = e^{\frac{2\pi i}{p} 0 \cdot t} = 1 \in \mathbb{T} \tag{2.1.8}$$

is a character, since $|I(t)| = |\exp((2\pi i/p) \cdot 0 \cdot t)| = 1$, and

$$I(t_1 + t_2) = I(t_1) I(t_2) = 1.$$

To determine all characters of the congruence class $\mathbb{Z}/p$, for each $t \in \mathbb{Z}/p$, set

$$\chi_0(t) = e^{\frac{2\pi i}{p} t}, \tag{2.1.9}$$

it satisfies $|\chi_0(t)| = \left| e^{\frac{2\pi i}{p} t} \right| = 1$; and if $t_1, t_2 \in \mathbb{Z}/p$, $0 \leqslant t_1 + t_2 < p$, then

$$\chi_0(t_1 + t_2) = e^{\frac{2\pi i}{p}(t_1+t_2)} = e^{\frac{2\pi i}{p} t_1} e^{\frac{2\pi i}{p} t_2} = \chi_0(t_1) \chi_0(t_2).$$

Moreover, $\forall k \in \mathbb{Z}/p$, we define

$$\chi(k) \equiv \{\chi_0(t)\}^k = \left(e^{\frac{2\pi i}{p} t}\right)^k = e^{\frac{2\pi i}{p} tk} = \chi_0(kt).$$

Since

$$\chi(kt) = \chi\underbrace{\left(t + t + \cdots + t\right)}_{k} = \underbrace{\chi(t)\chi(t)\cdots\chi(t)}_{k} = \{\chi(t)\}^k,$$

then, by the mod p operation, we have

$$\Gamma_{\mathbb{Z}/p} = \left\{ e^{\frac{2\pi i}{p} \cdot 0}, e^{\frac{2\pi i}{p} \cdot 1}, e^{\frac{2\pi i}{p} \cdot 2}, \cdots, e^{\frac{2\pi i}{p} \cdot (p-1)} \right\} \subset \mathbb{T}. \tag{2.1.10}$$

2.2 Character group Γ_p of K_p

We show properties and structures of a character χ in Γ_p, and properties of character group $\Gamma_p \equiv \Gamma_{K_p^+}$ for the addition group K_p^+ of a local field K_p (for simply, the character group Γ_p of K_p). Main reference of this section is [100].

2.2.1 Properties of $\chi \in \Gamma_p$ and Γ_p

1. Properties of $\chi \in \Gamma_p$

Theorem 2.2.1 *Suppose that χ is a non-trivial character of K_p^+, $\chi \in \Gamma_p$, $\chi \neq 1$. Then there exists an integer $k \in \mathbb{Z}$, s.t. χ is trivial on B^k, i.e., $\chi|_{B^k} = 1$.*

Proof. By the continuity of χ, and by the base $\{B^k : k \in \mathbb{Z}\}$ of zero $0 \in K_p$, then for $\varepsilon = \dfrac{\sqrt{2}}{2}$, $\exists k \in \mathbb{Z}$, s.t.

$$|\chi(x) - \chi(0)| = |\chi(x) - 1| < \frac{\sqrt{2}}{2}, \quad \forall x \in B^k.$$

On the other hand, $\forall l \in \mathbb{N}$ and $x \in B^k$, since

$$|x| \leqslant p^{-k} \Rightarrow |lx| = \Big|\underbrace{x + \cdots + x}_{l}\Big| \leqslant \max\{|x|, \cdots, |x|\} = |x| \leqslant p^k \Rightarrow lx \in B^k,$$

thus

$$lx = \underbrace{x + \cdots + x}_{l} \in B^k,$$

hence $\chi(lx) = \chi\Big(\underbrace{x + \cdots + x}_{l}\Big) = \chi^l(x)$, and holds

$$\left|(\chi(x))^l - 1\right| = |\chi(lx) - 1| < \frac{\sqrt{2}}{2}, \quad \forall l \in \mathbb{N}. \tag{2.2.1}$$

We claim that $\chi(x)|_{x \in B^k} = 1$. If not so, we will have the contradiction as follows.

In fact, if $\chi|_{B^k} \neq 1$, by $\chi(x) \in \{z \in \mathbb{C} : |z| = 1\}$, we would suppose

$$\chi(x) = e^{i\theta}, \quad \theta \in \left(0, \frac{\pi}{2}\right).$$

Thus, $\exists n \in \mathbb{N}$, s.t. $\dfrac{2\pi}{3} < n\theta < \dfrac{4\pi}{3}$, and

$$\left|e^{in\theta} - 1\right| \geqslant \sqrt{3}. \tag{2.2.2}$$

(2.2.2) holds since (2.2.2) is equivalent to

$$\left(e^{in\theta} - 1\right)\left(e^{-in\theta}\right) \geqslant 3,$$

or

$$2\cos n\theta \leqslant -1.$$

Moreover, if $\dfrac{2\pi}{3} < n\theta < \dfrac{4\pi}{3}$, then $-1 \leqslant \cos n\theta \leqslant -\dfrac{1}{2}$, thus (2.2.2) holds clearly. This contradicts (2.2.1). Hence
$$\chi(x)|_{x\in B^k} = 1.$$
The theorem is proved.

Definition 2.2.1 (Ramification, ramified degree) *The property of a non-trivial character $\chi \in \Gamma_p$ of K_p^+ "there exists $k \in \mathbb{Z}$, such that $\chi|_{B^k} = 1$" is said to be the **ramified property** of χ; or the **ramification property of characters**; the smallest integer of the $k \in \mathbb{Z}$ is said to be the **ramified degree** of χ.*

The ramification property of characters plays role in harmonic analysis over an addition group K_p^+ of K_p. However, the ramification property of a character $\pi \in \Gamma_{K^*}$ in the character group Γ_{K^*} of multiplication group K_p^* of local field K_p, will play very important role, it is connected with the Mellin-Fourier transform, and with lots of interesting open problems.

The addition group K_p^+ of a local field K_p is isomorphic to its character group Γ_p by Pontryagin dual theorem. We would take a character $\chi \in \Gamma_p$ of K_p, such that it is trivial on $D = \{x \in K_p : |x| \leqslant 1\}$, i.e., $\chi|_D \equiv 1$; and is non-trivial on $B^{-1} = \{x \in K_p : |x| \leqslant p\}$, i.e., $\chi|_{B^{-1}} \neq 1$. The Theorem 2.2.1 tells us this kind of character $\chi \in \Gamma_p$ exists, and χ is said to be "a **basic character**".

Theorem 2.2.2 *If $\chi \in \Gamma_p$ is trivial in B^k for some $k \in \mathbb{Z}$, i.e., $\chi|_{B^k} \equiv 1$, then χ is constant on all cosets*
$$y + B^k, \quad y \in K_p \backslash B^k.$$

Proof. In fact, $\forall x \in y + B^k$, it has
$$x = y + z, \quad y \in K_p\backslash B^k, \quad z \in B^k,$$
thus,
$$\chi(x) = \chi(y+z) = \chi(y)\chi(z) = \chi(y), \quad y \in K_p\backslash B^k, \quad z \in B^k,$$
since $\chi(z) = 1, z \in B^k$. The proof is complete.

The following theorem establishes the one-one corresponding relationship $\Gamma_p \leftrightarrow K_p$.

Theorem 2.2.3 *Let χ be a non-trivial character $\chi \in \Gamma_p$ of K_p^+, i.e., $\chi \neq 1$. Then the corresponding relationship*
$$\lambda \in K_p \longleftrightarrow \chi_\lambda \in \Gamma_p \qquad (2.2.3)$$

is determined by
$$\chi_\lambda(x) = \chi(\lambda x), \qquad (2.2.4)$$
and the topological isomorphic is established for K_p and Γ_p; moreover, we have
$$\Gamma_p = \{\chi_\lambda : \lambda \in K_p\}.$$

Proof. Firstly, we prove $\chi_\lambda : x \in K_p \to \chi_\lambda(x) = \chi(\lambda x)$ determined in (2.2.4) is a character of K_p, since

1° $\forall \lambda \in K_p$, $\forall x \in K_p \Rightarrow \lambda x \in K_p \Rightarrow \chi_\lambda(x) = \chi(\lambda x) \in \mathbb{T}$;

2° $\forall x_1, x_2 \in K_p \Rightarrow \chi_\lambda(x_1 + x_2) = \chi(\lambda(x_1+x_2)) = \chi(\lambda x_1 + \lambda x_2)$
$$= \chi(\lambda x_1)\chi(\lambda x_2) = \chi_\lambda(x_1)\chi_\lambda(x_2);$$

3° the continuity of χ_λ and $|\chi_\lambda(x)| = |\chi(\lambda x)| = 1$ are clear.

Secondly, we have
$$\{\chi_\lambda : \lambda \in K_p\} \subset \Gamma_p, \qquad (2.2.5)$$
and $\{\chi_\lambda : \lambda \in K_p\} \leftrightarrow K_p$ implies $\{\chi_\lambda : \lambda \in K_p\}$ is an one dimension subspace of Γ_p.

Then, we prove that the character group Γ_p of K_p^+ is one dimension also.

In fact, for $\chi_1, \chi_2 \in \Gamma_p$,
$$(\chi_1, \chi_2) \to \chi_1\chi_2$$
determined by $(\chi_1\chi_2)(x) = \chi_1(x)\chi_2(x)$;
$$(\lambda, \chi) \to \chi_\lambda$$
determined by $\chi_\lambda(x) = \chi(\lambda x)$.

Denote $\Gamma_p \equiv (K_p)^\wedge$, it follows by Pontryagin dual theorem that
$$K_p \leftrightarrow (K_p)^\wedge \leftrightarrow ((K_p)^\wedge)^\wedge. \qquad (2.2.6)$$
If $\Gamma_p = (K_p)^\wedge \equiv A^d$ is $d \in \mathbb{N}$ dimensional, and A is one dimensional. Then (2.2.6) gives
$$((K_p)^\wedge)^\wedge \equiv (A^d)^\wedge = \underbrace{(A \times \cdots \times A)}_{d}^\wedge = \underbrace{A^\wedge \times \cdots \times A^\wedge}_{d} = (A^\wedge)^d.$$

Thus, $A^d = \Gamma_p$ and (2.2.6) give
$$A^d = (K_p)^\wedge \leftrightarrow ((K_p)^\wedge)^\wedge = A^{d^2}.$$
Hence, $d = d^2 \Rightarrow d = 1$.

Finially, since (2.2.5), we get $\Gamma_p = \{\chi_\lambda : \lambda \in K_p\}$. The proof of theorem is complete.

2. Properties of Γ_p

To endow topology of the character group Γ_p, we introduce the concept of "annihilator".

Definition 2.2.2 (Annihilator) *The subset in the character group Γ_p of addition group K_p^+*

$$\Gamma^k = \{\chi \in \Gamma_p : \forall x \in B^k \Rightarrow \chi(x) = 1\}$$

is said to be an **annihilator** *of* $B^k = \{x \in K_p : |x| \leqslant p^{-k}\}$.

Theorem 2.2.4 *For the character group Γ_p, we have the following:*

(i) $\Gamma^k \subset \Gamma^{k+1}$, $k \in \mathbb{Z}$, *are increasing, open and closed, as well as compact subset sequence in Γ_p;*

(ii) $\bigcup\limits_{k=-\infty}^{+\infty} \Gamma^k = \Gamma_p$ *and* $\bigcap\limits_{k=-\infty}^{+\infty} \Gamma^k = \{I\}$, *$I$ is the unit of Γ_p;*

(iii) $\{\Gamma^k\}_{k=-\infty}^{+\infty}$ *is a base of the unit I of character group Γ_p, such that Γ_p is isomorphic to K_p^+, and it is a locally compact group;*

(iv) *The locally compact group Γ_p can be endowed the non-Archimedean valued norm*

$$\chi_\lambda \in \Gamma_p \to |\lambda| \in p^k, \quad k \in \mathbb{Z},$$

such that

$$\Gamma^k = \{\chi_\lambda \in \Gamma_p : |\lambda| \leqslant p^k\}, \quad k \in \mathbb{Z}.$$

The proofs are left for exercises.

In the next section, we will show the representatives of the character groups of p-series field S_p and p-aidc filed A_p, respectively.

2.2.2 Character group of p-series field S_p

Suppose that the characteristic number of S_p is a prime $p \geqslant 2$.

1. Determine a character $\chi \in \Gamma_{S_p}$ of p-series field S_p

For $\chi \in \Gamma_{S_p}$, $\forall x \in S_p \Rightarrow |\chi(x)| = 1$, and for each $j \in \mathbb{N}$, it follows

$$\chi(p\beta^{-j}) = \chi\Big(\underbrace{\beta^{-j} + \cdots + \beta^{-j}}_{p}\Big) = \chi\left(0 \cdot \beta^{-j}\right) \equiv 1. \tag{2.2.7}$$

On the other hand, by

$$\chi(p\beta^{-j}) = \chi\Big(\underbrace{\beta^{-j} + \cdots + \beta^{-j}}_{p}\Big) = \chi\left(\beta^{-j}\right) \cdot \cdots \cdot \chi\left(\beta^{-j}\right) = \chi\left(\beta^{-j}\right)^{p},$$

thus, by (2.2.7),
$$\chi(\beta^{-j})^{p} = 1, \tag{2.2.8}$$

this tells that, $\chi(\beta^{-j})$ is a root of equation $z^{p} = 1$, then $\chi(\beta^{-j})$, $j \in \mathbb{Z}$, takes values in the set

$$\left\{1, e^{\frac{2\pi i}{p} \cdot 1}, e^{\frac{2\pi i}{p} \cdot 2}, \cdots, e^{\frac{2\pi i}{p} \cdot (p-1)}\right\}. \tag{2.2.9}$$

Take a basic character by Theorem 2.2.1 as

$$\chi\left(\beta^{-j}\right) = \begin{cases} e^{\frac{-2\pi i}{p}}, & j = -1, \\ 1, & j \neq -1. \end{cases} \tag{2.2.10}$$

By (2.2.10), we now determine $\chi \in \Gamma_{S_p}$.

If $x \in S_p$, when $x \in B^{s_0} \subset S_p$, $s_0 \leqslant 0$, by the Section 1.2.6,

$$x = \sum_{s=s_0}^{+\infty} x_s \beta^s = x_{s_0}\beta^{s_0} + x_{s_0+1}\beta^{s_0+1} + \cdots + x_{-1}\beta^{-1} + x_0\beta^0 + x_1\beta^1 + \cdots, \tag{2.2.11}$$

where
$$x_s \in \{0, 1, 2, \cdots, p-1\}, \quad s = s_0, s_0+1, \cdots. \tag{2.2.12}$$

Then
$$\chi(x) = \chi\left(\sum_{s=s_0}^{+\infty} x_s\beta^s\right) = \chi\left(\sum_{s=s_0}^{-2} x_s\beta^s + x_{-1}\beta^{-1} + \sum_{s=0}^{+\infty} x_s\beta^s\right)$$
$$= \chi\left(x_{-1}\beta^{-1}\right) \prod_{s \neq -1} \chi(x_s\beta^s) = \chi\left(x_{-1}\beta^{-1}\right) \cdot 1$$
$$= \chi\Big(\underbrace{\beta^{-1} + \cdots + \beta^{-1}}_{x_{-1}}\Big) = \chi\left(\beta^{-1}\right)^{x_{-1}} = e^{\frac{2\pi i}{p}x_{-1}}. \tag{2.2.13}$$

Let
$$\omega = e^{\frac{2\pi i}{p}},$$

(2.2.13) can be rewrite as

$$\chi(x) = \chi\left(\sum_{s=s_0}^{+\infty} x_s\beta^s\right) = \omega^{x_{-1}}, \quad x_{-1} \in \{0, 1, \cdots, p-1\}. \tag{2.2.14}$$

This is a character of p-series field S_p.

The next example is a character of 2-series field S_2.

Take $x \in S_2$, then
$$x = x_{-s}\beta^{-s} + x_{-s+1}\beta^{-s+1} + \cdots + x_{-1}\beta^{-1} + x_0\beta^0 + x_1\beta^1 + \cdots$$
$$= \sum_{j=-s}^{-1} x_j\beta^j + z, \qquad (2.2.15)$$
where $x_{-s} = 1$, $x_j \in \{0,1\}$, $j = -s+1, \cdots, -1$, $|\beta| = 2^{-1}$, and $z \in D$, where $D = \{x \in K_2 : |x| \leqslant 1\}$.

Define a basic character:
$$\chi(\beta^k) = \begin{cases} -1, & k = -1, \\ 1, & k \neq -1. \end{cases}$$

Then
$$\chi(x) = \chi\left(\sum_{j=-s}^{-1} x_j\beta^j + z\right) = \chi(z)\chi\left(\sum_{j=-s}^{-1} x_j\beta^j\right)$$
$$= \chi(x_{-s}\beta^{-s})\chi(x_{-s+1}\beta^{-s+1})\cdots\chi(x_{-1}\beta^{-1})$$
$$= (1)^{x_{-s}} \cdot (1)^{x_{-s+1}} \cdots (1)^{x_{-2}} \cdot (-1)^{x_{-1}} = (-1)^{x_{-1}}.$$

For S_2, we have $\omega = e^{\frac{2\pi i}{p}} = e^{\pi i} = -1$. Then (2.2.14) becomes
$$\chi(x) = (-1)^{x_{-1}}.$$

This is a character of S_2.

2. Determine the character group Γ_{S_p} of p-series field S_p

Let
$$\lambda = \lambda_{-l}\beta^{-l} + \lambda_{-l+1}\beta^{-l+1} + \cdots + \lambda_{-1}\beta^{-1} + \lambda_0\beta^0 + \lambda_1\beta^1 + \cdots, \quad l \geqslant 0,$$
$$x = x_{-s}\beta^{-s} + x_{-s+1}\beta^{-s+1} + \cdots + x_{-1}\beta^{-1} + x_0\beta^0 + x_1\beta^1 + \cdots, \quad s \geqslant 0,$$
with $x_j, \lambda_k \in \{0, 1, \cdots, p-1\} = GF(p)$, $j \geqslant -s$, $k \geqslant -l$.

On the other hand, since we have $K_p \leftrightarrow \Gamma_p = \{\chi_\lambda : \lambda \in K_p\}$ by Theorem 2.2.3, and the corresponding relation $\varphi : \lambda \in K_p \leftrightarrow \chi_\lambda \in \Gamma_p$, as well as $\chi_\lambda(x) = \chi(\lambda x)$, it follows
$$\chi_\lambda(x) = \chi(\lambda x) = e^{(\lambda x)_{-1}}, \qquad (2.2.16)$$
where $(\lambda x)_{-1}$ is the coefficient of β^{-1}, determined by
$$\left(\sum_{j=-l}^{+\infty} \lambda_j\beta^j\right)\left(\sum_{j=-s}^{+\infty} x_j\beta^j\right)$$

$$= \left(\lambda_{-l}\beta^{-l} + \lambda_{-l+1}\beta^{-l+1} + \cdots + \lambda_{-1}\beta^{-1} + \lambda_0\beta^0 + \lambda_1\beta^1 + \cdots\right)$$
$$\cdot \left(x_{-s}\beta^{-s} + x_{-s+1}\beta^{-s+1} + \cdots + x_{-1}\beta^{-1} + x_0\beta^0 + x_1\beta^1 + \cdots\right).$$
(2.2.17)

The sum in (2.2.17) is a finite one, and the coefficient of β^{-1} is $(\lambda x)_{-1} = \sum \lambda_{-1-j}x_j$.

Then, the character group Γ_{S_p} of S_p is

$$\Gamma_{S_p} = \left\{\chi_\lambda : \chi_\lambda(x) = e^{\frac{2\pi i}{p}(\lambda x)_{-1}}, \lambda, x \in S_p\right\}. \quad (2.2.18)$$

2.2.3 Character groups of p-adic field A_p

For $K_p = A_p$, the characteristic number is 0 (i.e., it is ∞).

Since the expressions of elements in p-adic filed A_p have similar expressions of those in p-series field S_p, they are both formal power series on $GF(p)$, however, operations on two kind of fields are quite different, the former one, in A_p, is term by term, mod$p$, carrying from left to right, and the latter one, in S_p, is term by term, mod$p$, no carrying.

1. Determine a character $\chi \in \Gamma_{A_p}$ of p-adic field A_p

For $x \in A_p$, let $x \in B^{s_0} \subset A_p$, $s_0 \leqslant 0$, then

$$x = \sum_{s=s_0}^{+\infty} x_s\beta^s = x_{s_0}\beta^{s_0} + x_{s_0+1}\beta^{s_0+1} + \cdots + x_{-1}\beta^{-1} + x_0\beta^0 + x_1\beta^1 + \cdots$$
$$= x_{s_0}\beta^{s_0} + x_{s_0+1}\beta^{s_0+1} + \cdots + x_{-1}\beta^{-1} + z, \quad z \in D, \quad (2.2.19)$$

where $x_{s_0}, x_{s_0+1}, \cdots \in \{0, 1, \cdots, p-1\}$. Hence,

$$\chi(x) = \chi\left(x_{s_0}\beta^{s_0} + x_{s_0+1}\beta^{s_0+1} + \cdots + x_{-1}\beta^{-1} + z\right)$$
$$= \chi\left(x_{s_0}\beta^{s_0}\right) \cdot \chi\left(x_{s_0+1}\beta^{s_0+1}\right) \cdots \cdot \chi\left(x_{-1}\beta^{-1}\right), \quad (2.2.20)$$

since $\chi(z) = 1$ in (2.2.20).

The addition operation in A_p is term by term, mod p, carrying from left to right, thus

$$\chi(p\beta^{-j}) = \chi\Big(\underbrace{\beta^{-j} + \cdots + \beta^{-j}}_{p}\Big) = \chi\left(0 \cdot \beta^{-j} + \beta^{-j+1}\right) = \chi\left(\beta^{-j+1}\right),$$
(2.2.21)

and for $s = 0, 1, \cdots, p-1$, it follows

$$\chi(s\beta^{-j}) = \chi\Big(\underbrace{\beta^{-j} + \cdots + \beta^{-j}}_{s}\Big) = \chi\left(\beta^{-j}\right) \cdots \chi\left(\beta^{-j}\right) = \chi\left(\beta^{-j}\right)^s.$$
(2.2.22)

So that
$$\chi(p\beta^{-j}) = \chi\left(\beta^{-j+1}\right),$$
(2.2.23)

this shows that the p-power of $\chi(\beta^{-j})$ is from "$-j$", to "$-j+1$", so we have to consider so called **different levers of a basic character**.

(i) **First lever** Choose a basic character
$$\chi\left(\beta^{-j}\right) = \begin{cases} e^{\frac{-2\pi i}{p}}, & j = -1, \\ 1, & j \neq -1. \end{cases}$$

Then
$$\left\{1, e^{\frac{2\pi i}{p} \cdot 1}, e^{\frac{2\pi i}{p} \cdot 2}, \cdots, e^{\frac{2\pi i}{p} \cdot (p-1)}\right\} \subset \Gamma_{A_p}$$

constructs the first lever character subgroup of Γ_{A_p}. And $\forall x \in B^{-1}$,
$$x = x_{-1}\beta^{-1} + z, \quad z \in D,$$

it has
$$\chi(x) = e^{\frac{2\pi i}{p} x_{-1}}, \quad x_{-1} \in \{0, 1, \cdots, p-1\}.$$
(2.2.24)

(ii) **Second lever** Choose a basic character
$$\chi\left(\beta^{-j}\right) = \begin{cases} e^{\frac{-2\pi i}{p^2}}, & j = -2, \\ 1, & j \neq -2. \end{cases}$$

Then
$$\left\{1, e^{\frac{2\pi i}{p^2} \cdot 1}, e^{\frac{2\pi i}{p^2} \cdot 2}, \cdots, e^{\frac{2\pi i}{p^2} \cdot (p^2 - 1)}\right\} \subset \Gamma_{A_p}$$

constructs the second lever character subgroup of Γ_{A_p}. And $\forall x \in B^{-2}$,
$$x = x_{-2}\beta^{-2} + x_{-1}\beta^{-1} + z, \quad z \in D,$$

it has
$$\chi(x) = e^{\frac{2\pi i}{p^2}(x_{-1} + x_{-2})}, \quad x_{-1}, x_{-2} \in \{0, 1, \cdots, p-1\}.$$
(2.2.25)

(iii) **k-th lever** Choose a basic character for k, $k > 0$
$$\chi\left(\beta^{-j}\right) = \begin{cases} e^{\frac{-2\pi i}{p^k}}, & j = -k, \\ 1, & j \neq -k. \end{cases}$$

Then
$$\left\{1, e^{\frac{2\pi i}{p^k} \cdot 1}, e^{\frac{2\pi i}{p^k} \cdot 2}, \cdots, e^{\frac{2\pi i}{p^k} \cdot (p^k - 1)}\right\} \subset \Gamma_{A_p}$$

constructs the k-th lever character subgroup of Γ_{A_p}. And $\forall x \in B^{-k}$,
$$x = x_{-k}\beta^{-k} + x_{-k+1}\beta^{-k+1} + \cdots + x_{-1}\beta^{-1} + z, \quad z \in D,$$
it has
$$\chi(x) = e^{\frac{2\pi i}{p^k}(x_{-1}+x_{-2}+\cdots+x_{-k})}, \quad x_{-1},\cdots,x_{-k} \in \{0,1,\cdots,p-1\}. \tag{2.2.26}$$

This is a character of p-adic field A_p (compare with (2.2.14)).

The following example is a characters of 2-adic field A_2.

Take $x \in A_2$, then
$$x = x_{-s}\beta^{-s} + x_{-s+1}\beta^{-s+1} + \cdots + x_{-1}\beta^{-1} + x_0\beta^0 + x_1\beta^1 + \cdots$$
$$= \sum_{j=-k}^{-1} x_j\beta^j + z, \tag{2.2.27}$$
with $x_{-k} = 1, x_j \in \{0,1\}, j = -k+1,\cdots,-1, k > 0, z \in D, |\beta| = 2^{-1}$.
Thus
$$\chi(x) = \chi\left(\sum_{j=-k}^{-1} x_j\beta^j + z\right) = \chi(z)\chi\left(\sum_{j=-k}^{-1} x_j\beta^j\right)$$
$$= \chi\left(x_{-k}\beta^{-k}\right)\chi\left(x_{-k+1}\beta^{-k+1}\right)\cdots\chi\left(x_{-1}\beta^{-1}\right)$$
$$= (-1)^{x_{-k}+x_{-k+1}+\cdots+x_{-1}}.$$

This is a character of A_2.

2. Determine the character group Γ_{A_p} of p-adic field A_p

Similarly, we may determine the character group Γ_{A_p} of A_p:
$$\Gamma_{A_p} = \left\{\chi_\lambda : \chi_\lambda(x) = e^{\frac{2\pi i}{p^k}(\lambda \odot x)}, \lambda, x \in A_p, k \in \mathbb{N}\right\}, \tag{2.2.28}$$
where
$$\lambda \odot x = \sum_{n=l}^{-k-1} x_n \left(\sum_{s=n}^{-k-1} \frac{\lambda_{-s}}{p^{s-n+1}}\right),$$
with
$$\lambda = \sum_{j=s}^{+\infty} \lambda_j\beta^j, \quad \lambda_j \in GF(p), \quad j = s, s+1, \cdots,$$
$$x = \sum_{j=l}^{+\infty} x_j\beta^j, \quad x_j \in GF(p), \quad j = l, l+1, \cdots.$$

2.3 Some formulas in local fields

There are some important formulas that play key roles in the analysis over local fields[100].

2.3.1 *Haar measures of certain important sets in K_p*

Let $E \subset K_p$ be a Haar measurable set in K_p, its characteristic function be denoted by

$$\Phi_E(x) = \begin{cases} 1, & x \in E, \\ 0, & x \in K_p \backslash E. \end{cases} \tag{2.3.1}$$

Then the Haar measure of E is

$$|E| = \int_E dx = \int_{K_p} \Phi_E(x) dx. \tag{2.3.2}$$

(i) $|D| = \int_D dx = \int_{K_p} \Phi_D(x) dx = 1.$

Since $D = \{x \in K_p : |x| \leqslant 1\}$ in K_p is an open, closed and compact subgroup of K_p in the topology endowed by non-Archimedean valued norm $|x|$ for which we may suppose the Haar measure is normalized, so the Haar measure of D can be taken as $|D| = 1$.

(ii) $|B| = \int_B dx = \int_{K_p} \Phi_B(x) dx = p^{-1}.$

By (1.2.13), it follows

$$D/B = \{0 \cdot \beta^0 + B, 1 \cdot \beta^0 + B, \cdots, (p-1) \cdot \beta^0 + B\}$$

with $B \equiv B^1 = \{x \in K_p : |x| \leqslant p^{-1}\}$, and $D = \bigcup_{j=0}^{p-1} (j \cdot \beta^0 + B)$. Since each coset

$$j \cdot \beta^0 + B, \quad j = 0, 1, \cdots, p-1$$

has same Haar measure, then $p|B| = |D|$ gives $|B| = p^{-1}$.

(iii) $|B^k| = \int_{B^k} dx = p^{-k}, k \in \mathbb{Z}.$

By $B^k = \{x \in K_p : |x| \leqslant p^{-k}\}$ and $B^k/B^{k+1} = \bigcup_{j=0}^{p-1} (j \cdot \beta^k + B^{k+1})$, as well as (ii).

(iv) $\int_{|x|=1} dx = p(1-p^{-1})$.

By $\{x \in K_p : |x| = 1\} = D \backslash B$, where $D \backslash B$ is the difference of sets D and B.

(v) $\int_{|x|=p^{-k}} dx = p^{-k}(1-p^{-1})$, $k \in \mathbb{Z}$.

By $\{x \in K_p : |x| = p^{-k}\} = B^k \backslash B^{k+1}$.

2.3.2 Integrals for characters in K_p

Choose a basic character χ

$$\chi(x) \begin{cases} \equiv 1, & x \in D, \\ \neq 1, & x \in K_p \backslash D, \end{cases} \quad (2.3.3)$$

then each character χ_y of K_p in Γ_p can be expressed by the basic character χ, see the formulas (2.2.18) and formula (2.2.28).

(i) $\int_D \chi(x) dx = 1$.

Since the basic character in (2.3.3) satisfies $\chi|_D = 1$, thus

$$\int_D \chi(x) dx = \int_D dx = 1.$$

Moreover, $\forall \psi \in \Gamma_p$, if $\psi|_D$ is not trivial in D, then $\int_D \psi(x) dx = 0$, since if $\psi|_D$ is not trivial in D, we have

$\Rightarrow \exists a \in D$, s.t. $\psi(a) \neq 1$

$\Rightarrow \int_D \psi(x) dx = \int_D \psi(x+a) dx = \int_D \psi(x) \psi(a) dx = \psi(a) \int_D \psi(x) dx$

$\Rightarrow (\psi(a) - 1) \int_D \psi(x) dx = 0 \Rightarrow \int_D \psi(x) dx = 0$.

(ii) $\int_{B^{-k}} \chi(x) dx = \int_{|x| \leq p^k} \chi(x) dx = \begin{cases} p^k, & k \leq 0, \\ 0, & k > 0. \end{cases}$ (2.3.4)

Since $B^{-k} = \{x \in K_p : |x| \leq p^k\}$ is a subgroup of K_p, thus $B^{-k} = \beta^{-k} D$, and

$$k \leq 0 \Rightarrow B^{-k} = \beta^{-k} D \subseteq D \Rightarrow \chi|_{B^{-k}} \equiv 1,$$

$$k > 0 \Rightarrow B^{-k} = \beta^{-k} D \supsetneq D \Rightarrow \chi|_{B^{-k}} \neq 1,$$

hence
$$\int_{B^{-k}} \chi(x)dx = \int_D \chi\left(\beta^{-k}t\right)\left|\beta^{-k}\right|dt = \begin{cases} p^k, & k \leqslant 0, \\ 0, & k > 0. \end{cases}$$

(iii) $\int_{|x|=p^k} \chi(x)dx = \begin{cases} p^k\left(1-p^{-1}\right), & k \leqslant 0, \\ -1, & k = 1, \\ 0, & k > 1, \end{cases}$ (2.3.5)

and

$$\int_{|x|=p^k} \overline{\chi}(x)dx = \begin{cases} p^k\left(1-p^{-1}\right), & k \leqslant 0, \\ -1, & k = 1, \\ 0, & k > 1. \end{cases}$$ (2.3.6)

Since $\int_{|x|=p^k}\chi(x)dx = \int_{|x|\leqslant p^k}\chi(x)dx - \int_{|x|\leqslant p^{k-1}}\chi(x)dx$, by the result in (2.3.4), it follows

$$\int_{B^{-k}}\chi(x)dx = \int_{|x|\leqslant p^k}\chi(x)dx = \begin{cases} p^k, & k \leqslant 0, \\ 0, & k > 0. \end{cases}$$

The equality (2.3.6) can be deduced since the right hand side in (2.3.5) is real value.

(iv) $\int_{|x|=p^k}\chi_\xi(x)dx = \begin{cases} p^k\left(1-p^{-1}\right), & k \leqslant -l, \\ -p^{-l}, & k = -l+1, \text{ with } |\xi| = p^l, \\ 0, & k > -l+1 \end{cases}$

and

$$\int_{|x|=p^k}\overline{\chi}_\xi(x)dx = \begin{cases} p^k\left(1-p^{-1}\right), & k \leqslant -l, \\ -p^{-l}, & k = -l+1, \text{ with } |\xi| = p^l. \\ 0, & k > -l+1 \end{cases}$$

Since

$$\int_{|x|=p^k}\chi_\xi(x)dx = \int_{|x|\leqslant p^k}\chi_\xi(x)dx - \int_{|x|\leqslant p^{k-1}}\chi_\xi(x)dx,$$

take substitution of variables, it follows

$$\int_{|x\xi|\leqslant p^{k+l}}\chi(x\xi)|\xi|dx - \int_{|x\xi|\leqslant p^{k+l-1}}\chi(x\xi)|\xi|dx$$

$$=p^l\left\{\int_{|\eta|\leqslant p^{k+l}}\chi(\eta)dx - \int_{|\eta|\leqslant p^{k+l-1}}\chi(\eta)dx\right\},$$

then we get formulas in (iv) by (2.3.4).

2.3.3 Integrals for some functions in K_p

We compute Haar integrals of some functions in this section.

(i) Let $f(x) = |x|^\alpha, x \in K_p$, evaluate the integral $\int_D f(x)dx = \int_{|x| \leqslant 1} f(x)dx$.

$$\int_{|x| \leqslant 1} |x|^\alpha dx = \sum_{j=0}^{+\infty} \int_{|x|=p^{-j}} |x|^\alpha dx$$

$$= \int_{|x|=1} |x|^\alpha dx + \int_{|x|=p^{-1}} |x|^\alpha dx + \cdots + \int_{|x|=p^{-j}} |x|^\alpha dx + \cdots$$

$$= \int_{|x|=1} 1 dx + \int_{|x|=p^{-1}} p^{-\alpha} dx + \cdots + \int_{|x|=p^{-j}} p^{-j\alpha} dx + \cdots$$

$$= \int_{|x|=1} dx + p^{-\alpha} \int_{|x|=p^{-1}} dx + \cdots + p^{-j\alpha} \int_{|x|=p^{-j}} dx + \cdots$$

$$= (1-p^{-1}) + p^{-\alpha} p^{-1}(1-p^{-1}) + \cdots + p^{-j\alpha} p^{-j}(1-p^{-1}) + \cdots$$

$$= (1-p^{-1})\left\{1 + p^{-\alpha-1} + \cdots + p^{-j\alpha-j} + \cdots\right\}$$

$$= (1-p^{-1})\left\{1 + p^{-(\alpha+1)} + \cdots + p^{-j(\alpha+1)} + \cdots\right\},$$

if $\alpha > -1$, then

$$\int_D |x|^\alpha dx = (1-p^{-1}) \frac{1}{1-p^{-(\alpha+1)}};$$

if $\alpha \leqslant -1$, then the integral dose not exist.

Similarly, evaluate $\int_{|x| \geqslant 1} |x|^\alpha dx$:

$$\int_{|x| \geqslant 1} |x|^\alpha dx = \sum_{j=0}^{+\infty} \int_{|x|=p^j} |x|^\alpha dx$$

$$= \int_{|x|=1} |x|^\alpha dx + \int_{|x|=p} |x|^\alpha dx + \cdots + \int_{|x|=p^j} |x|^\alpha dx + \cdots$$

$$= \int_{|x|=1} 1 dx + \int_{|x|=p} p^\alpha dx + \cdots + \int_{|x|=p^j} p^{j\alpha} dx + \cdots$$

$$= \int_{|x|=1} dx + p^\alpha \int_{|x|=p^1} dx + \cdots + p^{j\alpha} \int_{|x|=p^j} dx + \cdots$$

$$= (1-p^{-1}) + p^\alpha p^1 (1-p^{-1}) + \cdots + p^{j\alpha} p^j (1-p^{-1}) + \cdots$$

$$= (1-p^{-1})\left\{1 + p^{\alpha+1} + \cdots + p^{j(\alpha+1)} + \cdots\right\},$$

if $\alpha < -1$, then
$$\int_{|x|\geqslant 1} |x|^\alpha dx = (1-p^{-1})\frac{1}{1-p^{(\alpha+1)}};$$
if $\alpha \geqslant -1$, the integral dose not exist.

(ii) Let $f(x) = \ln\frac{1}{|x|}, x \in K_p\setminus\{0\}$, evaluate $\int_D f(x)dx = \int_{|x|\leqslant 1} f(x)dx$.

$$\int_{|x|\leqslant 1} \ln\frac{1}{|x|}dx = \sum_{j=0}^{+\infty} \int_{|x|=p^{-j}} \ln\frac{1}{|x|}dx = \sum_{j=1}^{+\infty} \int_{|x|=p^{-j}} \ln p^j dx$$
$$= \sum_{j=1}^{+\infty} (\ln p) \int_{|x|=p^{-j}} j dx = (\ln p) \sum_{j=1}^{+\infty} j \int_{|x|=p^{-j}} dx$$
$$= (\ln p) \sum_{j=1}^{+\infty} jp^{-j}(1-p^{-1}) = (\ln p)(1-p^{-1}) \sum_{j=1}^{+\infty} jp^{-j}$$
$$= (\ln p)p^{-1}(1-p^{-1})^{-1} = \frac{\ln p}{p(1-p^{-1})}.$$

Exercises

1. Prove Theorem 2.1.1, the property theorem of character group.
2. Consider the characters and character group of K_p^*, the multiplication group of local field K_p.
3. Consider the characters and character group of finite algebraic extension field $K_q, q = p^c, c \in \mathbb{N}$ of a local field K_p.
4. Prove Theorem 2.2.4.
5. Consider the Haar measures on finite algebraic extension field $K_q, q = p^c, c \in \mathbb{N}$ of K_p.
6. Compute the following integrals on K_q:

(i) Let $f(x) = |x|^\alpha, x \in K_q$, evaluate $\int_D f(x)dx = \int_{|x|\leqslant 1} f(x)dx$;

(ii) Let $f(x) = \ln\frac{1}{|x|}, x \in K_q\setminus\{0\}$, evaluate $\int_D f(x)dx = \int_{|x|\leqslant 1} f(x)dx$.

Chapter 3

Harmonic Analysis on Local Fields

Firstly, we introduce the Fourier analysis theory on local fields in this chapter, it is one of the main parts of harmonic analysis and fractal analysis on local fields. Then, a new kind calculus on local fields — p-type calculus is introduced by virtue of pseudo-differential operators. Finally, the other important content — operator theory and construction theory of function defined on local fields are presented.

3.1 Fourier analysis on a local field K_p

Let $f: K_p \to \mathbb{C}$ be a Haar measurable real or complex function on K_p.

$C \equiv C(K_p) = \{f : K_p \to \mathbb{C}, f \text{ is bounded and continuous}\}$;

$C_0 \equiv C_0(K_p) = \left\{f \in C(K_p) : \lim_{|x| \to +\infty} f(x) = 0\right\}$;

$C_C \equiv C_C(K_p) = \{f \in C(K_p) : \operatorname{supp} f \text{ is compact in } K_p\}$;

$L^r \equiv L^r(K_p), 1 \leqslant r \leqslant +\infty$, the set of r-power Haar integrable functions

$$L^r(K_p) = \left\{f : K_p \to \mathbb{C}, \int_{K_p} |f(x)|^r \, dx < +\infty\right\}, \quad 1 \leqslant r < +\infty,$$

$$L^\infty(K_p) = \left\{f : K_p \to \mathbb{C}, \operatorname*{esssup}_{x \in K_p} |f(x)| < +\infty\right\}.$$

Similar to that of $\mathbb{R}^n$, we may endow the linear operations, topological structures, such that they are complete normed linear spaces, for example, $L^r \equiv L^r(K_p), 1 \leqslant r \leqslant +\infty$, with norm

$$\|f\|_r \equiv \|f\|_{L^r(K_p)} = \begin{cases} \left\{\int_{K_p} |f(x)|^r dx < +\infty\right\}^{\frac{1}{r}}, & 1 \leqslant r < +\infty, \\ \operatorname*{esssup}_{x \in K_p} |f(x)|, & r = +\infty, \end{cases}$$

with $\operatorname*{esssup}_{x \in K_p} |f(x)| = \inf_{\mu e = 0} \sup_{x \in K_p \backslash e} |f(x)|$, then $L^r(K_p), 1 \leqslant r \leqslant +\infty$ becomes a Banach space.

3.1.1 L^1-theory

The Fourier analysis of L^1-theory[1],[9],[100] is introduced in this section.

Let K_p be a local field, and $\Gamma_p = \{\chi_\xi : \xi \in K_p\}$ the character group of K_p. By the Section 2.2, $\Gamma_p \xleftrightarrow{\text{isomorphic}} K_p$.

1. Fourier transformation of L^1-function

For $f \in L^1(K_p)$, the Fourier transformation of f is defined in abstract harmonic analysis as

$$(Ff)(\xi) \equiv (Ff)(\chi_\xi) = \int_{K_p} f(x)\overline{\chi}_\xi(x) dx, \quad \chi_\xi \in \Gamma_p,$$

where $Ff(\chi_\xi)$ is defined on character group Γ_p. However, as in the Chapter 2, we know that

$$\chi_\xi \in \Gamma_p \xleftrightarrow{\text{one-one, isom.}} \xi \in K_p,$$

thus we can substitute the character $\chi_\xi \in \Gamma_p$ by $\xi \in K_p$, and regard as $\xi \in \Gamma_p$, so the Fourier transformation of $f \in L^1(K_p)$ is defined as:

Definition 3.1.1 (Fourier transformation of L^1-function) The mapping $F : f \to f^\wedge$ for $f \in L^1(K_p)$ is defined by

$$f^\wedge(\xi) \equiv (Ff)(\xi) = \int_{K_p} f(x)\overline{\chi}_\xi(x) dx, \quad \xi \in \Gamma_p \qquad (3.1.1)$$

and $f^\wedge$ is said to be a **Fourier transformation of f**.

Example 3.1.1 Let $\alpha > -1$, evaluate the Fourier transformation $f^\wedge(\xi)$ of function

$$f(x) = \begin{cases} |x|^\alpha, & |x| \leqslant 1, \\ 0, & |x| > 1, \end{cases} \quad x \in K_p.$$

Solution. Let $|\xi| = p^l$, we have

$$f^\wedge(\xi) = \int_{K_p} f(x)\overline{\chi_\xi}(x)dx = \int_{|x|\leqslant 1}^{0} |x|^\alpha \overline{\chi_\xi}(x)dx$$

$$= \sum_{k=-\infty}^{0} \int_{|x|=p^k} |x|^\alpha \overline{\chi_\xi}(x)dx = \sum_{k=-\infty}^{0} p^{k\alpha} \int_{|x|=p^k} \overline{\chi_\xi}(x)dx = \sum_{k=-\infty}^{0} p^{k\alpha} I_k$$

with

$$I_k = \begin{cases} p^k(1-p^{-1}), & k \leqslant -l, \\ -p^{-l}, & k = -l+1, \\ 0, & k > -l+1, \end{cases}$$

then

$$f^\wedge(\xi) = \begin{cases} \dfrac{1-p^{-1}}{1-p^{-(1+\alpha)}}, & |\xi| \leqslant 1, \\ \dfrac{1-p^\alpha}{1-p^{-(1+\alpha)}} \cdot \dfrac{1}{|\xi|^{1+\alpha}}, & |\xi| > 1, \end{cases} \quad \xi \in \Gamma_p.$$

Example 3.1.2 Evaluate the Fourier transformation $f^\wedge(\xi)$ of function

$$f(x) = \begin{cases} \ln \dfrac{1}{|x|}, & |x| \leqslant 1, \\ 0, & |x| > 1, \end{cases} \quad x \in K_p.$$

Solution. By definition, it follows that for $\xi \in \Gamma_p$

$$f^\wedge(\xi) = \int_{K_p} f(x)\overline{\chi_\xi}(x)dx = \int_{|x|\leqslant 1} \ln\dfrac{1}{|x|}\overline{\chi_\xi}(x)dx = \begin{cases} \dfrac{\ln p}{1-p^{-1}} p^{-1}, & |\xi| \leqslant 1, \\ \dfrac{\ln p}{1-p^{-1}} \cdot \dfrac{1}{|\xi|}, & |\xi| > 1. \end{cases}$$

Example 3.1.3 Let K_p be a local field, $B^k = \{x \in K_p : |x| \leqslant p^{-k}\}$ the ball in K_p with center $0 \in K_p$ and radius p^{-k}, $k \in \mathbb{Z}$, as well as the characteristic function of B^k:

$$\Phi_{B^k}(x) = \begin{cases} 1, & x \in B^k, \\ 0, & x \in K_p \backslash B^k. \end{cases} \tag{3.1.2}$$

Evaluate $(\Phi_{B^k})^\wedge(\xi)$.

Solution. By formulas in the Section 2.3, we get for $k \in \mathbb{Z}$,

$$(\Phi_{B^k})^\wedge(\xi) = \int_{K_p} \Phi_{B^k}(x)\overline{\chi_\xi}(x)dx = \int_{B^k} \overline{\chi_\xi}(x)dx$$

$$= \int_{|x|\leqslant p^{-k}} \overline{\chi}(\xi x)\,dx = \int_{|\beta^{-k}x|\leqslant 1} \chi(-\xi x)\,dx$$

$$= p^{-k} \int_{|\eta| \leq 1} \chi\left(-\xi \beta^k \eta\right) d\eta = \begin{cases} p^{-k}, & \left|-\xi \beta^k \eta\right| \leq 1, \\ 0, & \left|-\xi \beta^k \eta\right| > 1. \end{cases}$$

Since $\left|\xi \beta^k \eta\right| = p^{-k} |\xi| |\eta| \leq 1 \Leftrightarrow |\xi| |\eta| \leq p^k$, thus

If $|\xi| \leq p^k$, then $|\xi| |\eta| \leq p^k$, so that $\chi\left(-\xi \beta^k \eta\right)$ is trivial when $\left|-\xi \beta^k \eta\right| \leq 1$, it follows $\int_{|\eta| \leq 1} \chi\left(-\xi \beta^k \eta\right) d\eta = 1$, hence $|\xi| \leq p^k \Rightarrow \left(\Phi_{B^k}\right)^\wedge (\xi) = p^{-k}$.

If $|\xi| > p^k$, then there exists $\eta_0 \in K_p$ with $|\eta_0| < 1$, such that $p^{-k} |\xi| |\eta_0| > 1$, and $\chi\left(-\xi \beta^k \eta\right)$ is non-trivial, hence the integral equals 0, thus, implies that

$$|\xi| > p^k \Rightarrow \left(\Phi_{B^k}\right)^\wedge (\xi) = 0.$$

Then we have

$$\left(\Phi_{B^k}\right)^\wedge (\xi) = \begin{cases} p^{-k}, & |\xi| \leq p^k \\ 0, & |\xi| > p^k \end{cases} = p^{-k} \Phi_{\Gamma^k}(\xi), \quad \xi \in \Gamma_p, k \in \mathbb{Z}.$$

Definition 3.1.2 (Translation, dilation, reflection) *For a complex function $f : K_p \to \mathbb{C}$ defined on a local field K_p, then*
(i) *Translation operator $\tau_h : f \to \tau_h f$, $h \in K_p$, defined by*
$$(\tau_h f)(x) = f(x - h), \quad x \in K_p.$$
(ii) *Dilation operator $\rho_s : f \to \rho_s f$, $s \in \mathbb{Z}$, defined by*
$$(\rho_s f)(x) = f(\beta^s x), \quad x \in K_p.$$
(iii) *Reflection operator $\sim : f \to \tilde{f}$, defined by*
$$\tilde{f}(x) = f(-x), \quad x \in K_p;$$
in the above definitions, the operations $x - h$, $\beta^s x$, $-x$ are those of in local field K_p.

Theorem 3.1.1 *For $f \in L^1(K_p)$, the Fourier transformations of translation τ_h, dilation ρ_s, reflection $\sim$ of f, the following formulas hold:*
(i) $(\tau_h f)^\wedge (\xi) = \overline{\chi}_h(\xi) f^\wedge(\xi)$, $\xi \in \Gamma_p$, $h \in K_p$;
$(\chi_h(\cdot) f(\cdot))^\wedge (\xi) = \tau_h f^\wedge(\xi)$, $\xi \in \Gamma_p$, $h \in K_p$;
(ii) $(\rho_s f)^\wedge (\xi) = p^{-s} \rho_{-s} f^\wedge(\xi)$, $\xi \in \Gamma_p$, $s \in \mathbb{Z}$;
(iii) $\left(\overline{\tilde{f}}\right)^\wedge (\xi) = \overline{f^\wedge}(\xi)$, $\left(\tilde{f}\right)^\wedge (\xi) = \widetilde{(f)^\wedge}(\xi)$, $\xi \in \Gamma_p$.

Proof. Prove the first one in (i):
$$(\tau_h f)^\wedge (\xi) = \int_{K_p} \tau_h f(x) \overline{\chi}_\xi(x) dx = \int_{K_p} f(x - h) \overline{\chi}_\xi(x) dx$$

$$= \int_{K_p} f(y)\overline{\chi_\xi}(y+h)\,dy = \int_{K_p} f(y)\overline{\chi_\xi}(y)\overline{\chi_\xi}(h)\,dy$$

$$= \overline{\chi_\xi}(h) \int_{K_p} f(y)\overline{\chi_\xi}(y)dy = \overline{\chi_\xi}(h) f^\wedge(\xi) = \overline{\chi_h}(\xi) f^\wedge(\xi).$$

The proof of the second one in (i):

$$(\chi_h(\cdot)f(\cdot))^\wedge(\xi) = \int_{K_p} \chi_h(x)f(x)\overline{\chi_\xi}(x)dx = \int_{K_p} f(x)\chi(-\xi x)\chi(hx)\,dx$$

$$= \int_{K_p} f(x)\chi(hx-\xi x)\,dx = \int_{K_p} f(x)\chi((h-\xi)x)\,dx$$

$$= \int_{K_p} f(x)\overline{\chi_{\xi-h}}(x)dx = f^\wedge(\xi - h) = (\tau_h f^\wedge)(\xi).$$

Next is to prove (ii), for $\forall s \in \mathbb{Z}$,

$$(\rho_s f)^\wedge(\xi) = \int_{K_p} (\rho_s f)(x)\overline{\chi_\xi}(x)dx = \int_{K_p} f(\beta^s x)\overline{\chi_\xi}(x)dx$$

$$= \int_{K_p} f(\beta^s x)\chi(-\xi x)\,dy = \int_{K_p} f(\eta)\chi\left(-\xi \beta^{-s}\eta\right) p^s d\eta$$

$$= p^s \int_{K_p} f(\eta)\overline{\chi_{\beta^{-s}\xi}}(\eta)\,d\eta = p^s f^\wedge(\beta^{-s}\xi) = p^s \rho_{-s} f^\wedge(\xi).$$

The proof of (iii) is left to exercise.

The characteristic function $\Phi_{B^k}(x)$ in Example 3.1.3 plays important role in the Fourier analysis on local fields, we study its properties now.

Theorem 3.1.2 *The translation* $\tau_h \Phi_{B^k}(x) = \Phi_{h+B^k}(x)$ *of* $\Phi_{B^k}(x)$ *is continuous on* K_p.

Proof. $\Phi_{h+B^k}(x)$ is the characteristic function of the ball $h + B^k$ in K_p with center $h \in K_p$, radius $p^{-k}, k \in \mathbb{Z}$, because

$$\Phi_{h+B^k}(x) = \Phi_{B^k}(x-h) = \tau_h \Phi_{B^k}(x) \equiv \tau_h \Phi_k(x), \tag{3.1.3}$$

and

$$\tau_h \Phi_{B^k}(x) = \Phi_{h+B^k}(x) = \begin{cases} 1, & x \in h + B^k, \\ 0, & x \notin h + B^k. \end{cases}$$

If $x_0 \in h + B^k$, by the Theorem 1.2.5 of positions of two balls in local fields, it follows that

$$x_0 + B^k \subset h + B^k.$$

Thus, if $x \in x_0 + B^k \subset h + B^k$, then $\forall \varepsilon > 0$,

$$|\tau_h \Phi_{B^k}(x) - \tau_h \Phi_{B^k}(x_0)| = |1 - 1| = 0 < \varepsilon;$$

if $x_0 \notin h + B^k$, by the Theorem 1.2.5, $\exists x_0 + B^s$, such that
$$x_0 + B^s \cap h + B^k = \varnothing,$$
hence, $\tau_h \Phi_{B^k}(x_0) = 0$. This shows that $\forall \varepsilon > 0$, $\forall x \in x_0 + B^s$, it follows that
$$|\tau_h \Phi_{B^k}(x) - \tau_h \Phi_{B^k}(x_0)| = |0 - 0| = 0 < \varepsilon.$$
The continuity of $\tau_h \Phi_{B^k}(x) = \Phi_{h+B^k}(x)$ is proved.

Theorem 3.1.3 *The Fourier transformation formulas of* $\Phi_{B^k}(x)$, $\tau_h \Phi_{B^k}$ *and* $\rho_s \Phi_{B^k}$ *hold*:

(i) $(\Phi_{B^k})^\wedge (\xi) = p^{-k} \Phi_{\Gamma^k}(\xi)$, $\xi \in \Gamma_p$, $k \in \mathbb{Z}$;

(ii) $(\tau_h \Phi_{B^k})^\wedge (\xi) = p^{-k} \overline{\chi_h}(\xi) \Phi_{\Gamma^k}(\xi)$, $\xi \in \Gamma_p$, $h \in K_p$, $k \in \mathbb{Z}$;

(iii) $(\rho_s \Phi_{B^k})^\wedge (\xi) = p^s \rho_{-s} (\Phi_{B^k})^\wedge (\xi)$, $\xi \in \Gamma_p$, $s \in \mathbb{Z}$, $k \in \mathbb{Z}$.

2. Test function class $\mathbb{S}(K_p)$

An important function class on local field K_p — the test function class $\mathbb{S}(K_p)$, is introduced[88],[100].

Definition 3.1.3 (Test function class $\mathbb{S}(K_p)$ on local field K_p)
The set
$$\mathbb{S}(K_p) = \left\{ \varphi : K_p \to \mathbb{C}, \varphi(x) = \sum_{j=1}^n c_j \tau_{h_j} \Phi_{B^{k_j}}(x), c_j \in \mathbb{C}, h_j \in K_p, k_j \in \mathbb{Z}, 1 \leqslant j \leqslant n \right\}$$

(3.1.4)

*is said to be a **test function class on local field** K_p, and an element $\varphi(x) \in \mathbb{S}(K_p)$ is said to be a **test function**.*

In fact, $\mathbb{S}(K_p)$ is the space of finite linear combinations of functions in the form $\tau_h \Phi_{B^k}$, $h \in K_p$, $k \in \mathbb{Z}$.

Similarly, the test function class $\mathbb{S}(\Gamma_p)$ on Γ_p can be defined also. However, since K_p is isomorphic to Γ_p, so $\mathbb{S}(K_p)$ and $\mathbb{S}(\Gamma_p)$ can be regarded as equivalent.

Theorem 3.1.4 *The test function class $\mathbb{S}(K_p)$ has the following properties*:

(i) $\mathbb{S}(K_p)$ *is an algebra that consists of finite linear combinations of those continuous functions in form* $\tau_h \Phi_{B^k}(x)$, $h \in K_p$, $k \in \mathbb{Z}$, *on K_p with compact support; and $\mathbb{S}(K_p)$ separates points of K_p;*

(ii) $\mathbb{S}(K_p)$ *is dense in $C_0(K_p)$ and in $L^r(K_p)$, $1 \leqslant r < +\infty$.*

Proof. For (i). Firstly, prove: $\forall \varphi(x) = \tau_h \Phi_{B^k}(x)$ is continuous with compact support.

In fact, for $k \in \mathbb{Z}$ by (3.1.2),
$$\Phi_{B^k}(x) = \begin{cases} 1, & x \in B^k, \\ 0, & x \in K_p \backslash B^k, \end{cases}$$
it has a compact support B^k.

Then, take $h \in K_p$: if $h \in B^k$, then $h + B^k = B^k$, so
$$\varphi(x) = \tau_h \Phi_{B^k}(x) = \Phi_{B^k}(x),$$
its support is B^k; if $h \in K_p \backslash B^k$, then
$$\tau_h \Phi_{B^k}(x) = \Phi_{B^k}(x - h) = \begin{cases} 1, & x \in h + B^k, \\ 0, & x \notin h + B^k, \end{cases}$$
it has compact support, clearly. The continuity of functions is obvious by Theorem 3.1.2.

Secondly, prove: $\mathbb{S}(K_p)$ separates points of K_p[3], i.e.,
$$\forall x_1, x_2 \in K_p, x_1 \neq x_2 \Rightarrow \exists \varphi \in \mathbb{S}(K_p), \text{ s.t. } \varphi(x_1) \neq \varphi(x_2).$$

In fact, $\forall x_1, x_2 \in K_p, x_1 \neq x_2, \exists k \in \mathbb{Z}$ by Theorem 1.2.5, such that
$$(x_1 + B^k) \cap (x_2 + B^k) = \varnothing.$$
This $\varphi = \tau_{x_1} \Phi_{B^k}(x)$ is a needed function(note, this φ is not unique).

Then, prove: $\mathbb{S}(K_p)$ is an algebra. i.e., the 3 operations on $\mathbb{S}(K_p)$, the addition $+$, the number product $\alpha\cdot, \alpha \in \mathbb{C}$, the multiplication $\times$, satisfy

① $\mathbb{S}(K_p)$ is a linear space in $+, \alpha\cdot$.
② $\mathbb{S}(K_p)$ is a commutative ring with unit in $+, \times$.
③ The combination law, $\alpha \cdot (\varphi_1 \times \varphi_2) = (\alpha \cdot \varphi_1) \times \varphi_2 = \varphi_1 \times (\alpha \cdot \varphi_2)$ in $\alpha\cdot, \times$ holds. These can be verified easily.

For (ii). By Stone–Weierstrass Theorem[3](*Suppose that X is a T_2-type compact topological space, and $C_C(X)$ is a Banach algebra of continuous functions with compact support, and unit $I \in C_C(X)$. Moreover, $A \subset C_C(X)$ is a sub-algebra and $I \in A$. Then $\overline{A} = C_C(X)$ if and only if A separates points of X*), take $A = \mathbb{S}(K_p)$, it follows that $\overline{\mathbb{S}(K_p)} = C_C(K_p)$.

Hence, we have
$$\overline{\mathbb{S}(K_p)} = C_0(K_p) \quad \text{and} \quad \overline{\mathbb{S}(K_p)} = L^r(K_p), \quad 1 \leqslant r < +\infty,$$
by $\overline{C_C(K_p)} = C_0(K_p)$ and $\overline{C_0(K_p)} = L^r(K_p)$.

Theorem 3.1.5 *The equivalent property holds for the test function class* $\mathbb{S}(K_p)$: $\forall \varphi \in \mathbb{S}(K_p)$ *if and only if* $\exists! (k,l) \in \mathbb{Z} \times \mathbb{Z}$ *(exists and unique), such that φ is constants on cosets of* B^k, *and* $\text{supp } \varphi = B^l$.

Proof. Necessity. Suppose that $\varphi \in \mathbb{S}(K_p)$ is

$$\varphi(x) = \sum_{j=1}^{n} c_j \tau_{h_j} \Phi_{B^{k_j}}(x) = \sum_{j=1}^{n} c_j \Phi_{B^{k_j}}(x - h_j),$$

where $c_j \in \mathbb{R}, h_j \in K_p, k_j \in \mathbb{Z}, j = 1, 2, \cdots, n$.

Let $h_j \in B^{r_j}$, and $|h_j| = p^{-r_j}$, $j = 1, 2, \cdots, n$. Thus, let $l = \min\{r_1, r_2, \cdots, r_n\}$, then

$$\text{supp } \varphi = B^l.$$

On the other hand, let $k = \max\{k_1, k_2, \cdots, k_n\}$, then

$$h_j + B^k \subset h_j + B^{k_j} \subset B^l, \qquad (*)$$

the first $\subset$ in the $(*)$ is since $k > h_j$; and the second one in $(*)$ is because $h_j \in B^{r_j} \subset B^l$.

Note, $(*)$ implies $k \geqslant l$. If we take the set of all cosets that B^k in B^l, i.e., $\{B^l/B^k\}$, then $(*)$ shows that $h_j + B^k \in \{B^l/B^k\}$, and the test function

$$\varphi(x) = \sum_{j=1}^{n} c_j \tau_{h_j} \Phi_{B^{k_j}}(x)$$

takes constants on these cosets.

Sufficiency. Suppose the support of $\varphi(x)$ is B^l, $\text{supp } \varphi = B^l$, and it is constants on the cosets of B^k. Clearly, $k \geqslant l$. We have

$$\{B^l/B^k\} = \{B^k, \beta^{k-1}+B^k, \cdots, (p-1)\beta^{k-1}+B^k; \cdots; \beta^l+B^k, \cdots, (p-1)\beta^l+B^k\},$$

and the coset $A_j \in \{B^l/B^k\}$ is

$$A_j = b_j + B^k, \quad j = 1, 2, \cdots, m$$

with $\varphi|_{A_j} = \alpha_j$, $j = 1, 2, \cdots, m$; m is a finite integer, $|b_j| = p^{-r_j}$, $b_j \in B^{r_j}$, $k \geqslant r_j \geqslant l$.

Thus, $\varphi(x)$ has form $\varphi(x) = \sum_{j=1}^{m} \alpha_j \Phi_{b_j+B^k}(x)$, the proof is complete.

Definition 3.1.4 (Index pair of $\varphi \in \mathbb{S}(K_p)$) *The integer index pair (k,l) in Theorem 3.1.5 is said to be an **index pair** of $\varphi \in \mathbb{S}(K_p)$.*

We discuss the operation structure and topological structure of $\mathbb{S}(K_p)$.

Endow the **addition operation** $+$:

$$\varphi, \psi \in \mathbb{S}(K_p) \Rightarrow (\varphi + \psi)(x) = \varphi(x) + \psi(x), \quad x \in K_p,$$

the **number product operation** $\alpha\cdot$:

$$\varphi \in \mathbb{S}(K_p),\ \alpha \in \mathbb{C} \Rightarrow (\alpha \cdot \varphi)(x) = \alpha\varphi(x),\quad x \in K_p,$$

then $\mathbb{S}(K_p)$ is a linear space.

Definition 3.1.5 (Topological structure on $\mathbb{S}(K_p)$) Endow the following topology τ to $\mathbb{S}(K_p)$: a sequence $\{\varphi_n\}_{n=1}^{+\infty} \subset \mathbb{S}(K_p)$ in $\mathbb{S}(K_p)$ is said to be **a null sequence** if
  (i) There exists the index pair (k, l) for all $\varphi_n(x) \in \mathbb{S}(K_p)$;
  (ii) $\lim_{n \to +\infty} \varphi_n(x) = 0$, uniformly for all $x \in K_p$.

Thus, $\mathbb{S}(K_p)$ is a topological linear space under above operations $+$, $\alpha \cdot$, and topology τ.

Theorem 3.1.6 *The test function class $\mathbb{S}(K_p)$ is a complete, T_2-type, separable, topological linear space under the above operations $+$, $\alpha \cdot$, and the topology τ.*

Proof. Complete property and T_2-type are clear.

Separable property: the subset of $\mathbb{S}(K_p)$:

$$\mathbb{S}_r(K_p) = \left\{ \varphi \in \mathbb{S}(K_p) : \varphi(x) = \sum_{j=1}^{n} c_j \tau_{h_j} \Phi_{B^{k_j}}(x),\ c_j \in \mathbb{Q} \right\}$$

is a countable set with coefficients c_j in rational number field $\mathbb{Q}$; and $\mathbb{S}_r(K_p)$ is dense in the space $\mathbb{S}(K_p)$, so $\mathbb{S}(K_p)$ is separable.

The following is about the Fourier transformation properties of $\mathbb{S}(K_p)$.

$\mathbb{S}(K_p) \subset L^1(K_p)$ by Theorem 3.1.4, thus the Fourier transformation of $\varphi \in \mathbb{S}(K_p)$ is defined by (3.1.1):

$$\varphi^{\wedge}(\xi) = \int_{K_p} \varphi(x) \overline{\chi}_\xi(x) dx,\quad \xi \in \Gamma_p.$$

Theorem 3.1.7 (i) *The Fourier transformation $\varphi^{\wedge}(\xi)$ of a test function $\varphi \in \mathbb{S}(K_p)$ is in the test function class $\mathbb{S}(\Gamma_p)$. And the Fourier transformation operator $F : \mathbb{S}(K_p) \to \mathbb{S}(\Gamma_p)$ is an one-one topological isomorphism from $\mathbb{S}(K_p)$ onto $\mathbb{S}(\Gamma_p)$.*

(ii) *If the index pair of $\varphi \in \mathbb{S}(K_p)$ is (k, l), then the index pair of $\varphi^{\wedge}$ is $(l, k) \in \mathbb{Z} \times \mathbb{Z}$, that is, $\varphi^{\wedge}$ is constant on cosets of $\Gamma^l(\leftrightarrow B^{-l})$, and is supported on Γ^k, $\mathrm{supp}\,\varphi^{\wedge} = \Gamma^k(\leftrightarrow B^{-k})$.*

Proof. For (i), we have $(\tau_h \Phi_{B^k})^{\wedge}(\xi) \in \mathbb{S}(\Gamma_p)$ by Theorem 3.1.3 (ii).

The mapping $F : \mathbb{S}(K_p) \to \mathbb{S}(\Gamma_p)$ is continuous by Theorem 3.1.2; it is one-one by Theorem 3.1.3 (i) and (ii).

For (ii), let the index pair of $\varphi \in \mathbb{S}(K_p)$ be (k,l). Firstly, if $\varphi(x) = \Phi_{B^k}(x)$, by Theorem 3.1.3 (i) and (ii), we have

$$(\Phi_{B^k})^{\wedge}(\xi) = p^{-k}\Phi_{\Gamma^k}(\xi)$$

and

$$(\tau_h \Phi_{B^k})^{\wedge}(\xi) = p^{-k}\overline{\chi}_h(\xi)\Phi_{\Gamma^k}(\xi),$$

the support of them is Γ^k. Then, the character $\overline{\chi}_h(\xi)$ is constants on $\{\Gamma^k/\Gamma^l\}$ by definition of character $\overline{\chi}_h(\xi) = \overline{\chi}(h\xi)$ and by Theorem 2.2.2.

Secondly, if $\varphi(x) = \sum_{j=1}^{n} c_j \tau_{h_j} \Phi_{B^{k_j}}(x)$, we get $\varphi^{\wedge} \in \mathbb{S}(\Gamma_p)$ since φ is a finite linear combinations of $\tau_{h_j} \Phi_{B^{k_j}}(x)$. Then take $k = \max\{k_1, k_2, \cdots, k_n\}$ and $l = \min\{r_1, r_2, \cdots, r_n\}$ with r_j, $|h_j| = p^{-r_j}$, $1 \leqslant j \leqslant n$, thus, the index of $\varphi^{\wedge}$ is (l, k). The proof is complete.

This Theorem can be diagramed by

$$F : \begin{bmatrix} \varphi \in \mathbb{S}(K_p) \\ (k,l) \end{bmatrix} \to \begin{bmatrix} \varphi^{\wedge} \in \mathbb{S}(\Gamma_p) \\ (l,k) \end{bmatrix}. \qquad (3.1.5)$$

3. Convolution operators, k-cutout operators

The convolution operator has important application background: suppose a linear system has an impulse response $h(t)$, and the input signal is $f(t)$, then what is the output signal $g(t)$?

$$f(t) \longrightarrow \boxed{h(t)} \longrightarrow g(t)$$

Since the input signal $f(t)$ has time-delay $t - \tau$, thus the output signal is

$$g(t) = f * h(t) = \int f(t - \tau) h(\tau) d\tau,$$

thus the idea of convolution plays an essential role in signal processing. Similar to the classical case, we define convolution operator on a local field K_p.

Definition 3.1.6 (Convolution operator) *Let $f, g : K_p \to \mathbb{C}$ be Haar measurable functions. The integral*

$$(f * g)(x) = \int_{K_p} f(x - z) g(z) dz = \int_{K_p} f(z) g(x - z) dz \qquad (3.1.6)$$

*is said to be a **convolution** of f and g.*

Note. The definition of convolution on K_p is similar to that of $\mathbb{R}^n$, but the operation "$-$" in (3.1.6) is the inverse operation of $+$ in K_p, and the integral $\int_{K_p}$ is the Haar integral on K_p.

For the convolution and its Fourier transformation, we has the following properties:

Theorem 3.1.8 Let $f \in L^1(K_p)$, $g \in L^r(K_p)$, $1 \leqslant r < +\infty$. Then

(i) $f * g \in L^r(K_p)$, and $\|f * g\|_{L^r(K_p)} \leqslant \|f\|_{L^r(K_p)} \|g\|_{L^1(K_p)}$;
(*closed property*)

(ii) $(f * g) * h = f * (g * h)$; (*associativity*)

(iii) $f * g = g * f$; (*commutation*)

(iv) $(f * g)^\wedge = f^\wedge g^\wedge$, for $r = 1$; (FT *of convolution*)

(v) $\int_{K_p} f(x) g^\wedge(x) dx = \int_{K_p} f^\wedge(x) g(x) dx$, $r = 1$. (*Parseval formula*)

Proof. For (i)$\sim$(iv), the proofs are clear by the technique of real analysis. For (v), it can be obtained by the Fubini Theorem on K_p.

Definition 3.1.7 (*k-cutout operator*) For $f \in L_{\text{loc}}(K_p)$ and $k \in \mathbb{Z}$, the operator Λ_k defined by

$$\Lambda_k f = \int_{K_p} f(x) \Phi_{B^{-k}}(x) dx = \int_{B^{-k}} f(x) dx, \qquad (3.1.7)$$

is said to be the *k*-**cutout operator**, simply, **cutout operator** Λ_k : $L_{\text{loc}}(K_p) \to \mathbb{C}$.

The motivation of k-cutout operator is: it provides some approximation properties of integrals of $g \in L_{\text{loc}}(K_p)$. It is similar to the Fredrich operators in the classical case.

Theorem 3.1.9 Let $f \in L^1(K_p)$. Then

(i) $\lim\limits_{k \to +\infty} \Lambda_k f = \int_{K_p} f(x) dx$;

(ii) $\Lambda_k (f^\wedge \chi_x) = \dfrac{1}{|B^k|} \int_{x+B^k} f(t) dt = p^k \int_{x+B^k} f(t) dt \equiv f_k(x)$;

(iii) $\lim\limits_{k \to +\infty} \Lambda_k (f^\wedge \chi_x) = f(x)$, a.e. $x \in K_p$.

Proof. For (i), by definition,

$$\Lambda_k f = \int_{K_p} f(x) \Phi_{B^{-k}}(x) dx = \int_{B^{-k}} f(x) dx,$$

it follows that
$$\lim_{k\to+\infty} \Lambda_k f = \lim_{k\to+\infty} \int_{B^{-k}} f(x)dx = \int_{K_p} f(x)dx.$$
For (ii), by
$$\Lambda_k\left(f^\wedge \chi_x\right) = \int_{K_p} f^\wedge(\xi)\chi_x(\xi)\Phi_{B^{-k}}(\xi)d\xi = \int_{K_p} f^\wedge(\xi)\left\{\chi_x(\xi)\Phi_{B^{-k}}(\xi)\right\}d\xi$$
and Theorem 3.1.8 (v), we have
$$\Lambda_k\left(f^\wedge \chi_x\right) = \int_{\Gamma_p} f(\xi)\left\{\chi_x(\cdot)\Phi_{B^{-k}}(\cdot)\right\}^\wedge(\xi)d\xi. \qquad (*)$$

Let $g(\eta) \equiv \chi_x(\eta)\Phi_{B^{-k}}(\eta)$. Then by the formula $(\chi_h(\cdot)f(\cdot))^\wedge(\xi) = (\tau_h f^\wedge)(\xi)$ in Theorem 3.1.1, and $(\Phi_{B^{-k}}(\cdot))^\wedge(\xi) = p^k\Phi_{\Gamma^{-k}}(\xi)$ in Theorem 3.1.3 (i), we have
$$(g(\cdot))^\wedge(\xi) = \left\{\chi_x(\cdot)\Phi_{B^{-k}}(\cdot)\right\}^\wedge(\xi) = \left(\tau_x\left\{\Phi_{B^{-k}}(\cdot)\right\}^\wedge\right)(\xi)$$
$$= \tau_x\left\{p^k\Phi_{\Gamma^{-k}}(\xi)\right\} = p^k\Phi_{\Gamma^{-k}}(\xi - x).$$

Substitute it into $(*)$, and by $\Phi_{\Gamma^{-k}} = \Phi_{B^k}$, it follows that
$$\Lambda_k\left(f^\wedge \chi_x\right) = \int_{K_p} f(\xi)p^k\Phi_{\Gamma^{-k}}(\xi - x)d\xi = p^k \int_{x+B^k} f(t)dt \equiv f_k(x).$$
(ii) is proved.

The line of proof of (iii) is: let
$$f_k(x) = p^k \int_{x+B^k} f(t)dt, \quad \varphi_k(x) = p^k \int_{x+B^k} \varphi(t)dt.$$
We have

① If $\varphi \in \mathbb{S}(K_p)$, then $\varphi_k(x) = \varphi(x)$ for some large $k \in \mathbb{N}$.

② If $f \in L_{\text{loc}}(K_p)$, then we only need to prove for $f \in L^1(K_p)$. Let
$$f_1(t) = (\tau_x\Phi_0)(t)f(t) = \Phi_0(t-x)f(t),$$
then $f_1 \in L^1(K_p)$.

③ For $f_1 \in L^1(K_p)$, $\forall \varepsilon > 0$, there exists $g \in \mathbb{S}(K_p)$, such that $\|f_1 - g\|_{L^1(K_p)} < \varepsilon$.

④ For $f \in L^1(K_p)$, then $f - f_k = (f-g) - (f-g)_k$, and
$$0 \leqslant \limsup|f(x) - f_k(x)| \leqslant |f(x) - g(x)| + \limsup|(f-g)_k|.$$

⑤ $\forall \delta > 0$, let
$$E = \{x \in K_p : \limsup|f(x) - f_k(x)| > \delta\},$$
then

$$E \subset \left\{ x \in K_p : |f(x) - g(x)| > \frac{\delta}{2} \right\} \cup \left\{ x \in K_p : \limsup |(f-g)_k(x)| > \frac{\delta}{2} \right\}$$
$$\equiv E_1 \cup E_2.$$

⑥ $|E_1| = \int_{K_p} \Phi_{E_1}(x) dx = \int_{E_1} 1 \cdot dx \leqslant \int_{K_p} \frac{2}{\delta} |f(x) - g(x)| dx \leqslant \frac{2}{\delta} \|f - g\|_{L^1(K_p)};$

$$|E_2| = \left| \left\{ x \in K_p : \limsup |(f-g)_k(x)| > \frac{\delta}{2} \right\} \right|$$
$$= \left| \left\{ x \in K_p : \limsup \left| p^k \int_{|x-z| \leqslant p^{-k}} (f-g)(z) dz \right| > \frac{\delta}{2} \right\} \right|$$
$$\leqslant \left| \left\{ x \in K_p : \sup_{k \in \mathbb{Z}} \left| p^k \int_{|x-z| \leqslant p^{-k}} (f-g)(z) dx \right| > \frac{\delta}{2} \right\} \right|$$
$$\leqslant \left| \left\{ x \in K_p : M(f-g)(x) > \frac{\delta}{2} \right\} \right| \leqslant \left(\frac{\delta}{2} \right)^{-1} \|f-g\|_{L^1(K_p)}$$
$$= \frac{2}{\delta} \|f - g\|_{L^1(K_p)},$$

then we get
$$|E| \leqslant |E_1| + |E_2| \leqslant \frac{4}{\delta} \|f - g\|_{L^1(K_p)} < \frac{4}{\delta} \varepsilon.$$

Fix δ, and let $\varepsilon \to 0$, it follows that $|E| = 0$, i.e., $\lim_{k \to +\infty} f_k(x) = f(x)$, a.e. $x \in K_p$; or
$$\lim_{k \to +\infty} \Lambda_k (f^\wedge \chi_x)(x) = f(x), \quad \text{a.e. } x \in K_p.$$

After proofs of the above 6 items, the proof of Theorem 3.1.9 (iii) is complete.

4. L^1-Fourier transformation and its inverse Fourier transformation

Theorem 3.1.10 *For L^1-Fourier transformation, the following analysis properties hold:*

(i) *The mapping $F : f \to f^\wedge$ is a linear bounded non-increasing norm transformation from $L^1(K_p)$ onto $L^\infty(K_p)$, that is,*

$(f + g)^\wedge \to f^\wedge + g^\wedge; \quad (\alpha f)^\wedge \to \alpha f^\wedge, \quad \alpha \in \mathbb{C};$ *(linearity)*

$\|f^\wedge\|_{L^\infty(\Gamma_p)} \leqslant \|f\|_{L^1(K_p)};$ *(boundedness, non-increasing norm)*

(ii) $f \in L^1(K_p)$, *then $f^\wedge(\xi)$ is an uniform continuous function on $\xi \in \Gamma_p$;* *(uniform continuity)*

(iii) $f \in L^1(K_p)$, then $\lim\limits_{|\xi| \to +\infty} f^\wedge(\xi) = 0$; (*Riemann–Lebesgue lemma*)

(iv) $f \in L^1(K_p), f^\wedge \in L^1(\Gamma_p)$, then
$$f(x) = \int_{K_p} f^\wedge(\xi) \chi_x(\xi) d\xi, \text{ a.e. } x \in K_p; \quad (\textit{inverse formula})$$

(v) $f, g \in L^1(K_p)$, $f^\wedge, g^\wedge \in L^1(\Gamma_p)$, then
$$\int_{K_p} f(x) \overline{g}(x) dx = \int_{\Gamma_p} f^\wedge(\xi) \overline{g^\wedge}(\xi) d\xi. \quad (\textit{multiplication formula})$$

Proof. Linear, bounded, non-increasing properties are clear.

To prove (ii), take $h \in \Gamma_p$, since
$$f^\wedge(\xi + h) - f^\wedge(\xi) = \int_{K_p} f(x) \overline{\chi}_{\xi+h}(x) dx - \int_{K_p} f(x) \overline{\chi}_\xi(x) dx$$
$$= \int_{K_p} f(x) \overline{\chi}_\xi(x) \overline{\chi}_h(x) dx - \int_{K_p} f(x) \overline{\chi}_\xi(x) dx$$
$$= \int_{K_p} f(x) \overline{\chi}_\xi(x) [\overline{\chi}_h(x) - 1] dx,$$

by virtue of $f \in L^1(k_p)$, $|\overline{\chi}_\xi(x)| = 1$, the continuity of $\overline{\chi}_h(x) = \overline{\chi}(hx)$, $\lim\limits_{h \to 0} \overline{\chi}_h(x) = 1$ and the Lebesgue dominated convergence theorem, it follows that for any $\xi \in \Gamma_p$,
$$\lim_{h \to 0} \{f^\wedge(\xi + h) - f^\wedge(\xi)\} = 0$$
uniformly holds. The proof of (ii) is complete.

For (iii), firstly, for $\varphi \in \mathbb{S}(K_p) \subset L^1(K_p)$, then $\varphi^\wedge \in \mathbb{S}(\Gamma_p)$, and $\varphi^\wedge$ has compact support, so that $\lim\limits_{|\xi| \to +\infty} \varphi^\wedge(\xi) = 0$.

Secondly, for $f \in L^1(K_p)$, then for each $\varepsilon > 0$, there exists $g_\varepsilon \in \mathbb{S}(K_p)$ by density, such that
$$\|f - g_\varepsilon\|_{L^1(K_p)} < \varepsilon.$$

Note that $\operatorname{supp}(\varphi_\varepsilon)^\wedge$ is compact, to prove $\lim\limits_{|\xi| \to +\infty} f^\wedge(\xi) = 0$ only need to consider for the case $x \notin \operatorname{supp} \varphi_\varepsilon$, thus
$$|f^\wedge(\xi)| = |(f - g_\varepsilon)^\wedge(\xi)| \leqslant \|(f - g_\varepsilon)^\wedge\|_{L^\infty(\Gamma_p)} \leqslant \|f - g_\varepsilon\|_{L^1(K_p)} \leqslant \varepsilon,$$
hence $\lim\limits_{|\xi| \to +\infty} f^\wedge(\xi) = 0$.

For (iv), since $f^\wedge \in L^1(\Gamma_p)$, for k-cutout operator

$$\Lambda_k\left(f^\wedge \chi_x\right)(x) = \int_{\Gamma_p} f^\wedge(\xi)\chi_x(\xi)\Phi_{\Gamma^k}(\xi)d\xi$$
$$= \overline{\int_{\Gamma_p} \overline{f^\wedge(\xi)}\overline{\Phi_{\Gamma^k}(\xi)}\overline{\chi_x}(\xi)d\xi},$$

let

$$G_k(\xi) \equiv \overline{f^\wedge(\xi)}\overline{\Phi_{\Gamma^k}(\xi)} = \overline{f^\wedge(\xi)}\Phi_{\Gamma^k}(\xi) \in L^1(\Gamma_p),$$

then we have

$$\Lambda_k\left(f^\wedge \chi_x\right)(x) = \overline{G_k^\wedge(x)}.$$

Thus, it follows that: for each $k \in \mathbb{N}$, the function $\Lambda_k\left(f^\wedge \chi_x\right)$ is uniformly continuous on $x \in K_p$ by (i), so $\lim_{k\to+\infty} \Lambda_k\left(f^\wedge \chi_x\right)(x)$ exists and continuous, and

$$\lim_{k\to+\infty} \Lambda_k\left(f^\wedge \chi_x\right)(x) = \lim_{k\to+\infty} \int_{\Gamma_p} f^\wedge(\xi)\chi_x(\xi)\Phi_{\Gamma^k}(\xi)d\xi$$
$$= \lim_{k\to+\infty} \int_{\Gamma^k} f^\wedge(\xi)\chi_x(\xi)d\xi = \int_{\Gamma_p} f^\wedge(\xi)\chi_x(\xi)d\xi.$$

On the other hand, if $f \in L^1(K_p)$, by Theorem 3.1.9 (iii), then

$$\lim_{k\to+\infty} \Lambda_k\left(f^\wedge \chi_x\right)(x) = f(x), \quad \text{a.e.,} \quad x \in K_p.$$

This proves (iv):

$$f(x) = \int_{\Gamma_p} f^\wedge(\xi)\chi_x(\xi)d\xi, \quad \text{a.e.,} \quad x \in K_p.$$

To prove (v), by Fubini theorem and (iv),

$$\int_{\Gamma_p} f^\wedge(\xi)\overline{g^\wedge}(\xi)dx = \int_{\Gamma_p} \left\{\int_{K_p} f(x)\overline{\chi_\xi}(x)dx\right\} \overline{g^\wedge}(x)dx$$
$$= \int_{K_p} f(x)\left[\overline{\int_{\Gamma_p} g^\wedge(\xi)\chi_x(\xi)d\xi}\right]dx = \int_{K_p} f(x)\overline{g}(x)dx.$$

The proof is complete.

Since $\mathbb{S}(K_p) \subset L^1(K_p)$, Theorem 3.1.10 holds for $\varphi \in \mathbb{S}(K_p)$.

To discuss inverse transformation of L^1-function, we consider functions in the test function class $\mathbb{S}(\Gamma_p)$ first.

Definition 3.1.8 (Inverse Fourier transformation of $\mathbb{S}(\Gamma_p)$ functions) *For $\psi \in \mathbb{S}(\Gamma_p)$,*

$$\psi^\vee(x) \equiv F^{-1}\psi(x) = \int_{\Gamma_p} \psi(\xi)\chi_x(\xi)d\xi, \quad x \in K_p \qquad (3.1.8)$$

is said to be the **inverse Fourier transformation** of ψ.

It follows that
$$F : \varphi \in \mathbb{S}(K_p) \to \varphi^\wedge \in \mathbb{S}(\Gamma_p)$$
and
$$F^{-1} : \psi \in \mathbb{S}(\Gamma_p) \to \psi^\vee \in \mathbb{S}(K_p).$$

Since $K_p \overset{\text{iso.}}{\longleftrightarrow} \Gamma_p$, so the Fourier transformation operator F and the inverse Fourier transformation operator F^{-1} both can be regarded from $\mathbb{S}(K_p)$ onto $\mathbb{S}(K_p)$. Thus
$$\varphi \in \mathbb{S}(K_p) \underset{F^{-1}}{\overset{F}{\rightleftarrows}} \varphi^\wedge \in \mathbb{S}(\Gamma_p) \overset{\text{iso.}}{\longleftrightarrow} \varphi^\wedge \in \mathbb{S}(K_p),$$
and
$$(\varphi^\wedge)^\vee = \varphi = (\varphi^\vee)^\wedge, \quad \forall \varphi \in \mathbb{S}(K_p).$$

Example 3.1.4 Let $\Phi_{\Gamma^k}(\xi) = \begin{cases} 1, & \xi \in \Gamma^k \\ 0, & \xi \in K_p \backslash \Gamma^k \end{cases}$. Evaluate $(\Phi_{\Gamma^k})^\vee(x)$.

By
$$(\Phi_{\Gamma^k})^\vee(x) = \int_{\Gamma_p} \Phi_{\Gamma^k}(\xi)\chi_x(\xi)d\xi = \overline{\int_{\Gamma_p} \overline{\Phi_{\Gamma^k}(\xi)}\overline{\chi_x(\xi)}dx}$$
$$= \overline{\int_{\Gamma^k} \overline{\chi_x(\xi)}d\xi} = \overline{p^k \Phi_{B^k}(x)} = p^k \Phi_{B^k}(x),$$

it follows that
$$(\Phi_{\Gamma^k})^\vee(x) = p^k \Phi_{B^k}(x), \quad x \in K_p.$$

Recall Theorem 3.1.7, the Fourier transformation $F : \mathbb{S}(K_p) \to \mathbb{S}(\Gamma_p)$ is an one-one, onto, isomorphic mapping. Then by the definitions of $F : \varphi \to \varphi^\wedge$ and $F^{-1} : \psi \to \psi^\vee$, the Theorem 3.1.7, the relationships $\varphi \in \mathbb{S}(K_p) \subset L^1(K_p)$ and $\varphi^\wedge \in \mathbb{S}(\Gamma_p) \subset L^1(\Gamma_p)$, as well as Theorem 3.1.10 (iv), it follows that
$$\varphi(x) = \int_{\Gamma_p} \varphi^\wedge(\xi)\chi_x(\xi)d\xi, \quad x \in K_p.$$

Moreover, the above inverse formula holds everywhere by the continuity of $\varphi(x)$. Then

$$\begin{bmatrix} \varphi \in \mathbb{S}(K_p) \\ (k,l) \end{bmatrix} \xrightarrow{F} \begin{bmatrix} \varphi^\wedge \in \mathbb{S}(\Gamma_p) \\ (l,k) \end{bmatrix} \xrightarrow{F^{-1}} \begin{bmatrix} \varphi = (\varphi^\wedge)^\vee \in \mathbb{S}(K_p) \\ (k,l) \end{bmatrix}. \quad (3.1.9)$$

Definition 3.1.9 (Inverse Fourier transformation of L^1-functions) For $g \in L^1(\Gamma_p)$,
$$g^\vee(x) \equiv F^{-1}g(x) = \int_{\Gamma_p} g(\xi)\chi_x(\xi)d\xi, \quad x \in K_p$$
is said to be the **inverse Fourier transformation** of g.

Note that, $K_p \underset{\text{iso.}}{\overset{1-1}{\longleftrightarrow}} \Gamma_p$, the inverse Fourier transformation can be rewritten as
$$g^\vee(x) = \int_{K_p} g(\xi)\chi_x(\xi)d\xi, \quad x \in K_p. \tag{3.1.10}$$

Example 3.1.5 Evaluate the inverse Fourier transformation of
$$g(\xi) = \begin{cases} \ln\frac{1}{|\xi|}, & 0 < |\xi| \leqslant 1, \\ 0, & |\xi| > 1, \end{cases} \quad \xi \in \Gamma_p.$$

Solution. Find $g^\vee(x)$ by Definition 3.1.9:
$$g^\vee(x) = \int_{\Gamma_p} g(\xi)\chi_x(\xi)d\xi = \int_{0<|\xi|\leqslant 1} \ln\frac{1}{|\xi|}\chi_x(\xi)d\xi$$
$$= \begin{cases} \dfrac{\ln p}{1-p^{-1}}\dfrac{1}{p}, & |x| \leqslant 1, \\ \dfrac{\ln p}{1-p^{-1}}\dfrac{1}{|x|}, & |x| > 1, \end{cases} \quad x \in K_p.$$

Some inverse formulas for the Fourier transformation are in the following theorem.

Theorem 3.1.11 *We have*

(i) $f(x) \in L^1(K_p)$ *implies that there exists* $g \in C(K_p)$, *such that*
$$f(x) = g(x), \quad a.e. \ x \in K_p;$$
that is, $f(x) \in L^1(K_p)$ *equals a continuous function, for a.e.* $x \in K_p$;

(ii) $f \in L^1(K_p)$ *and* $f^\wedge \in L^1(\Gamma_p)$ *imply that the inverse formula of Fourier transformation holds*:
$$f(x) = \int_{K_p} f^\wedge(\xi)\chi_x(\xi)d\xi, \quad a.e. \ x \in K_p;$$
that is, $f, f^\wedge \in L^1 \Rightarrow (f^\wedge)^\vee = f$; *specially*, $\varphi, \varphi^\wedge \in \mathbb{S}(K_p) \Rightarrow (\varphi^\wedge)^\vee = \varphi$;

(iii) $f \in L^1(K_p), f^\wedge \geqslant 0$, *and* f *is continuous at* $x = 0$, *then* $f^\wedge \in L^1(\Gamma_p)$; *thus the inverse formula of Fourier transformation holds*:
$$f(x) = \int_{\Gamma_p} f^\wedge(\xi)\chi_x(\xi)d\xi, \quad a.e. \ x \in K_p;$$

in particular, $f(0) = \int_{\Gamma_p} f^\wedge(\xi)d\xi$.

Proof. (i) By $\overline{C(K_p)} = L^1(K_p)(C(K_p)$ is dense in $L^1(K_p))$; (ii) is the Theorem 3.1.10 (iv).

To prove (iii), we claim that: "$f \in L^1(K_p)$, $f^\wedge \geqslant 0$, and f is continuous at $x = 0$" implies "$f^\wedge \in L^1(\Gamma_p)$".

In fact, $f \in L^1(K_p) \Rightarrow f^\wedge \in C(\Gamma_p)$ by Theorem 3.1.10 (i), and $f^\wedge$ is continuous uniformly on K_p. Moreover, since $f^\wedge \geqslant 0$, by the Fatou lemma, Theorem 3.1.9 (iii), and by the continuity at $x = 0$ of $f(x)$, it can be got

$$0 \leqslant \int_{\Gamma_p} |f^\wedge(\xi)| d\xi = \int_{\Gamma_p} f^\wedge(\xi)d\xi = \int_{\Gamma_p} \lim_{k \to +\infty} f^\wedge(\xi)\Phi_{\Gamma^k}(\xi)d\xi$$
$$\leqslant \lim_{k \to +\infty} \int_{\Gamma_p} f^\wedge(\xi)\Phi_{\Gamma^k}(\xi)d\xi = \lim_{k \to +\infty} \int_{\Gamma_p} f^\wedge(\xi)\chi_0(\xi)\Phi_{\Gamma^k}(\xi)d\xi,$$

where $\chi_0(\xi) \equiv 1$. Then, by the definition of the cutout operator Λ_k

$$\int_{\Gamma_p} f^\wedge(\xi)\chi_0(\xi)\Phi_{\Gamma^k}(\xi)d\xi = \Lambda_k(f^\wedge\chi_0),$$

hence,

$$0 \leqslant \int_{\Gamma_p} |f^\wedge(\xi)| d\xi \leqslant \lim_{k \to +\infty} \Lambda_k(f^\wedge\chi_0) = \lim_{k \to +\infty} \Lambda_k(f^\wedge\chi_0) = f(0) < +\infty,$$

this implies $f^\wedge \in L^1(\Gamma_p)$. Then, (iii) is proved by Theorem 3.1.10 (iv).

Thus, the proof of Theorem 3.1.11 is complete.

The uniqueness theorem of the Fourier transformation is as follows.

Theorem 3.1.12 *If $f, g \in L^1(K_p)$, and $f^\wedge = g^\wedge$, then $f = g$, a.e.*

Proof. We deduce by Theorem 3.1.11:

$$f^\wedge = g^\wedge \Rightarrow (f - g)^\wedge = 0,$$

then,

$$f(x) - g(x) = \int_{\Gamma_p} (f^\wedge - g^\wedge)(\xi)\chi_x(\xi)d\xi$$
$$= \int_{\Gamma_p} (f - g)^\wedge(\xi)\chi_x(\xi)d\xi = 0, \quad \text{a.e.}$$

The proof is complete.

5. H–L maximal operator M

Definiton 3.1.10 (Hardy–Littlewood maximal operator M) *If $g \in L_{\text{loc}}(K_p)$, the operator M is defined by*

$$Mg(x) = \sup_{k \in \mathbb{Z}} \frac{1}{p^{-k}} \int_{x+B^k} |g(z)| \, dz = \sup_{\substack{x \in S \\ S: \text{ball in } K_p}} \frac{1}{|S|} \int_S |g(t)| \, dt,$$

and $M : g \to Mg$ is said to be the **Hardy–Littlewood maximal operator**.

We display a covering lemma without proof.

Theorem 3.1.13[100] **(Wiener covering lemma)** If $E \subset K_p$ is a measurable subset with finite Haar measure, and $\{S_\alpha\}$ is a covering of E by spheres, that is, $\bigcup_\alpha S_\alpha \supset E$, where S_α are balls in K_p, then for any given $\eta, 0 < \eta < 1$, there exists a finite sub-collection $\{S_k\}_{k=1}^N$ with mutually disjoint spheres, such that $\sum_{k=1}^N |S_k| > \eta |E|$.

Hardy-Littlewood maximal operators have the following property.

Theorem 3.1.14 If $f \in L^1(K_p)$, $\lambda > 0$, then

$$|\{x \in K_p : Mf(x) > \lambda\}| \leq \frac{1}{\lambda} \|f\|_{L^1(K_p)}, \qquad (3.1.11)$$

that is, the H–L maximal operator M is a **w-(1, 1) type operator**.

Proof. For $\lambda > 0$, let $E = \{x \in K_p : Mf(x) > \lambda\}$. Choose any measurable subset E_1 of E with finite Haar measure, i.e., $E_1 \subset E$, $|E_1| < +\infty$. $\forall x \in E_1$, there exists S_x with $x \in S_x$, and by $f \in L^1(K_p)$, $\lambda > 0$, we get

$$\int_{K_p} |f(t)| \, dt > \lambda |S_x|. \qquad (3.1.12)$$

Thus, $\forall x \in E_1$, we have $Mf(x) = \sup\limits_{\substack{x \in S \\ S: \text{ball in } K_p}} \frac{1}{|S|} \int_S |f(t)| \, dt > \lambda$. So there exists S_x, such that

$$\frac{1}{|S_x|} \int_{S_x} |f(t)| \, dt > \lambda. \qquad (3.1.13)$$

Then, the set $\{S_x\}_{x \in E_1}$ satisfies the conditions of the covering Theorem 3.1.14, $\bigcup_{x \in E_1} S_x \supset E_1$.

Hence it follows that, given $\eta > 0$, $0 < \eta < 1$, there exists a finite sub-collection

$$\{S_1, S_2, \cdots, S_N\} \subset \{S_x\}_{x \in E_1}$$

with mutually disjoint spheres and

$$\sum_{k=1}^N |S_k| > \eta |E_1|.$$

Combining (3.1.12) and (3.1.13), it follows

$$\eta |E_1| < \sum_{k=1}^{N} |S_k| < \sum_{k=1}^{N} \frac{1}{\lambda} \int_{S_k} |f| \leq \frac{\|f\|_{L^1(K_y)}}{\lambda}.$$

Let $\eta \to 1$, then

$$|E_1| \leq \frac{\|f\|_{L^1(K_y)}}{\lambda}. \tag{3.1.14}$$

To prove (3.1.11), for $f \in L^1(K_p)$, suppose there exists $\lambda_0 > 0$, such that

$$|E_0| = |\{x \in K_p : Mf(x) > \lambda_0\}| = +\infty.$$

Then, $\forall \lambda > \lambda_0$, the relation $\{x \in K_p : Mf(x) > \lambda\} \supset \{x \in K_p : Mf(x) > \lambda_0\}$ implies

$$|\{x \in K_p : Mf(x) > \lambda\}| \geq |\{x \in K_p : Mf(x) > \lambda_0\}| = +\infty.$$

On the other hand, $\forall N_0 > 0$, there exists a ball S_x large enough with $x \in E_0$ and $|S_x| > N_0$, as well as $|f(x)| > \lambda_0$ on S_x (otherwise, it would be $\forall x \in E_0$ implies $|f(x)| \leq \lambda_0$ so that $M|f|(x) \leq \lambda_0$, and it contradicts $x \in E_0$), hence

$$\|f\|_1 > \int_{S_x} |f| > \lambda_0 |S_x| > \lambda_0 N_0,$$

contradicts $f \in L^1(K_p)$. Then

$$|E| = |\{x \in K_p : Mf(x) > \lambda\}| < +\infty.$$

Since $E_1 \subset E$ in (3.1.14) is arbitrary subset of E with finite measure, so that (3.1.14) holds for $E_1 = E$, thus $|E| \leq \dfrac{\|f\|_{L^1(K_y)}}{\lambda}$, (3.1.11) is proved.

3.1.2 L^2-theory

We introduce L^2-theory of Fourier analysis[1],[9],[100] in this section.

1. Fourier transformation of L^2-function

Theorem 3.1.15 If $f \in L^1(K_p) \cap L^2(K_p)$, then $\|f^\wedge\|_{L^2(\Gamma_p)} = \|f\|_{L^2(K_p)}$.

Proof. For $f \in L^1(K_p) \cap L^2(K_p)$, set

$$g(x) = \overline{f(-x)},$$

then $g \in L^1(K_p) \cap L^2(K_p)$, and

$$g^\wedge(\xi) = \int_{K_p} \overline{f(-x)} \overline{\chi_\xi}(x)dx = \overline{\int_{K_p} f(-x) \chi_\xi(x)dx}$$
$$= \overline{\int_{K_p} f(t)\chi_\xi(-t)dt} = \overline{\int_{K_p} f(t)\overline{\chi_\xi}(t)dt} = \overline{f^\wedge(\xi)}.$$

Thus, by Theorem 3.1.8 (i) and (iv), we have
$$f, g \in L^1(K_p) \Rightarrow f * g \in L^1(K_p)$$
$$\Rightarrow (f * g)^\wedge = f^\wedge g^\wedge = f^\wedge \overline{f^\wedge} = |f^\wedge|^2 \geq 0. \quad (3.1.15)$$

On the other hand, we assert that
$$f, g \in L^2(K_p) \Rightarrow f * g \in C(K_p). \quad (3.1.16)$$

In fact, by the Holder inequality,
$$|f * g(x+y) - f * g(x)|$$
$$= \left| \int_{K_p} [f(x+y-z) - f(x-z)] g(z)dz \right|$$
$$\leq \left\{ \int_{K_p} |f(x+y-z) - f(x-z)|^2 dz \right\}^{\frac{1}{2}} \left\{ \int_{K_p} |g(z)|^2 dz \right\}^{\frac{1}{2}}$$
$$= \|f(\cdot + y) - f(\cdot)\|_2 \|g\|_2.$$

Then, $\|f(\cdot + y) - f(\cdot)\|_{L^2(K_p)} \|g\|_{L^2(K_p)} = o(1)$, $y \to 0$, by the L^2-continuity. This implies that
$$\lim_{y \to 0} \{f * g(x+y) - f * g(x)\} = 0,$$
thus the continuity of $f * g$ is proved.

By virtue of (3.1.15) and (3.1.16), $f * g$ has the following properties
① $(f * g)^\wedge = |f^\wedge|^2 \geq 0$;
② $f * g \in L^1(K_p)$;
③ $f * g$ is continuous.

Then, $(f * g)^\wedge \in L^1(\Gamma_p)$ by Theorem 3.1.12 (iii), and
$$(f * g)(0) = \int_{K_p} g(0-z) f(z) dz = \int_{K_p} \overline{f(-(0-z))} f(z) dz$$
$$= \int_{K_p} \overline{f(z)} f(z) dz = \int_{K_p} |f(z)|^2 dz.$$

Since $g^\wedge(z) = \overline{f^\wedge(z)}$, then

$$(f * g)(0) = \int_{\Gamma_p} (f * g)^\wedge (z) dz = \int_{\Gamma_p} f^\wedge(z) g^\wedge(z) dz$$
$$= \int_{\Gamma_p} f^\wedge(z) \overline{f^\wedge(z)} dz = \int_{\Gamma_p} |f^\wedge(z)|^2 dz,$$

that is $\|f^\wedge\|_2 = \|f\|_2$, the proof of theorem is complete.

We define the Fourier transformation of $f \in L^2(K_p)$.

Definition 3.1.11 (Fourier transformation of L^2-function) Let $f \in L^2(K_p)$, and let

$$F_k(x) = f(x) \Phi_{B^{-k}}(x).$$

Then $F_k(x)$ is said to be k-*cut function* of $f(x)$, $k \in \mathbb{Z}$.

Clearly, $F_k \in L^1(K_p) \cap L^2(K_p)$. Then

$$(F_k)^\wedge(\xi) = \int_{K_p} F_k(x) \overline{\chi_\xi}(x) dx = \int_{|x| \leqslant p^k} f(x) \overline{\chi_\xi}(x) dx.$$

If there exists a function $F \in L^2(\Gamma_p)$, such that

$$\lim_{k \to +\infty} \left\{ \int_{\Gamma_p} |F(\xi) - (F_k)^\wedge(\xi)|^2 d\xi \right\}^{\frac{1}{2}} = 0,$$

denoted by

$$F(\xi) = \underset{k \to +\infty}{\overset{(2)}{\text{l.i.m.}}} (F_k)^\wedge(\xi),$$

where $\underset{k \to +\infty}{\overset{(2)}{\text{l.i.m.}}}$ is the L^2-*limit in mean* as $k \to +\infty$, then F is said to be the **Fourier transformation** of $f \in L^2(K_p)$, denoted by $F = f^\wedge$, that is,

$$f^\wedge(\xi) = \underset{k \to +\infty}{\overset{(2)}{\text{l.i.m.}}} (F_k)^\wedge(\xi) = \underset{k \to +\infty}{\overset{(2)}{\text{l.i.m.}}} \int_{|x| \leqslant p^k} f(x) \overline{\chi_\xi}(x) dx, \quad x \in K_p \quad (3.1.17)$$

is the Fourier transformation of $f \in L^2(K_p)$. Clearly, the Fourier transformation $f^\wedge \in L^2(\Gamma_p)$ of $f \in L^2(K_p)$ satisfies

$$\lim_{k \to +\infty} \left\| f^\wedge(\xi) - \int_{|x| \leqslant p^k} f(x) \overline{\chi_\xi}(x) dx \right\|_{L^2(\Gamma_p)} = 0.$$

Theorem 3.1.16 *The L^2-Fourier transformation of $f \in L^2(K_p)$ has the following operator properties:*

(i) The mapping $F: f \to f^\wedge$ is linear from $L^2(K_p)$ to $L^2(\Gamma_p)$, i.e.,
$$(f+g)^\wedge \to f^\wedge + g^\wedge; \quad (\alpha f)^\wedge \to \alpha f^\wedge, \quad \alpha \in \mathbb{C}.$$

(ii) The L^2-Fourier transformation of $f \in L^2(K_p)$ has the norm preserved property, i.e.,
$$\|f^\wedge\|_{L^2(\Gamma_p)} = \|f\|_{L^2(K_p)}.$$

(iii) The mapping $F: f \to f^\wedge$ is isomorphism from $L^2(K_p)$ to $L^2(\Gamma_p)$, i.e., onto, one-one, bi-continuous.

(iv) If $f, g \in L^2(K_p)$, then the **Parseval formula** holds:
$$\int_{K_p} f^\wedge(x) g(x) dx = \int_{K_p} f(x) g^\wedge(x) dx.$$

(v) The L^2-Fourier transformation on $L^2(K_p)$ is unitary, i.e., the **multiplication formula**
$$\int_{\Gamma_p} f^\wedge(\xi) \overline{g^\wedge(\xi)} d\xi = \int_{K_p} f(x) \overline{g(x)} dx.$$

holds, in which the left side and the right side are the inner products of the spaces $L^2(\Gamma_p)$ and $L^2(K_p)$, respectively, $(f^\wedge, g^\wedge) = (f, g)$.

Proof. For (i), the linear property of $L^2(K_p)$-Fourier transformation is clear.

To prove (ii), for the norm-preserved of L^2-Fourier transformation $F: f \to f^\wedge$ on K_p, by Theorem 3.1.15 for F_k, we have
$$\|(F_k)^\wedge\|_{L^2(\Gamma_p)} = \|F_k\|_{L^2(K_p)}.$$

Moreover,
$$\|(F_k)^\wedge - f^\wedge\|_{L^2(\Gamma_p)} \to 0 \quad \text{and} \quad \|F_k - f\|_{L^2(K_p)} \to 0,$$

hence, $\|f^\wedge\|_{L^2(\Gamma_p)} = \|f\|_{L^2(K_p)}$.

The proof of (iii), we prove $F: f \to f^\wedge$ is one-one, firstly, i.e., prove
$$\left(L^2(K_p)\right)^\wedge = \{f^\wedge : f \in L^2(K_p)\}.$$

① The relationship $\left(L^2(K_p)\right)^\wedge \subset L^2(\Gamma_p)$ is clear by the definition of L^2-Fourier transformation;

② To prove $\left(L^2(K_p)\right)^\wedge \supset L^2(\Gamma_p)$, let $\left(L^2(K_p)\right)^\wedge \subsetneqq L^2(\Gamma_p)$, then we assert that: "$\exists g \in L^2(\Gamma_p) \setminus \left(L^2(K_p)\right)^\wedge$, $g \neq 0$, $\|g\|_{L^2(K_p)} \neq 0$, but no any $f \in L^2(K_p)$ such that $g = f^\wedge$ holds, then $\int_{\Gamma_p} \overline{g} f^\wedge = 0$ holds for $\forall f \in L^2(K_p)$".

Otherwise, $\exists f \in L^2(K_p)$, such that $g = f^\wedge$, then $0 \neq \int_{\Gamma_p} g\bar{g} = \int_{\Gamma_p} f^\wedge \overline{f^\wedge} = \|f^\wedge\|^2_{L^2(\Gamma_p)}$. However, $0 = \int_{\Gamma_p} \bar{g} f^\wedge = \int_{\Gamma_p} \bar{g} g$, this contradiction gives the assertion.

On the other hand, since $0 = \int_{\Gamma_p} \bar{g} f^\wedge = \int_{K_p} (\bar{g})^\wedge f$ implies
$$\int_{K_p} (\bar{g})^\wedge f = 0, \quad \forall f \in L^2(K_p).$$

By the uniqueness of the Haar integral, we have $\overline{g^\wedge} = (\bar{g})^\wedge = 0$. However, this contradicts $\|g^\wedge\|_{L^2(\Gamma_p)} = \|g\|_{L^2(K_p)} \neq 0$, thus
$$\left(L^2(K_p)\right)^\wedge = L^2(\Gamma_p),$$

then $F : f \to f^\wedge$ is onto.

The mapping $F : f \to f^\wedge$ is one-one, bi-continuous, see Exercise.

(iv) $f, g \in L^2(K_p)$ implies that: as $k \to +\infty$, we have
$$\|F_k - f\|_{L^2(K_p)} = \left(\int_{K_p} |f(x)\Phi_{B^{-k}}(x) - f(x)|^2 \, dx\right)^{\frac{1}{2}} = \left(\int_{|x| > p^k} |f(x)|^2 \, dx\right)^{\frac{1}{2}} \to 0;$$

and as $j \to +\infty$,
$$\|G_j - g\|_{L^2(K_p)} = \left(\int_{K_p} |g(x)\Phi_{B^{-j}}(x) - g(x)|^2 \, dx\right)^{\frac{1}{2}} = \left(\int_{|x| \triangleright p^j} |g(x)|^2 \, dx\right)^{\frac{1}{2}} \to 0,$$

where F_k and G_j are the k-cut and j-cut functions of f and g, respectively. Moreover,
$$F_k \in L^1(K_p) \cap L^2(K_p) \Rightarrow (F_k)^\wedge \in L^2(\Gamma_p) \Rightarrow \left\|(F_k)^\wedge - f^\wedge\right\|_{L^2(\Gamma_p)} \to 0,$$
$$G_j \in L^1(K_p) \cap L^2(K_p) \Rightarrow (G_j)^\wedge \in L^2(\Gamma_p) \Rightarrow \left\|(G_j)^\wedge - g^\wedge\right\|_{L^2(\Gamma_p)} \to 0.$$

By Theorem 3.1.8 (v),
$$\int_{K_p} F_k(x)(G_j)^\wedge(x) dx = \int_{K_p} (F_k)^\wedge(x) G_j(x) dx. \tag{3.1.18}$$

Then, fix k in (3.1.18), for the left side
$$\left|\int_{K_p} F_k(x)(G_j)^\wedge(x) dx - \int_{K_p} F_k(x) g^\wedge(x) dx\right|$$
$$= \left|\int_{K_p} F_k(x) \left[(G_j)^\wedge(x) - g^\wedge(x)\right] dx\right|$$

$$\leqslant \int_{K_p} |F_k(x)\,[(G_j)^\wedge(x) - g^\wedge(x)]|\,dx$$

$$\leqslant \|F_k\|_{L^2(K_p)} \|(G_j)^\wedge - g^\wedge\|_{L^2(\Gamma_p)},$$

and let $j \to +\infty$,

$$\lim_{j\to+\infty} \int_{K_p} F_k(x)\,(G_j)^\wedge(x)\,dx = \int_{K_p} F_k(x) g^\wedge(x)\,dx; \qquad (3.1.19)$$

Moreover, for the right side of (3.1.18), as $j \to +\infty$, we have

$$\lim_{j\to+\infty} \int_{K_p} (F_k)^\wedge(x) G_j(x)\,dx = \int_{K_p} (F_k)^\wedge(x) g(x)\,dx; \qquad (3.1.20)$$

(3.1.19) and (3.1.20) imply that

$$\int_{K_p} F_k(x) g^\wedge(x)\,dx = \int_{K_p} (F_k)^\wedge(x) g(x)\,dx.$$

Let $k \to +\infty$, we get

$$\int_{K_p} f(x) g^\wedge(x)\,dx = \int_{K_p} f^\wedge(x) g(x)\,dx.$$

The proof of (iv) is complete.

Note that, the proof of (iv) can be done by $f, g \in L^2(K_p) \Rightarrow f \cdot g \in L^1(K_p)$ and by the Fubini Theorem.

(v) L^2-Fourier transformation is unitary, since by $\left(L^2(K_p)\right)^\wedge = L^2(\Gamma_p)$, we can take $g \in L^2(K_p)$ with $\overline{g} = (g_1)^\wedge$, then

$$(f, g) = \int_{K_p} f\overline{g} = \int_{K_p} f(g_1)^\wedge = \int_{K_p} f^\wedge g_1 = \int_{\Gamma_p} f^\wedge \overline{g^\wedge} = (f^\wedge, g^\wedge).$$

(Here we use the relationship "$\overline{g} = (g_1)^\wedge \Rightarrow g_1 = \overline{g^\wedge}$", see Exercise 6). The proof is complete.

2. Inverse Fourier transformation of L^2-function

Definition 3.1.12 (Inverse Fourier transformation of L^2-function)
For $f \in L^2(K_p)$, then $f^\wedge \in L^2(\Gamma_p)$, moreover, the mapping

$$F : f \to f^\wedge \qquad (3.1.21)$$

is an one-one isomorphism from $L^2(K_p)$ onto $L^2(\Gamma_p)$; and the inverse mapping F^{-1} of F exists for $g \in L^2(\Gamma_p)$

$$F^{-1} : g \to g^\vee, \qquad (3.1.22)$$

where $g^\vee \in L^2(K_p)$ is said to be the **inverse Fourier transformation** of $g \in L^2(\Gamma_p)$ for $g \in L^2(\Gamma_p)$.

Thus, except the properties of operators of L^2-Fourier transformation in Theorem 3.1.16, we have the following operation properties of L^2-Fourier transformation:

Theorem 3.1.17 *The L^2-Fourier transformation of $f \in L^2(K_p)$ has the following properties*:

(i) $(\tau_h f)^\wedge(\xi) = \overline{\chi}_h(\xi) f^\wedge(\xi)$, $(\chi_h f)^\wedge(\xi) = \tau_h f^\wedge(\xi)$, $\xi \in \Gamma_p$, $h \in K_p$;

(ii) $(\rho_s f)^\wedge(\xi) = p^{-s} \rho_{-s} f^\wedge(\xi)$, $\xi \in \Gamma_p$, $s \in \mathbb{Z}$;

(iii) $\left(\overline{\tilde{f}}\right)^\wedge(\xi) = \overline{f^\wedge(\xi)}$, $\overline{\left(\tilde{f}\right)^\wedge}(\xi) = (\overline{f})^\wedge(\xi), \xi \in \Gamma_p$;

(iv) $f \in L^2(K_p)$, $g \in L^1(K_p)$, then $(f * g)^\wedge(\xi) = f^\wedge(\xi) g^\wedge(\xi), \xi \in \Gamma_p$;

(v) $f \in L^2(K_p)$, $f^\wedge \in L^1(\Gamma_p)$, then $f(x) = \int_{\Gamma_p} f^\wedge(\xi) \chi_x(\xi) d\xi$, a.e., $x \in K_p$.

The proofs of the Theorem 3.1.17 are left as exercise.

Theorem 3.1.18 *The inverse Fourier transformation of $f \in L^2(K_p)$ has the form*:

$$f^\wedge(\xi) = \lim_{k \to +\infty} \int_{|x| \leqslant p^k} f(x) \overline{\chi}_\xi(x) dx, \ \xi \in \Gamma_p, \quad a.e., \quad (3.1.23)$$

or

$$f^\wedge(\xi) = \lim_{k \to +\infty} \int_{K_p} f(x) \Phi_{B^{-k}}(x) \overline{\chi}_\xi(x) dx, \ \xi \in \Gamma_p, \quad a.e. \quad (3.1.24)$$

Proof. Denote $\Phi_{-k} \equiv \Phi_{B^{-k}}$, evaluate for $\xi \in \Gamma_p$ as

$$\int_{|x| \leqslant p^k} f(x) \overline{\chi}_\xi(x) dx$$

$$= \int_{K_p} f(x) \Phi_{-k}(x) \overline{\chi}_\xi(x) dx$$

$$= \int_{K_p} f(x) \overline{\Phi_{-k}(x) \chi_\xi(x)} dx$$

$$= \int_{\Gamma_p} f^\wedge(\eta) \overline{[\Phi_{-k}(\cdot) \chi_\xi(\cdot)]^\wedge(\eta)} d\eta$$

(by Theorem 3.1.16 (v))

$$= \int_{\Gamma_p} f^\wedge(\eta) \overline{[\tau_\xi \Phi^\wedge_{-k}(\eta)]} d\eta \qquad (\Phi^\wedge_{-k}(\eta) = p^k \Phi_k(\eta))$$

$$= p^k \int_{\Gamma_p} f^\wedge(\eta) \overline{[\tau_\xi \Phi_k(\eta)]} d\eta \qquad (\overline{\tau_\xi \Phi_k(\eta)} = \tau_\xi \Phi_k(\eta))$$

$$= p^k \int_{\Gamma_p} f^\wedge(\eta) \tau_\xi \Phi_k(\eta) d\eta$$

$$= p^k \int_{\Gamma_p} f^\wedge(\eta) \Phi_k(\eta - \xi) d\eta$$

$$= p^k \int_{|\eta - \xi| \leq p^{-k}} f^\wedge(\eta) d\eta = (f^\wedge)_k(\xi).$$

Since $f \in L^2(K_p) \Rightarrow f^\wedge \in L^2(\Gamma_p) \Rightarrow f^\wedge \in L_{\mathrm{loc}}(\Gamma_p)$, so by definition

$$(f^\wedge)_k(\xi) \to f^\wedge(\xi), \quad \xi \in \Gamma_p, \quad \text{a.e.}$$

we get (3.1.23), (3.1.24).

3.1.3 L^r-Theory, $1 < r < 2$

We introduce the theory of the Fourier analysis of $L^r(K_p)$, $1 < r < 2$.

1. The Fourier transformation of $L^r(K_p)(1 < r < 2)$-function

For $f \in L^r(K_p)$, $1 < r < 2$, the Fourier transformation theory is based on the theory of Fourier transformation of $r = 1$, $r = 2$.

For any $f \in L^r(K_p)$, take a constant $\alpha > 0$, let

$$f^\alpha = \begin{cases} f, & |f| > \alpha, \\ 0, & |f| \leq \alpha, \end{cases} \quad f_\alpha = \begin{cases} 0, & |f| > \alpha, \\ f, & |f| \leq \alpha. \end{cases} \tag{3.1.25}$$

Then, $f = f^\alpha + f_\alpha$, and $f^\alpha \in L^1(K_p) \cap L^r(K_p)$, $f_\alpha \in L^2(K_p) \cap L^r(K_p)$, since

$$f = f^\alpha + f_\alpha, \quad |f|^r = |f_\alpha|^r + |f^\alpha|^r.$$

Moreover,

$$|f^\alpha| \leq |f| \Rightarrow f^\alpha \in L^r(K_p), \quad |f^\alpha| \leq (\alpha)^{-(r-1)}|f|^r \Rightarrow f^\alpha \in L^1(K_p);$$

$$|f_\alpha| \leq |f| \Rightarrow f_\alpha \in L^r(K_p), \quad |f_\alpha|^2 \leq \alpha^{2-r}|f_\alpha|^r \Rightarrow f_\alpha \in L^2(K_p).$$

Thus, we have the decomposition for an $f \in L^r(K_p)$ as $f = f^\alpha + f_\alpha$, and let

$$L^r(K_p) = L^1(K_p) + L^2(K_p)$$
$$= \{f \in L^r(K_p) : f = f^\alpha + f_\alpha, f^\alpha \in L^1(K_p), f_\alpha \in L^2(K_p)\},$$

then we may define the L^r-Fourier transformation.

Definition 3.1.13 (Fourier transformation of L^r-function) For $f \in L^r(K_p)$, $1 < r < 2$, take functions in (3.1.25), $f^\alpha \in L^1(K_p)$, $f_\alpha \in L^2(K_p)$, such that $f = f^\alpha + f_\alpha$, and define the Fourier transformation of an L^r-function f as

$$f^\wedge = (f^\alpha)^\wedge + (f_\alpha)^\wedge, \qquad (3.1.26)$$

where $(f^\alpha)^\wedge$ is in L^1-FT sense, and $(f_\alpha)^\wedge$ is in L^2-FT sense.

Theorem 3.1.19 *The L^r-Fourier transformation $F : f \to f^\wedge$ of $f \in L^r(K_p)$ is a linear, non-increase norm operator from $L^r(K_p)$ to $L^{r'}(\Gamma_p)$, with $\dfrac{1}{r} + \dfrac{1}{r'} = 1$, $1 < r < 2$.*

Proof. ① L^r-Fourier transformation is determined uniquely. Let

$$f = f_1 + f_2, \quad f_1 \in L^1(K_p), \quad f_2 \in L^2(K_p); \qquad (3.1.27)$$
$$f = g_1 + g_2, \quad g_1 \in L^1(K_p), \quad g_2 \in L^2(K_p). \qquad (3.1.28)$$

By (3.1.27) and (3.1.28), we have

$$f_1 + f_2 = f = g_1 + g_2,$$

that is $f_1 - g_1 = g_2 - f_2$. Then,

$$f_1 - g_1 \in L^1(K_p), \quad g_2 - f_2 \in L^2(K_p),$$

this shows that $f_1 - g_1 \in L^1(K_p) \cap L^2(K_p)$, $g_2 - f_2 \in L^1(K_p) \cap L^2(K_p)$. Clearly,

$$(f_1 - g_1)^\wedge \in L^2(K_p), \quad (g_2 - f_2)^\wedge \in L^2(K_p).$$

The uniqueness of $f^\wedge$ can be obtained by

$$f_1 - g_1 = g_2 - f_2 \Rightarrow (f_1 - g_1)^\wedge = (g_2 - f_2)^\wedge$$
$$\Rightarrow (f_1)^\wedge - (g_1)^\wedge = (g_2)^\wedge - (f_2)^\wedge$$
$$\Rightarrow (f_1)^\wedge + (f_2)^\wedge = (g_2)^\wedge + (g_1)^\wedge$$
$$\Rightarrow (f_1 + f_2)^\wedge = (g_1 + g_2)^\wedge.$$

$F : f \to f^\wedge$ is linear, clearly.

② L^r-Fourier transformation has a non-increasing norm. By

$$f \in L^1(K_p) \Rightarrow f^\wedge \in L^\infty(\Gamma_p) \Rightarrow \|f^\wedge\|_{L^\infty(\Gamma_p)} \leqslant \|f\|_{L^1(K_p)};$$
$$f \in L^2(K_p) \Rightarrow f^\wedge \in L^2(\Gamma_p) \Rightarrow \|f^\wedge\|_{L^2(\Gamma_p)} = \|f\|_{L^2(K_p)}.$$

Thus, the operator $F : f \to f^\wedge$ is $L^1(K_p) \to L^\infty(K_p)$, $L^2(K_p) \to L^2(\Gamma_p)$, continuous linear operator, and is s-$(1, \infty)$ type, s-$(2, 2)$ type, respectively, so by Riesz-Thorin convex theorem[65], it follows that $F : f \to f^\wedge$ is $L^r(K_p) \to L^{r'}(\Gamma_p)$, non-increasing norm: for $1 < r < 2$,

$$f \in L^r(K_p) \Rightarrow f^\wedge \in L^{r'}(\Gamma_p) \Rightarrow \|f^\wedge\|_{L^{r'}(\Gamma_p)} \leqslant \|f\|_{L^r(K_p)}. \qquad (3.1.29)$$

The proof is complete.

Remark. The definition L^r-Fourier transformation $f^\wedge$ of $f \in L^r(K_p)$ in Definition 3.1.13 is well-defined by Theorem 3.1.19. And the formula (3.1.29) holds for $1 \leqslant r < 2$.

2. Properties of $L^r(K_p)(1 < r < 2)$ Fourier transformation

Theorem 3.1.20 *The Fourier transformation of $f \in L^r(K_p)$, $1 < r < 2$, has the following properties*:

(i) $(\tau_h f)^\wedge(\xi) = \overline{\chi_h} f^\wedge(\xi)$, $(\chi_h f)^\wedge(\xi) = \tau_h f^\wedge(\xi)$, $\xi \in \Gamma_p$, $h \in K_p$;

(ii) $(\rho_s f)^\wedge(\xi) = p^{-s} \rho_{-k} f^\wedge(\xi)$, $\xi \in \Gamma_p$, $s \in \mathbb{Z}$;

(iii) $\left(\bar{\tilde{f}}\right)^\wedge(\xi) = \overline{f^\wedge(\xi)}$, $\left(\tilde{f}\right)^\wedge(\xi) = \overline{(\bar{f})^\wedge(\xi)}$, $\xi \in \Gamma_p$;

(iv) $f \in L^r(K_p)$, $g \in L^1(K_p)$, then $(f * g)^\wedge(\xi) = f^\wedge(\xi) g^\wedge(\xi)$, $\xi \in \Gamma_p$;

(v) $f, g \in L^r(K_p)$, then $\int_{K_p} f g^\wedge dx = \int_{K_p} f^\wedge g dx$;

(vi) $f \in L^r(K_p)$, $g \in L^1(K_p)$, $g^\wedge \in L^1(\Gamma_p)$, then $\int_{K_p} f \bar{g} = \int_{\Gamma_p} f^\wedge \overline{g^\wedge}$;

(vii) $f \in L^r(K_p)$, $f^\wedge \in L^1(\Gamma_p)$, then $f(x) = \int_{\Gamma_p} f^\wedge(\xi) \chi_x(\xi) d\xi$, a.e. $x \in K_p$.

The proofs of the Theorem 3.1.20 are left as exercise.

3.1.4 Distribution theory on K_p

1. Basic knowledge of distribution theory

To complete the Fourier analysis theory on local fields, we introduce the distribution theory on K_p[3],[88]. Similar to that of on $\mathbb{R}^n$, the distribution theory are indispensable part of Fourier analysis on local fields, because we have the properties of Fourier transformation $F : f \to f^\wedge$ for $L^r(K_p) \to L^{r'}(\Gamma_p)$ with $1 \leqslant r \leqslant 2$:

$$f \in L^r(K_p) \Rightarrow f^\wedge \in L^{r'}(\Gamma_p), \quad \frac{1}{r} + \frac{1}{r'} = 1,$$

$$\|f^\wedge\|_{L^{r'}(\Gamma_p)} \leqslant \|f\|_{L^r(K_p)},$$

but it is not true for $r > 2$, so that we need distribution theory for completing this gap, on one hand; and on the other hand, we have to deal with some signals appeared in physics and computer science, such as Dirac distribution δ, and have to evaluate "Fourier transformation" of δ.

To study the Distribution theory on K_p, we emphasize that: certain definitions and properties of the test function class $\mathbb{S}(K_p)$ are groundwork and basic knowledge of distribution theory, for example, definitions of Fourier transformation $\varphi^\wedge$ of $\varphi \in \mathbb{S}(K_p)$, inverse Fourier transformation $\psi^\vee$ of $\psi \in \mathbb{S}(\Gamma_p)$; Theorem 3.1.7 in which certain analytic properties of $\varphi \in \mathbb{S}(K_p)$ are listed. Following Theorem 3.1.21 shows some useful operations on $\mathbb{S}(K_p)$, for instance, translation, dilation, reflection, multiplication, convolution, and so on, they are closed operations onto $\mathbb{S}(K_p)$ itself. As well as the following summarized Theorem 3.1.22 of some important operation properties of test functions in $\mathbb{S}(K_p)$.

Theorem 3.1.21 *If $\varphi, \psi \in \mathbb{S}(K_p)$, then the translation $\tau_h \varphi$, dilation $\rho_s \varphi$, reflection $\tilde{\varphi}$, multiplication $\varphi \cdot \psi$, convolution $\varphi * \psi$, all belong to $\mathbb{S}(K_p)$. Moreover, in the test function class $\mathbb{S}(K_p)$, the mappings $\varphi \to \tau_h \varphi, \varphi \to \rho_s \varphi, \varphi \to \tilde{\varphi}$ are also isomorphic from $\mathbb{S}(K_p)$ onto $\mathbb{S}(K_p)$.*

Theorem 3.1.22 *If $\varphi \in \mathbb{S}(K_p)$, then we have*

(i) $(\tau_h \varphi(\cdot))^\wedge (\xi) = \overline{\chi}_h(\xi) \varphi^\wedge(\xi), \xi \in \Gamma_p, h \in K_p$,

where $(\tau_h \varphi)(x) = \varphi(x - h), x, h \in K_p$;   (*FT of translation*)

(ii) $\tau_h \varphi^\wedge(\xi) = [\chi_h(\cdot) \varphi(\cdot)]^\wedge (\xi), \xi \in \Gamma_p, h \in K_p$;   (*translation of FT*)

(iii) $[\rho_s \varphi(\cdot)]^\wedge (\xi) = p^{-s} [\rho_{-s} \varphi^\wedge](\xi), \xi \in K_p, s \in \mathbb{Z}$,

where $(\rho_s \varphi)(x) = \varphi(p^s x), x \in K_p$;   (*FT of dilation*)

(iv) $(\tilde{\varphi}(\cdot))^\wedge (\xi) = \varphi^\vee(\xi), \xi \in K_p$;   (*FT of reflection*)

(v) $(\overline{\varphi})^\wedge (\xi) = \overline{(\tilde{\varphi}(\cdot))^\wedge}(\xi), \xi \in K_p$;   (*FT of conjugate*)

(vi) $(\varphi^\wedge)^\vee (x) = \varphi(x), x \in K_p$;

$(\psi^\vee)^\wedge (\xi) = \psi(\xi), \xi \in \Gamma_p$;   (*IFT formula*)

(vii) $(\varphi * \psi(\cdot))^\wedge (\xi) = \varphi^\wedge(\xi) \psi^\wedge(\xi), \xi \in K_p$;   (*FT of convolution*)

(viii) $\int_{K_p} \varphi(x) \psi^\wedge(x) dx = \int_{K_p} \varphi^\wedge(x) \psi(x) dx$;   ($x \in K_p \leftrightarrow \xi \in \Gamma_p$)

(*Parseval formula*)

(ix) $\int_{K_p} \varphi(x) \overline{\psi}(x) dx = \int_{\Gamma_p} \varphi^\wedge(\xi) \overline{\psi^\wedge}(\xi) d\xi$.   (*multiplication formula*)

2. Distributions on $\mathbb{S}(K_p)$, distribution space $\mathbb{S}^*(K_p)$

Definition 3.1.14 (Distribution on $\mathbb{S}(K_p)$) *A continuous linear functional $T : \mathbb{S}(K_p) \to \mathbb{C}$ is said to be a **distribution** on $\mathbb{S}(K_p)$; The collection of all distributions on $\mathbb{S}(K_p)$ is said to be the **distribution space** on $\mathbb{S}(K_p)$, denoted by*

$$\mathbb{S}^*(K_p) = \{T : K_p \to \mathbb{C}, \ T \text{ is a continuous linear functional on } K_p\}.$$

In the distribution space $\mathbb{S}^*(K_p), \forall T \in \mathbb{S}^*(K_p)$ acts on $\varphi \in \mathbb{S}(K_p)$ denoted by

$$\langle T, \varphi \rangle, \tag{3.1.30}$$

where $\langle T, \varphi \rangle \in \mathbb{C}$.

Endow the w^*-topology on $\mathbb{S}^*(K_p)$: for $T_n \in \mathbb{S}^*(K_p)$,

$$T_n \xrightarrow{\mathbb{S}^*(K_p)} 0 \Leftrightarrow \langle T_n, \varphi \rangle \xrightarrow{\mathbb{C}} 0, \quad \forall \varphi \in \mathbb{S}(K_p). \tag{3.1.31}$$

Thus, the distribution space $\mathbb{S}^*(K_p)$ is a topological space under the w^*-topology.

Further, we define "addition" and "scalar product" on $\mathbb{S}^*(K_p)$. For $T, S \in \mathbb{S}^*(K_p), \alpha \in \mathbb{C}$, we have

(i) **Addition.** $T + S$ is defined as a distribution $T + S \in \mathbb{S}^*(K_p)$ by

$$\langle T + S, \varphi \rangle = \langle T, \varphi \rangle + \langle S, \varphi \rangle, \quad \forall \varphi \in \mathbb{S}(K_p).$$

(ii) **Scalar product.** αT is defined as a distribution $\alpha T \in \mathbb{S}^*(K_p)$ by

$$\langle \alpha T, \varphi \rangle = \alpha \langle T, \varphi \rangle = \langle T, \alpha \varphi \rangle, \quad \forall \varphi \in \mathbb{S}(K_p).$$

Hence, $\mathbb{S}^*(K_p)$ becomes a linear space on $\mathbb{C}$. And under the w^*-topology, the addition (i) and scalar product (ii) are continuous. Thus, $\mathbb{S}^*(K_p)$ is a topological linear space under the w^*-topology.

The following is examples in $\mathbb{S}^*(K_p)$.

Example 3.1.6 $L_{\text{loc}}(K_p) \subset \mathbb{S}^*(K_p)$, where $f \in L_{\text{loc}}(K_p)$ is a function which is Haar integrable at any Haar measurable set $E \subset K_p$ with finite Haar measure $\mu(E) < +\infty$.

Proof. $\forall f \in L_{\text{loc}}(K_p)$, the integral

$$\langle T, \varphi \rangle \equiv \langle f, \varphi \rangle = \int_{K_p} f(x)\varphi(x)dx, \quad \forall \varphi \in \mathbb{S}(K_p)$$

determines a linear functional. It is continuous, since take $\varphi_n \in \mathbb{S}(K_p)$ with
$$\varphi_n \xrightarrow{\mathbb{S}(K_p)} \varphi,$$
let $\sup_n |\varphi_n(x)| = \psi(x)$, then $\psi \in L^1(K_p)$. By Lebesgue dominated convergence theorem, it follows
$$\langle f, \varphi_n \rangle \xrightarrow{\mathbb{C}} \langle f, \varphi \rangle.$$
So that $f \in L_{\text{loc}}(K_p) \Rightarrow T = f \in \mathbb{S}^*(K_p)$, i.e., $L_{\text{loc}}(K_p) \subset \mathbb{S}^*(K_p)$.

Example 3.1.7 $L^r(K_p) \subset \mathbb{S}^*(K_p)$, $1 \leqslant r < +\infty$.

Proof. It is true since $L^r(K_p) \subset L_{\text{loc}}(K_p)$.

Example 3.1.8 Dirac distribution $\delta \in \mathbb{S}^*(K_p)$.

Proof. Dirac distribution δ is defined by
$$\langle \delta, \varphi \rangle = \varphi(0), \quad \forall \varphi \in \mathbb{S}(K_p). \tag{3.1.32}$$
It is clear that $\delta \in \mathbb{S}^*(K_p)$.

Definition 3.1.15 (Regular distribution and singular distribution) *A distribution $T \in \mathbb{S}^*(K_p)$ is said to be a **regular distribution**, if there exists a locally Haar integrable function $f \in L_{\text{loc}}(K_p)$ such that a distribution can be expressed by*
$$\langle T, \varphi \rangle = \int_{K_p} f(x) \varphi(x) dx, \quad \forall \varphi \in \mathbb{S}(K_p).$$
*Otherwise, $T \in \mathbb{S}^*(K_p)$ is said to be a **singular distribution**.*

Thus, the distribution $\delta \in \mathbb{S}^*(K_p)$ is a singular distribution.

3. Fourier transformation on the distribution space $\mathbb{S}^*(K_p)$

Definition 3.1.16 (Fourier transformation on $\mathbb{S}^*(K_p)$) *The Fourier transformation of a distribution $T \in \mathbb{S}^*(K_p)$ is defined as the distribution $T^\wedge \in \mathbb{S}^*(\Gamma_p)$ satisfying*
$$\langle T^\wedge, \varphi \rangle = \langle T, \varphi^\wedge \rangle, \quad \forall \varphi \in \mathbb{S}(\Gamma_p). \tag{3.1.33}$$
Or, equivalently, satisfying
$$\langle T^\wedge, \varphi^\vee \rangle = \langle T, \varphi \rangle, \quad \forall \varphi \in \mathbb{S}(K_p). \tag{3.1.34}$$

*The **inverse Fourier transformation** of $S \in \mathbb{S}^*(\Gamma_p)$ is defined as the distribution $S^\vee \in \mathbb{S}^*(K_p)$ satisfying*
$$\langle S^\vee, \psi \rangle = \langle S, \psi^\vee \rangle, \quad \forall \psi \in \mathbb{S}(K_p). \tag{3.1.35}$$

Or, equivalently, satisfying
$$\langle S^\vee, \psi^\wedge \rangle = \langle S, \psi \rangle, \quad \forall \psi \in \mathbb{S}(\Gamma_p). \tag{3.1.36}$$

Example 3.1.9 The Fourier transformation of Dirac $\delta \in \mathbb{S}^*(K_p)$.

Solution. The Fourier Transformation $\delta^\wedge$ of the Dirac distribution $\delta \in \mathbb{S}^*(K_p)$ can be evaluated: by (3.1.33), $\forall \varphi \in \mathbb{S}(K_p)$, we have

$$\langle \delta^\wedge, \varphi \rangle = \langle \delta, \varphi^\wedge \rangle = \varphi^\wedge(0) = \int_{K_p} \varphi(x)\overline{\chi_0}(x)dx = \int_{K_p} \varphi(x)dx = \langle 1, \varphi \rangle.$$

Thus $\delta^\wedge = 1$.

The Fourier transformation on $\mathbb{S}^*(K_p)$ has important analysis properties.

Theorem 3.1.23 *The Fourier transformation $F: T \to T^\wedge$ on $\mathbb{S}^*(K_p)$ and the inverse Fourier transformation $F^{-1}: S \to S^\vee$ on $\mathbb{S}^*(\Gamma_p)$ have the following analysis properties:*

(i) F and F^{-1} are one-one, linear mappings from $\mathbb{S}^*(K_p)$ onto $\mathbb{S}^*(\Gamma_p)$, and satisfying $(T^\wedge)^\vee = T = (T^\vee)^\wedge$.

(ii) The Fourier transformation $F: T \to T^\wedge$ and the inverse transform $F^{-1}: T \to T^\vee$ are homeomorphous (topological isomorphic) from $\mathbb{S}^*(K_p)$ onto $\mathbb{S}^*(\Gamma_p)$.

Proof. Since the Fourier transformation $F: \varphi \to \varphi^\wedge$ from test function class $\mathbb{S}(K_p)$ onto $\mathbb{S}(\Gamma_p)$ is homeomorphous (topological isomorphic) by Theorem 3.1.7, so does Fourier transformation $F: T \to T^\wedge$ from $\mathbb{S}^*(K_p)$ to $\mathbb{S}^*(\Gamma_p)$. Similar for F^{-1}. Moreover,

$$\left\langle (T^\wedge)^\vee, \varphi \right\rangle = \langle (T^\wedge), \varphi^\vee \rangle = \left\langle T, (\varphi^\vee)^\wedge \right\rangle = \langle T, \varphi \rangle;$$

$$\left\langle (T^\vee)^\wedge, \varphi \right\rangle = \langle (T^\vee), \varphi^\wedge \rangle = \left\langle T, (\varphi^\wedge)^\vee \right\rangle = \langle T, \varphi \rangle,$$

the conclusions of theorem are proved.

4. Operations on distribution space $\mathbb{S}^*(K_p)$

We define translation, dilation, reflection, multiplication with functions on distribution space $\mathbb{S}^*(K_p)$.

Definition 3.1.17 (Operations on $\mathbb{S}^*(K_p)$) (i) **Translation.** For $T \in \mathbb{S}^*(K_p)$, $h \in K_p$, the translation $\tau_h T$ of T is defined as a distribution $\tau_h T \in \mathbb{S}^*(K_p)$ satisfying

$$\langle \tau_h T, \varphi \rangle = \langle T, \tau_{-h}\varphi \rangle, \quad \forall \varphi \in \mathbb{S}(K_p),$$

where $\tau_{-h}\varphi(x) = \varphi(x+h)$ is the translation of φ.

(ii) **Dilation.** For $T \in \mathbb{S}^*(K_p)$, the dilation $\rho_s T$, $s \in \mathbb{N}$, of T is defined as a distribution $\rho_s T \in \mathbb{S}^*(K_p)$ satisfying

$$\langle \rho_s T, \varphi \rangle = \langle T, p^{-s} \rho_{-s} \varphi \rangle, \quad \forall \varphi \in \mathbb{S}(K_p),$$

where $(\rho_s \varphi)(x) = \varphi(p^s x)$ is the dilation of φ;

(iii) **Reflection.** For $T \in \mathbb{S}^*(K_p)$, the reflection $\tilde{T}$ is defined as a distribution $\tilde{T} \in \mathbb{S}^*(K_p)$ satisfying

$$\langle \tilde{T}, \varphi \rangle = \langle T, \tilde{\varphi} \rangle, \quad \forall \varphi \in \mathbb{S}(K_p),$$

where $\tilde{\varphi}(x) = \varphi(-x)$ is the reflection of φ;

(iv) **Multiplication with function.** For $T \in \mathbb{S}^*(K_p)$, the multiplication gT with function $g \in C(K_p)$ is defined as a distribution $gT \in \mathbb{S}^*(K_p)$ satisfying

$$\langle gT, \varphi \rangle = \langle T, g\varphi \rangle, \quad \forall \varphi \in \mathbb{S}(K_p),$$

where $(g\varphi)(x) = g(x)\varphi(x)$.

Theorem 3.1.24 *The Fourier transformations on $\mathbb{S}^*(K_p)$ have the following properties: let $T \in \mathbb{S}^*(K_p)$, then*

(i) $(\tau_h T)^\wedge = \overline{\chi}_h T^\wedge$, $h \in K_p$;   (*FT of translation $\tau_h T$*)

(ii) $\tau_h T^\wedge = (\chi_h T)^\wedge$, $h \in K_p$;   (*translation of FT $T^\wedge$*)

(iii) $(\rho_s T)^\wedge = p^{-s} \rho_{-s} T^\wedge$, $s \in \mathbb{Z}$;   (*FT of dilation $\rho_s T$*)

(iv) $\left(\tilde{T}\right)^\wedge = T^\vee$, $\overline{\left(\tilde{T}\right)^\wedge} = \overline{T^\wedge}$, $\overline{\left(\tilde{T}\right)}^\wedge = \left(\overline{T}\right)^\wedge$;   (*FT of reflection $\tilde{T}$*)

(v) $(T^\wedge)^\sim = T^\vee = \left(\tilde{T}\right)^\wedge$;   (*reflection of FT $T^\wedge$*)

(vi) $(T^\wedge)^\vee = T = (T^\vee)^\wedge$.   (*inverse FT of FT $T^\wedge$*)

Proof. (i) For any $\varphi \in \mathbb{S}(K_p)$, and $\forall h \in K_p$, it follows that: the left side of (i), $(\tau_h T)^\wedge$ acts on $\varphi \in \mathbb{S}(K_p)$,

$$\langle (\tau_h T)^\wedge, \varphi \rangle = \langle \tau_h T, \varphi^\wedge \rangle = \langle T, \tau_{-h} \varphi^\wedge \rangle = \langle T, (\overline{\chi}_h \varphi)^\wedge \rangle;$$

the right side of (i), $\overline{\chi}_h T^\wedge$ acts on $\varphi \in \mathbb{S}(K_p)$,

$$\langle \overline{\chi}_h T^\wedge, \varphi \rangle = \langle T^\wedge, \overline{\chi}_h \varphi \rangle = \langle T, (\overline{\chi}_h \varphi)^\wedge \rangle,$$

thus (i) is proved.

(ii) Take any $\varphi \in \mathbb{S}(K_p)$, and $\forall h \in K_p$, it follows that:

the left side of (ii), $\tau_h T^\wedge$ acts on $\varphi \in \mathbb{S}(K_p)$,
$$\langle \tau_h T^\wedge, \varphi \rangle = \langle T^\wedge, \tau_{-h}\varphi \rangle = \langle T, (\tau_{-h}\varphi)^\wedge \rangle = \langle T, \chi_h \varphi^\wedge \rangle;$$
the right side of (ii), $(\chi_h T)^\wedge$ acts on $\varphi \in \mathbb{S}(K_p)$,
$$\langle (\chi_h T)^\wedge, \varphi \rangle = \langle \chi_h T, \varphi^\wedge \rangle = \langle T, \chi_h \varphi^\wedge \rangle,$$
then we have
$$\tau_h T^\wedge = (\chi_h T)^\wedge, \quad h \in K_p.$$
This is (ii).

(iii) Take any $\varphi \in \mathbb{S}(K_p)$, it follows that for $s \in \mathbb{Z}$:
the left side of (iii), $(\rho_s T)^\wedge$ acts on $\varphi \in \mathbb{S}(K_p)$
$$\langle (\rho_s T)^\wedge, \varphi \rangle = \langle \rho_s T, \varphi^\wedge \rangle = \langle T, p^{-s} \rho_{-s} \varphi^\wedge \rangle;$$
the right side of (iii), $p^{-s} \rho_{-s} T^\wedge$ acts on $\varphi \in \mathbb{S}(K_p)$
$$\langle p^{-s} \rho_{-s} T^\wedge, \varphi \rangle = p^{-s} \langle \rho_{-s} T^\wedge, \varphi \rangle = p^{-s} \langle T^\wedge, p^s \rho_s \varphi \rangle = p^{-s} \langle T, p^s (\rho_s \varphi)^\wedge \rangle$$
$$= p^{-s} p^s \langle T, (\rho_s \varphi)^\wedge \rangle = \langle T, p^{-s} (\rho_{-s} \varphi^\wedge) \rangle.$$
This is (iii).

The proofs of (iv)~(vi) are left as exercises.

The following is the **Parseval formula of distributions**.

Theorem 3.1.25 *For a distribution $T \in \mathbb{S}^*(K_p)$ and a function $g \in \mathbb{S}(K_p)$, then*

(i) $\langle T, \overline{g} \rangle = \langle T^\wedge, \overline{g^\wedge} \rangle$;

(ii) $\langle T, g \rangle = \langle T^\wedge, g^\vee \rangle$.

Proof. For (i), to prove $\langle T, \overline{g} \rangle = \langle T^\wedge, \overline{g^\wedge} \rangle$, take $g \in \mathbb{S}(K_p)$, we rewrite $\overline{g}$ as
$$\overline{g}(x) = \left[(\overline{g})^\vee\right]^\wedge(x) = \left[\int_{\Gamma_p} \overline{g}(\xi) \chi_x(\xi) d\xi\right]^\wedge \equiv h^\wedge(x),$$
then
$$\langle T, \overline{g} \rangle = \langle T, h^\wedge \rangle = \langle T^\wedge, h \rangle = \left\langle T^\wedge, \int_{\Gamma_p} \overline{g}(\xi) \chi_x(\xi) d\xi \right\rangle$$
$$= \left\langle T^\wedge, \overline{\int_{\Gamma_p} g(\xi) \overline{\chi}_x(\xi) d\xi} \right\rangle = \langle T^\wedge, \overline{g^\wedge} \rangle.$$

For (ii), by
$$\langle T^\wedge, g^\vee \rangle = \langle T, (g^\vee)^\wedge \rangle = \langle T, g \rangle,$$

we get $\langle T, g \rangle = \langle T^\wedge, g^\vee \rangle$. The proof is complete.

Example 3.1.10 Find the Fourier transformation of the Dirac distribution $\delta_{t_0} \in \mathbb{S}^*(K_p)$ supported at t_0.

Solution. The Dirac distribution δ_{t_0} supported at t_0 is defined as
$$\langle \delta_{t_0}, \varphi \rangle = \varphi(t_0), \quad \forall \varphi \in \mathbb{S}(K_p).$$
Thus, by Theorem 3.1.24 (i),
$$\langle \delta_{t_0}, \varphi \rangle \equiv \langle \tau_{t_0} \delta, \varphi \rangle = \langle \delta, \tau_{-t_0} \varphi \rangle = \langle \delta, \varphi(\xi + t_0) \rangle = \varphi(t_0), \quad \forall \varphi \in \mathbb{S}(K_p),$$
and
$$(\delta_{t_0})^\wedge = (\tau_{t_0} \delta)^\wedge = \overline{\chi}_{t_0}(\xi) \delta^\wedge = \overline{\chi}_{t_0}(\xi) \cdot 1 = \overline{\chi}_{t_0}(\xi), \quad \xi \in \Gamma_p.$$
Then, $(\delta_{t_0})^\wedge = \overline{\chi}_{t_0}(\xi), \xi \in \Gamma_p$.

Example 3.1.11 The Fourier transformation of 1.

Solution. Since $1 \in L_{\text{loc}}(K_p)$, so
$$\langle 1^\wedge, \varphi \rangle = \langle 1, \varphi^\wedge \rangle = \int_{\Gamma_p} \varphi^\wedge(\xi) d\xi = \int_{\Gamma_p} \varphi^\wedge(\xi) \overline{\chi}_0(\xi) d\xi$$
$$= \int_{\Gamma_p} \varphi^\wedge(\xi) \chi_0(\xi) d\xi = (\varphi^\wedge)^\vee(0) = \varphi(0) = \langle \delta, \varphi \rangle, \quad \forall \varphi \in \mathbb{S}(K_p),$$
we have
$$1^\wedge = \delta.$$

5. The convolution on distribution space $\mathbb{S}^*(K_p)$

We define the support set of a distribution $T \in \mathbb{S}^*(K_p)$, firstly.

Definition 3.1.18 (Support set of distribution $T \in \mathbb{S}^*(K_p)$) *Suppose that $\Omega \subset K_p$ is an open set in K_p.*

(i) *$T \in \mathbb{S}^*(K_p)$ is said to be a **zero distribution** on an open set $V \subset \Omega$, if*

"$\forall \varphi \in \mathbb{S}(K_p)$ with supp $\varphi \subset V$ implies $\langle T, \varphi \rangle = 0$".

Since $\forall \varphi \in \mathbb{S}(K_p) \leftrightarrow (k, l)$, with (k, l) the index pair of φ, then supp $\varphi = B^l$, thus $\forall \varphi \in \mathbb{S}(K_p)$ has an open, closed and compact support.

(ii) *If $T \in \mathbb{S}^*(K_p)$ is zero distribution on each open subset $V_j \subset V, j \in \Lambda$, then the union $V = \bigcup_{j \in \Lambda} V_j$ is the **biggest open set** such that $T \in \mathbb{S}^*(K_p)$ is a zero distribution.*

(iii) *If V is the biggest open set such that $T \in \mathbb{S}^*(K_p)$ is a zero distribution, then the complementary set V^C of V is said to be the **support of** T, denoted by* supp T, *or*
$$\operatorname{supp} T = \overline{\{U \subset K_p : T|_U \neq 0\}}.$$

If supp T *is a compact set in K_p, the distribution T is said to have a **compact support**.*

Recall that the convolution for functions on local fields is defined in the Definition 3.1.6. For $f, g \in L^1(K_p)$, the convolution of f and g is defined by
$$f * g(x) = \int_{K_p} f(x-t)g(t)dt = \int_{K_p} f(y)g(x-y)\,dy = \int_{K_p} f(y)\tilde{g}(y-x)\,dy$$
$$= \int_{K_p} f(y)\tau_x\tilde{g}(y)dy = \langle f, \tau_x\tilde{g}\rangle, \tag{*}$$
where $\tau_x\tilde{g}(y) = \tilde{g}(y-x) = g(x-y)$. For each $x \in K_p$,
$$f * g(x) = \langle f, \tau_x\tilde{g}\rangle$$
is a complex valued function of $x \in K_p$, and $f * g \in L^1(K_p)$ can be regarded as a distribution. Moreover, $f * g(x) = \langle f, \tau_x\tilde{g}\rangle$ can also be regarded as a distribution, thus, $\forall \varphi \in \mathbb{S}(K_p)$, it follows that
$$\langle f * g(x), \varphi\rangle = \langle\langle f, \tau_x\tilde{g}\rangle, \varphi\rangle = \left\langle \int_{K_p} f(y)\tilde{g}(y-x)\,dy, \varphi \right\rangle$$
$$= \int_{K_p} \left\{ \int_{K_p} f(y)g(x-y)\,dy \right\} \varphi(x)dx$$
$$= \int_{K_p} f(y) \left\{ \int_{K_p} g(x-y)\varphi(x)dx \right\} dy$$
$$= \int_{K_p} f(y) \left\{ \int_{K_p} g(t)\varphi(y+t)\,dt \right\} dy \qquad (x-y=t)$$
$$= \int_{K_p} f(y) \left\{ \int_{K_p} g(x)\varphi(x+y)\,dx \right\} dy \qquad (t \xrightarrow{\text{change}} x)$$
$$= \langle f_y, \langle g_x, \varphi(x+y)\rangle\rangle = \langle f_x, \langle g_y, \varphi(x+y)\rangle\rangle,$$
so the convolution $f * g \in L^1(K_p)$ can be regarded as a distribution satisfying the equality
$$\langle f * g(x), \varphi\rangle = \langle f_x, \langle g_y, \varphi(x+y)\rangle\rangle, \quad \forall \varphi \in \mathbb{S}(K_p).$$

This motivates us, for $T \in \mathbb{S}^*(K_p)$ and $\psi \in \mathbb{S}(K_p)$, the convolution $T * \psi$ may be defined as a distribution satisfying
$$\langle T * \psi(x), \varphi \rangle = \langle T_x, \langle \psi_y, \varphi(x+y) \rangle \rangle, \quad \forall \varphi \in \mathbb{S}(K_p).$$
Now we turn to define the convolutions on distribution space $\mathbb{S}^*(K_p)$.

Definition 3.1.19 (Convolution) For a distribution $T \in \mathbb{S}^*(K_p)$, we define

(i) **Convolution of a distribution with a function**

The convolution of a distribution $T \in \mathbb{S}^*(K_p)$ and a function $\psi \in \mathbb{S}(K_p)$ is defined as a distribution $T * \psi \in \mathbb{S}^*(K_p)$ satisfying
$$\langle T * \psi, \varphi \rangle = \langle T_x, \langle \psi_y, \varphi(x+y) \rangle \rangle, \quad \forall \varphi \in \mathbb{S}(K_p). \tag{3.1.37}$$

(ii) **Convolution of distributions**

For distributions $S, T \in \mathbb{S}^*(K_p)$, suppose that one of them, for example, S has a compact support, the convolution $S * T$ is defined as a distribution satisfying
$$\langle S * T, \varphi \rangle = \langle S_x \otimes T_y, \varphi(x+y) \rangle, \quad \forall \varphi \in \mathbb{S}(K_p), \tag{3.1.38}$$
where $S_x \otimes T_y$ is the tensor product of S and T determined by
$$\langle S_x \otimes T_y, u(x,y) \rangle = \langle S_x, \langle T_y, u(x,y) \rangle \rangle$$
$$= \langle T_x, \langle S_y, u(x,y) \rangle \rangle, \quad \forall u \in \mathbb{S}(K_p \times K_p),$$
and $\forall \varphi, \psi \in \mathbb{S}(K_p)$, have $\langle S_x \otimes T_y, \varphi(x)\psi(y) \rangle = \langle S, \varphi \rangle \langle T, \psi \rangle$.

Since supp S is compact, then (3.1.38) becomes the form
$$\langle S * T, \varphi \rangle = \langle S_x, \langle T_y, \varphi(x+y) \rangle \rangle, \quad \forall \varphi \in \mathbb{S}(K_p).$$

For the details, we omit and refer to [1].

Theorem 3.1.26 For $T \in \mathbb{S}^*(K_p)$ and $\psi \in \mathbb{S}(K_p)$, the convolution $T * \psi$ has a form
$$T * \psi(x) = \langle T_y, \psi(x-y) \rangle, \quad x \in K_p.$$

Proof. In fact, $\forall \varphi \in \mathbb{S}(K_p)$, we can deduce by definition
$\langle T * \psi, \varphi \rangle = \langle T_x, \langle \psi_y, \varphi(x+y) \rangle \rangle$
$\uparrow$
$$\langle \psi_y, \varphi(x+y) \rangle = \int_{K_p} \psi(y)\varphi(x+y)\, dy = \int_{K_p} \psi(t-x)\varphi(t)dt$$
$\uparrow$

$$= \left\langle T_x, \int_{K_p} \psi(t-x)\varphi(t)dt \right\rangle = \left\langle T_y, \int_{K_p} \psi(x-y)\varphi(x)dx \right\rangle$$

(since $\varphi, \psi \in \mathbb{S}(K_p)$)

$\uparrow$ change $x \longrightarrow y, t$ change $\longrightarrow x$

$$= \int_{K_p} \langle T_y, \psi(x-y) \rangle \varphi(x) dx = \langle \langle T_y, \psi(x-y) \rangle, \varphi(x) \rangle,$$

$\uparrow$
the proof of changing T_y with $\int_{K_p}$, we refer to [1]

this implies $T * \psi(x) = \langle T_y, \psi(x-y) \rangle, x \in K_p$.

Remark. By Theorem 3.1.26, we conclude that the commutative law

$$T * \psi = \psi * T$$

holds for $T \in \mathbb{S}^*(K_p)$ and $\psi \in \mathbb{S}(K_p)$.

The operation formulas for distribution convolutions are listed in the following:

Theorem 3.1.27 For $T \in \mathbb{S}^*(K_p)$, for $f, g \in \mathbb{S}^*(K_p)$ both with compact support, moreover, $\delta \in \mathbb{S}^*(K_p)$ is the Dirac distribution, then we have

(i) $T * (\alpha f + \beta g) = \alpha T * f + \beta T * g, \quad \alpha, \beta \in \mathbb{C}$;
(ii) $(T * f) * g = T * (f * g)$;
(iii) $T * f = f * T$;
(iv) $T * \delta = \delta * T = T$;
(v) $\tau_h (T * f) = (\tau_h T) * f = T * (\tau_h f); (\tau_h \delta) * T = \tau_h T$;
(vi) $(T * \varphi)^\wedge = T^\wedge \varphi^\wedge$.

Proof. (i) is clear. For (ii), take any $\varphi \in \mathbb{S}(K_p)$, the left side of (ii) is

$$\langle (T * f) * g, \varphi \rangle = \langle (T * f)_x, \langle g_y, \varphi(x+y) \rangle \rangle = \langle T_x, \langle f_y, \langle g_y, \varphi(x+y) \rangle \rangle \rangle$$

$\uparrow$ definition of $(T * f) * g$ $\uparrow$ definition of $(T * f)_x$

$$= \left\langle T_x, \left\langle (f * g)_y, \varphi(x+y) \right\rangle \right\rangle = \left\langle T_x * (f * g)_y, \varphi(x+y) \right\rangle$$
$$= \langle T * (f * g), \varphi \rangle,$$

(ii) is follows.

To prove (iii), take 2 steps.

① Let $T_1, T_2 \in \mathbb{S}^*(K_p)$, if $T_1 * \varphi = T_2 * \varphi, \forall \varphi \in \mathbb{S}(K_p)$, then $T_1 = T_2$.

Since for each $\varphi \in \mathbb{S}(K_p)$, $T_1 * \varphi = T_2 * \varphi$ by assumption, so Theorem 3.1.26 gives $\forall T \in \mathbb{S}^*(K_p)$ that

$$T * \varphi(x) = \langle T_y, \varphi(x-y)\rangle = \langle T_y, \tilde{\varphi}(y-x)\rangle, \quad x \in K_p.$$

Let $x = 0$ in the above, then $T * \varphi(0) = \langle T_y, \tilde{\varphi}(y)\rangle$, thus

$$T * \tilde{\varphi}(0) = \langle T_y, \varphi(y)\rangle = \langle T, \varphi\rangle.$$

Take T_1 and T_2 instead of T, we have

$$\langle T_1, \varphi\rangle = (T_1 * \tilde{\varphi})(0) = (T_2 * \tilde{\varphi})(0) = \langle T_2, \varphi\rangle, \quad \forall \varphi \in \mathbb{S}(K_p).$$

Hence $T_1 = T_2$.

② Let $T \in \mathbb{S}^*(K_p)$, $f \in \mathbb{S}^*(K_p)$, and supp$f$ is compact, then $T * f = f * T$.

Since for any $\varphi, \psi \in \mathbb{S}(K_p)$, we have $\varphi * \psi = \psi * \varphi$. Thus

$$(T * f) * (\varphi * \psi) = (T * f) * (\psi * \varphi) = T * (f * \psi * \varphi)$$
$$\uparrow \qquad\qquad\qquad \uparrow$$

commutation of function convolutions   (ii) in this Theorem

$$= T * (f * \psi) * \varphi = T * \varphi * (f * \psi) = (T * \varphi) * (f * \psi)$$
$$\uparrow$$

Remark in Theorem 3.1.25

$$= (f * \psi) * (T * \varphi) = f * \psi * (T * \varphi) = f * (T * \varphi) * \psi$$
$$= (f * T) * (\varphi * \psi).$$

Then, we have (iii) by ①.

To prove (iv), $T * \delta = \delta * T = T$, take 2 steps.

① Prove $\delta * f = f$ for $T = f \in \mathbb{S}(K_p)$.

Since by Theorem 3.1.26,

$$\delta * f(x) = \langle \delta_y, f(x-y)\rangle = f(x-y)|_{y=0} = f(x),$$

moreover, Dirac δ is in $\mathbb{S}^*(K_p)$, $f \in \mathbb{S}(K_p)$, by the Remark of Theorem 3.1.25, we have

$$f = \delta * f = f * \delta.$$

② Prove $T * \delta = T$ for $T \in \mathbb{S}^*(K_p)$.

Since

$$(T * \delta) * \varphi = T * (\delta * \varphi) = T * \varphi,$$

we get $T * \delta = T$.

The property (iv) can be reinterpreted: the Dirac distribution $\delta \in \mathbb{S}^*(K_p)$ is as the unit of convolution operation $*$ in $\mathbb{S}^*(K_p)$.

The proof of the second formula of (v), $(\tau_h \delta) * T = \tau_h T$: take any $\varphi \in \mathbb{S}(K_p)$, then

$$\langle (\tau_h \delta) * T, \varphi \rangle = \langle T * (\tau_h \delta), \varphi \rangle = \langle T_x, \langle (\tau_h \delta)_y, \varphi_{x+y} \rangle \rangle$$
$$= \langle T_x, \varphi_{x+h} \rangle = \langle T, \tau_{-h} \varphi \rangle = \langle \tau_h T, \varphi \rangle,$$

this shows that $(\tau_h \delta) * T = \tau_h T$.

The proof of the first formula of (v), $\tau_h (T * f) = (\tau_h T) * f = T * (\tau_h f)$: take any $\varphi \in \mathbb{S}(K_p)$, then

$$\langle \tau_h (T * f), \varphi \rangle = \langle T * f, \tau_{-h} \varphi \rangle = \langle (T * f)_x, \varphi(x+h) \rangle$$
$$= \langle T_x, \langle f_y, \varphi_{x+y+h} \rangle \rangle = \langle T_x, \langle f_y, \tau_{-h} \varphi_{x+y} \rangle \rangle = \langle T_x, \langle \tau_h f_y, \varphi_{x+y} \rangle \rangle$$
$$= \langle T_x * \tau_h f_y, \varphi_{x+y} \rangle = \langle T * \tau_h f, \varphi \rangle.$$

Furthermore, by the commutative law $\tau_h (T * f) = \tau_h (f * T)$ and the similar way we get

$$\tau_h (f * T) = (\tau_h f) * T = T * (\tau_h f).$$

This tells the first in (v) holds.

The proof of (vi) is left as an exercise.

Similar to the classical case, the space $L^1(K_p)$ is a Banach algebra without unit under the addition, scalar product, L^1-norm, convolution operations. However, the convolution operation has a unit δ, the Dirac distribution, since $f * \delta = \delta * f = f$, $\forall f \in L^1(K_p)$.

The implication of (v) is: the convolution has invariability of translation.

We discuss the distribution theory on a local field K_p, by using it, we have studied the $L^r(K_p)$- theory, $r > 2$, of the Fourier transformation on K_p, that is, $\forall f \in L^r(K_p)$, $r > 2$, can be regarded as a distribution $f \in \mathbb{S}^*(K_p)$, thus we complete the theory of Fourier analysis on local field K_p.

Exercises

1. What is the idea of the Fourier analysis on the addition group K_p^+ of a local field K_p?
2. Show the proofs of Theorem 3.1.1 (iii), Theorem 3.1.3.

3. What do we need to do to establish the theory of the Fourier analysis on the multiplication group K_p^*?
4. Compare the proofs of Theorem 3.1.12 with that of the classical case in $\mathbb{R}$.
5. Prove that: if $f, g \in L^1(K_p) \cap L^2(K_p)$, then

   $1°\ (f*g)^\wedge = |f^\wedge|^2 \geq 0$; $2°\ f*g \in L^1(K_p)$; $3°\ f*g$ is continuous.
6. If $g \in L^2(K_p)$, and $\bar{g} = (g_1)^\wedge$, prove that $g_1 = \overline{g^\wedge}$.
7. Show the proofs of Theorem 3.1.17 and Theorem 3.1.20.
8. For a local field K_p, what is the characters of the product group K_p^*, and what is the character group of K_p^*? And what is the Fourier transformation of $L^r(K_p^*)$, $1 \leq r \leq 2$?
9. Show the proofs of Theorem 3.1.24 (iv)$\sim$(vi); Theorem 3.1.27 (vi).
10. Establish the distribution theory on the group K_p^*, as well as the Fourier analysis theory.
11. For distributions $S, T \in \mathbb{S}^*(K_p)$, consider the definition of the convolution $S * T$? (see Definition 3.1.19 and [1]).

3.2 Pseudo-differential operators on local fields

The p-adic calculus on a local field K_p is based upon pseudo-differential operator theory on local fields. To introduce the p-adic calculus on K_p, we establish the pseudo-differential operator theory on local fields in this section. Based on the test function space $\mathbb{S}(K_p)$ and its distribution space $\mathbb{S}^*(K_p)$, we define the symbol class in the following Subsection 3.2.1[88].

3.2.1 Symbol class $S^\alpha_{\rho\delta}(K_p) \equiv S^\alpha_{\rho\delta}(K_p \times \Gamma_p)$

Definition 3.2.1 (Symbol class) *If a complex valued function $\sigma : K_p \times \Gamma_p \to \mathbb{C}$ satisfies the following: for real numbers $\alpha \in \mathbb{R}$, $\rho \geq 0$, $\delta \geq 0$*

(i) *There exists a constant $c > 0$, such that*

$$|\sigma(x, \xi)| \leq c \langle \xi \rangle^\alpha, \quad x \in K_p,\ \xi \in \Gamma_p,$$

where $\langle \xi \rangle = \max\{1, |\xi|\}$.

(ii) *For any pair $(\mu, \nu) \in \mathbb{P} \times \mathbb{P}$, there exists a constant $c_{\mu\nu} > 0$, such that*

$$\left|\Delta^x_h \Delta^\xi_\zeta \sigma(x, \xi)\right| \leq c_{\mu\nu} |h|^\mu |\zeta|^\nu \langle \xi \rangle^{\alpha+\delta\mu-\rho\nu}, \quad x, h \in K_p,\ \xi, \zeta \in \Gamma_p, \xi \neq 0;$$

and exist constants $c_\mu > 0$, $c_\nu > 0$, such that

$$|\Delta^x_h \sigma(x, \xi)| \leq c_\mu |h|^\mu \langle \xi \rangle^{\alpha+\delta\mu}, \quad x, h \in K_p,\ \xi \in \Gamma_p, \xi \neq 0;$$

$$\left|\Delta^\xi_\zeta \sigma(x, \xi)\right| \leq c_\nu |\zeta|^\nu \langle \xi \rangle^{\alpha-\rho\nu}, \quad x \in K_p,\ \xi, \zeta \in \Gamma_p, |\zeta| < \langle \xi \rangle, \xi \neq 0,$$

where Δ_h^x, Δ_ζ^ξ, $\Delta_h^x \Delta_\zeta^\xi$ are the first, second order differences about x and ξ, respectively. Then $\sigma(x, \xi)$ is said to be a **symbol** on a local field K_p. The set of all symbols on K_p is said to be a **symbol class**, denoted by $S_{\rho\delta}^\alpha(K_p \times \Gamma_p)$. Then endow certain algebraic and topological structures on $S_{\rho\delta}^\alpha(K_p \times \Gamma_p)$, it becomes a topological linear space, and is said to be **symbol class on local fields**.

We also denote by $S_{\rho\delta}^\alpha(K_p) \equiv S_{\rho\delta}^\alpha(K_p \times \Gamma_p)$, for abbreviation.

For studying p-adic calculus we need the following properties.

Theorem 3.2.1 For $\alpha \in \mathbb{R}, \rho \geqslant 0$ and $\delta \geqslant 0$,

$$\sigma(x,\xi) = \langle \xi \rangle^\alpha \in S_{\rho\delta}^\alpha(K_p).$$

Proof. By Definition 3.2.1.

The decomposition theorem of the symbol class:

Theorem 3.2.2 Suppose that $\sigma(x,\xi) \in S_{\rho\delta}^m(K_p)$, then for $m < 0, \rho \geqslant 1$; or $m \leqslant 0, \rho > 1$; or $m + 3(1-\rho) < 0$, the series

$$\sigma(x,\xi) = \sum_{k,j=0}^{+\infty} \omega_{kj}(x)\varphi_{kj}(\xi)$$

converges absolutely and uniformly, where

$$\omega_{kj}(x) = \begin{cases} \int_{\Gamma_p} \sigma(x,\xi) \Phi_{\Gamma^0}(\xi) \overline{\chi}_{v(k)}(\xi) d\xi, & j = 0, \\ \int_{\Gamma_p} \sigma(x,\eta) \Phi_{\Gamma^0 \setminus \Gamma^{-1}}(\xi) \overline{\chi}_{v(k)}(\xi) d\xi, & j > 0, \end{cases}$$

and $|\eta| = p^j |\xi|$,

$$\varphi_{kj}(\xi) = \begin{cases} \Phi_{\Gamma^0}(\xi) \chi_{v(k)}(\xi), & j = 0, \\ \Phi_{\Gamma^j \setminus \Gamma^{j-1}}(\xi) \chi_{v(k)}(\theta) = \Phi_{\Gamma^0 \setminus \Gamma^{-1}}(\theta) \chi_{v(k)}(\theta), & j > 0, \end{cases}$$

and $|\theta| = p^{-j} |\xi|$; the set $\{v(k)\}_{k=0}^{+\infty}$ is the complete set of cosets of compact group $K_0 \subset K_p$ in K_p, and

$$\{v(k)\}_{k=0}^{+\infty} \leftrightarrow \{\chi_{v(k)}\}_{k=0}^{+\infty},$$

where the character set $\{\chi_{v(k)}\}_{k=0}^{+\infty}$ is the complete orthogonal set of compact group $K_0 = D$ (see [100]).

Proof. We give the lines of the proof[84],[96].

(i) Four useful lemmas.

① Let $\gamma > 0$, then there exists a constant $c > 0$, such that for any $x \in K_p$ with $|x| > p^{-n}$ we have
$$\int_{|\xi| \leqslant p^n} \frac{|\chi_x(\xi) - 1|}{|\xi|^{\gamma+1}} d\xi \geqslant c|x|^\gamma.$$

② For $j > 0$, $|\eta| = p^j |\xi|$, $|\varsigma| < 1$,
$$\left|\chi_{v(k)}(\varsigma) - 1\right| |\omega_{kj}(x)| = \left|\int_{\Gamma_p} \Delta_\varsigma^\xi \sigma(x, \eta) \Phi_{\Gamma^j \setminus \Gamma^{j-1}}(\eta) \overline{\chi}_{v(k)}(\xi) d\xi\right|,$$

thus, $\sigma(x, \xi) \in S_{\rho\delta}^m(K_p)$ has the decomposition
$$\sigma(x, \xi) = \sum_{k,j=0}^{+\infty} \omega_{kj}(x) \varphi_j(\xi) \chi_{v(k)}(p^j \xi),$$
which converges uniformly, where
$$\varphi_j(\xi) = \begin{cases} \Phi_{\Gamma^0}(\xi), & j = 0, \\ \Phi_{\Gamma^j \setminus \Gamma^{j-1}}(\xi), & j > 0, \end{cases}$$

and
$$\omega_{kj}(x) = \begin{cases} \int_{\Gamma_p} \sigma(x, \xi) \Phi_{\Gamma^0}(\xi) \overline{\chi}_{v(k)}(\xi) d\xi, & j = 0, \\ \int_{\Gamma_p} \sigma(x, \eta) \Phi_{\Gamma^0 \setminus \Gamma^{-1}}(\xi) \overline{\chi}_{v(k)}(\xi) d\xi, & j > 0. \end{cases}$$

③ Assume that $h_{kj}(t) = \int_{\Gamma_p} \varphi_j(\xi) \chi_{p^j v(k)+t}(\xi) d\xi$, $k, j \in \mathbb{P}$, then
$$\|h_{kj}\|_{L^1(K_p)} = \begin{cases} 1, & j = 0, \ k \geqslant 0, \\ 2(1 - p^{-1}), & j \geqslant 1, \ k \geqslant 0. \end{cases}$$

④ For any $n \in \mathbb{P}$, there exists a constant $c > 0$, such that for any $x \in K_p$ and $\xi \in \Gamma_p$, holds
$$|\chi_\xi(x) - 1| \leqslant c|x|^n |\xi|^n.$$

The proofs of the above lemmas are very delicate and technical, and they give rise to some special methods for dealing with problems for local fields. We omit them, and refer to [88].

(ii) The proof of Theorem. By
$$|v(k)|^\gamma |\omega_{kj}(x)| \leqslant \left(\int_{|\varsigma|<1} \frac{|\chi_{v(k)}(\varsigma) - 1|}{|\varsigma|^{\gamma+1}} d\varsigma\right) |\omega_{kj}(x)|$$

$$\leqslant \int_{|\varsigma|<1} \frac{|\chi_{v(k)}(\varsigma) - 1| |\omega_{kj}(x)|}{|\varsigma|^{\gamma+1}} d\varsigma$$

$$= \int_{|\varsigma|<1} \frac{1}{|\varsigma|^{\gamma+1}} \left| \int_{\Gamma_p} \Delta_\varsigma^\xi \sigma(x,\eta) \Phi_{\Gamma^j \setminus \Gamma^{j-1}}(\eta) \overline{\chi}_{v(k)}(\xi) d\xi \right| d\varsigma$$

$$\leqslant \left(\int_{|\varsigma|<1} c_\gamma p^{j(m+(1-\rho)\gamma)} d\varsigma \right) \leqslant c_\gamma p^{j(m+(1-\rho)\gamma)},$$

this implies

$$|\omega_{kj}(x)| \leqslant c_\gamma p^{j(m+(1-\rho)\gamma)} |v(k)|^{-\gamma}.$$

Note that $|v(k)| \geqslant p$, we may choose suitable $\gamma \in \mathbb{N}$, such that for $m < 0$, $\rho \geqslant 1$; or $m \leqslant 0$, $\rho > 1$; or $m + 3(1-\rho) < 0$, the series

$$\sigma(x,\xi) = \sum_{k,j=0}^{+\infty} \omega_{kj}(x) \varphi_{kj}(\xi)$$

is absolutely and uniformly convergent. The proof is complete.

3.2.2 Pseudo-differential operator T_σ on local fields

Definition 3.2.2 (Pseudo-differential operator) For $\sigma(x,\xi) \in S_{\rho\delta}^\alpha(K_p)$,

$$T_\sigma f(x) = \int_{\Gamma_p} \left\{ \int_{K_p} \sigma(x,\xi) f(t) \overline{\chi}_\xi(t-x) dt \right\} d\xi \qquad (3.2.1)$$

is said to be a **pseudo-differential operator** with symbol $\sigma \in S_{\rho\delta}^\alpha(K_p)$ on a local field K_p, where $f: K_p \to \mathbb{C}$ is a Haar measurable function on K_p and $x \in K_p$.

We do not study general theory of pseudo-differential operators on K_p, since it is similar to that of on $\mathbb{R}^n$, but just establish the p-adic calculus on K_p by using that theory, and establish some basic theory of p-adic calculus.

By Theorem 3.2.1, we study the pseudo-differential operator with symbol $\sigma(x,\xi) = \langle \xi \rangle^\alpha$

$$T_\alpha f(x) \equiv \int_{\Gamma_p} \left\{ \int_{K_p} \langle \xi \rangle^\alpha f(t) \overline{\chi}_\xi(t-x) dt \right\} d\xi, \quad \alpha \in \mathbb{R}, x \in K_p. \qquad (3.2.2)$$

This operator plays a key role in p-adic calculus theory.

1. Properties of pseudo-differential operator T_α on $\mathbb{S}(K_p)$

Theorem 3.2.3 *For $\alpha \in \mathbb{R}$, the operator T_α is a homeomorphism from test function space $\mathbb{S}(K_p)$ onto $\mathbb{S}(K_p)$.*

Proof. By (3.2.2) and $\varphi \in \mathbb{S}(K_p)$, we have

$$T_\alpha \varphi(x) = (\langle \cdot \rangle^\alpha \varphi^\wedge(\cdot))^\vee (x), x \in K_p \qquad (3.2.3)$$

and the Fourier transformation $F: \varphi \to \varphi^\wedge$, the inverse Fourier transformation $F^{-1}: \varphi \to \varphi^\vee$, the translation $\tau_h: \varphi \to \tau_h \varphi$ all are homeomorphisms from $\mathbb{S}(K_p)$ onto $\mathbb{S}(K_p)$, so we only need to prove

$$\langle \xi \rangle^\alpha \Phi_{\Gamma^k}(\xi) \in \mathbb{S}(\Gamma_p).$$

However,

$$\langle \xi \rangle^\alpha \Phi_{\Gamma^k}(\xi) = \begin{cases} \Phi_{\Gamma^k}(\xi), & |\xi| \leqslant 1, \\ |\xi|^\alpha \Phi_{\Gamma^k}(\xi), & |\xi| > 1, \end{cases} \quad \alpha \in \mathbb{R},$$

it is in $\mathbb{S}(\Gamma_p)$, clearly. Theorem is proved.

There is another function class on a local field K_p, the Foundational function class.

Definition 3.2.3 (Foundational function class) *For a complex valued Haar measurable function $\psi: K_p \to \mathbb{C}$ on K_p, if it satisfies*

(i) *For any non-negative integer $N \in \mathbb{P}$, there exists a constant $c_N > 0$, such that*

$$|\psi(x)| \leqslant c_N \langle x \rangle^{-N}, \quad x \in K_p,$$

where $\langle x \rangle = \max\{1, |x|\}$;

(ii) *For any pair $(\mu, N) \in \mathbb{P} \times \mathbb{P}$, there exists a constant $c_{\mu N} > 0$, such that*

$$|\Delta_h^x \psi(x)| \leqslant c_{\mu N} |h|^\mu \langle x \rangle^{-N}, \quad x, h \in K_p,$$

*where $\Delta_h^x \psi(x) = \psi(x + h) - \psi(x)$, then $\psi(x)$ is said to be a **foundational function**. The set of all foundational functions on K_p, denoted by $\mathbb{B}(K_p)$, endowed certain linear operations and topological structure, $\mathbb{B}(K_p)$ becomes a topological linear space, and is said to be the foundational function class, or the **foundational function space**.*

Example 3.2.1 Let $\varphi(x) = \begin{cases} e^{-|x|}, & |x| > 1, \\ e^{\frac{-1}{x}}, & |x| \leqslant 1, \end{cases}$ it is easy to verify that

$\varphi \in \mathbb{B}(K_p)$.

Theorem 3.2.4 *For $\alpha \in \mathbb{R}$, the operator T_α is a homeomorphism on $\mathbb{B}(K_p)$.*

Proof. We give a line of the proof.

(i) State 2 lemmas:

① Let $\gamma > 0$, then there exists a constant $c > 0$, such that for any $x \in K_p$ with $|x| > p^{-n}$, holds

$$\int_{|\xi| \leqslant p^n} \frac{|\chi_x(\xi) - 1|}{|\xi|^{\gamma+1}} d\xi \geqslant c|x|^\gamma.$$

② For any $n \in \mathbb{P}$, there exists a constant $c > 0$, such that for $x \in K_p$, $\xi \in \Gamma_p$, holds

$$|\chi_\xi(x) - 1| \leqslant c|x|^n |\xi|^n.$$

(ii) The line of proof of Theorem:

$$\psi \in \mathbb{B}(K_p) \Rightarrow T_\alpha \psi \in \mathbb{B}(K_p). \tag{3.2.4}$$

Consider two cases: $|x| \leqslant 1$ and $|x| > 1$, by using some properties of the Fourier transformation and definition of operator T_α, definition of foundational function space, as well as ① and ② in the above (i), we have for any $N \in \mathbb{P}$, there exists a constant $c_N > 0$, such that

$$|T_\alpha \psi(x)| \leqslant c_N \langle x \rangle^{-N}, \quad x \in K_p.$$

And for any pair $(\mu, N) \in \mathbb{P} \times \mathbb{P}$, there exists a constant $c_{\mu N} > 0$, such that

$$|\Delta_h^x T_\alpha \psi(x)| \leqslant c_{\mu N} |h|^\mu \langle x \rangle^{-N}, \quad x, h \in K_p.$$

Combining the above (i) and (ii), we can prove the Theorem.

For the detail of the proof, we refer to [88].

Theorem 3.2.5 *The translation, reflection, Fourier transformation on local field K_p all are homeomorphism on the space $\mathbb{B}(K_p)$.*

2. Extension of pseudo-differential operator T_α to spaces $\mathbb{S}^*(K_p)$ and $\mathbb{B}^*(K_p)$

The operator T_α can be extended to the distribution spaces $\mathbb{S}^*(K_p)$ and $\mathbb{B}^*(K_p)$.

Definition 3.2.4 (Pseudo-differential operators on distribution spaces) For a distribution $S \in \mathbb{S}^*(K_p)$, if the following equality holds

$$\langle T_\alpha S, \varphi \rangle = \langle S, T_\alpha \varphi \rangle, \quad \forall \varphi \in \mathbb{S}(K_p),$$

then $T_\alpha S \in \mathbb{S}^*(K_p)$ is a distribution; and $T_\alpha S$ is said to be the **pseudo-differential operator** T_α acts on distribution $S \in \mathbb{S}^*(K_p)$. Thus, T_α becomes a pseudo-differential operator on the space $\mathbb{S}^*(K_p)$.

Similarly, define a pseudo-differential operator T_α acts on a distribution $S \in \mathbb{B}^*(K_p)$ satisfying

$$\langle T_\alpha S, \psi \rangle = \langle S, T_\alpha \psi \rangle, \quad \forall \psi \in \mathbb{B}(K_p).$$

It is easy to verify:

Theorem 3.2.6 *A pseudo-differential operator T_α is homeomorphic on space $\mathbb{S}^*(K_p)$, it is homeomorphic on $\mathbb{B}^*(K_p)$ also.*

3.3 p-type derivatives and p-type integrals on local fields

To distinguish, we call the derivative and integral on local fields the p-type derivative and p-type integral (motivation of this kind of derivatives and integrals is the Gibbs derivative, see Subsection 3.3.3) are based upon the so called pseudo-differential operator theory on local fields.

3.3.1 p-type calculus on local fields

We establish p-type calculus for functions which are defined on local fields, and discuss their properties.

Definition 3.3.1 (Point-wise p-type derivative and L^r-strong p-type derivative) *Let $\alpha > 0$, if for a complex Haar measurable function $f : K_p \to \mathbb{C}$ on K_p, the integral*

$$T_{\langle \cdot \rangle^\alpha} f(x) \equiv \int_{\Gamma_p} \left\{ \int_{K_p} \langle \xi \rangle^\alpha f(t) \overline{\chi}_\xi (t - x) \, dt \right\} d\xi \tag{3.3.1}$$

*exists at $x \in K_p$, then $T_{\langle \cdot \rangle^\alpha} f(x)$ is said to be a **point-wise α-order p-type derivative** of $f(x)$ at x, denoted by $f^{\langle \alpha \rangle}(x)$.*

To define L^r-strong p-type derivative, let

$$f_k(x) = \begin{cases} f(x), & |x| \leqslant p^k, \\ 0, & |x| > p^k, \end{cases} \quad k \in \mathbb{Z}. \tag{3.3.2}$$

If there exists $g \in L^r(K_p)$, $1 \leqslant r < +\infty$, such that

$$\lim_{k \to +\infty} \left\| g(\cdot) - T_{\langle \cdot \rangle^\alpha} f_k(\cdot) \right\|_{L^r(K_p)} = 0, \tag{3.3.3}$$

then $g \in L^r(K_p)$ is said to be an L^r-**strong α-order p-type derivative** of $f(x)$, denoted by $D^{\langle\alpha\rangle}f(x)$.

Similarly, the p-type integral is defined as

Definition 3.3.2 (Point-wise p-type integral and L^r-strong p-type integral) Let $\alpha > 0$, if for a complex Haar measurable function $f : K_p \to \mathbb{C}$ on K_p, the integral

$$T_{\langle\cdot\rangle^{-\alpha}}f(x) \equiv \int_{\Gamma_p}\left\{\int_{K_p} \langle\xi\rangle^{-\alpha} f(t)\overline{\chi}_\xi(t-x)\,dt\right\}d\xi \tag{3.3.4}$$

exists at $x \in K_p$, then $T_{\langle\cdot\rangle^{-\alpha}}f(x)$ is said to be a **point-wise α-order p-type integral** of $f(x)$ at x, denoted by $f_{\langle\alpha\rangle}(x)$.

If there exists $h \in L^r(K_p), 1 \leqslant r < +\infty$, such that

$$\lim_{k\to+\infty}\left\|h(\cdot) - T_{\langle\cdot\rangle^{-\alpha}}f_k(\cdot)\right\|_{L^r(K_p)} = 0, \tag{3.3.5}$$

then $h \in L^r(K_p)$ is said to be an L^r-**strong α-order p-type integral** of $f(x)$, denoted by $I_{\langle\alpha\rangle}f(x)$.

Remark. We suppose that $\alpha > 0$ in the above definitions, thus the order α of p-type derivatives and integrals can be any positive real numbers, thus, fractional order derivatives and fractional order integrals all are contained. Moreover, for $\alpha = 0$, we agree on for $x \in K_p$

$$f^{\langle 0\rangle}(x) = f(x) = f_{\langle 0\rangle}(x) \quad \text{and} \quad D^{\langle 0\rangle}f(x) = f(x) = I_{\langle 0\rangle}f(x),$$

and in fact, this agreement can be proved for $\varphi \in \mathbb{S}(K_p)$.

3.3.2 Properties of p-type derivatives and p-type integrals of $\varphi \in \mathbb{S}(K_p)$

It is reasonable to call $f^{\langle\alpha\rangle}(x)$ and $f_{\langle\alpha\rangle}(x)$ derivative and integral of f, respectively, since lots of properties of them show that they can play the role in p-type calculus on local fields similar to that of in the Newton calculus on Euclidean spaces.

Theorem 3.3.1 If $\varphi \in \mathbb{S}(K_p)$, then φ has any order point-wise p-type derivatives at $x \in K_p$; also has any order L^r-strong p-type derivatives; and for any $\alpha \geqslant 0$,

$$D^{\langle\alpha\rangle}\varphi(x) = \varphi^{\langle\alpha\rangle}(x) \in \mathbb{S}(K_p). \tag{3.3.6}$$

Moreover, φ has any order point-wise p-type integrals at $x \in K_p$; also has any order L^r-strong p-type integrals; and for any $\alpha \geqslant 0$,

$$I_{\langle\alpha\rangle}\varphi(x) = \varphi_{\langle\alpha\rangle}(x) \in \mathbb{S}(K_p). \tag{3.3.7}$$

Proof. We just prove the conclusion about p-type derivatives, for that of the integrals can be obtained similarly.

① $\varphi \in \mathbb{S}(K_p)$ **has any order point-wise p-type derivatives.** In fact, by Theorem 3.2.3, $\varphi^{\langle\alpha\rangle} \in \mathbb{S}(K_p)$, thus for any $\alpha > 0$, test function $\varphi \in \mathbb{S}(K_p)$ has point-wise α-order p-type derivatives at $x \in K_p$. And for $\alpha = 0$, clearly, we have

$$\varphi^{\langle 0 \rangle}(x) = T_{\langle\cdot\rangle^0}\varphi(x)$$

$$= \int_{\Gamma_p} \left\{ \int_{K_p} \langle\xi\rangle^0 \varphi(t)\overline{\chi_\xi}(t-x)\,dt \right\} d\xi = (\varphi^\wedge(\cdot))^\vee(x) = \varphi(x).$$

② $\varphi \in \mathbb{S}(K_p)$ **has any order L^r-strong p-type derivatives.** By definition of L^r-strong derivatives, let

$$\varphi_k(x) = \begin{cases} \varphi(x), & |x| \leqslant p^k, \\ 0, & |x| > p^k, \end{cases} \quad k \in \mathbb{Z},$$

then $\varphi_k \in \mathbb{S}(K_p)$, and $T_{\langle\cdot\rangle^\alpha}\varphi_k(x) \in \mathbb{S}(K_p)$ for $\alpha > 0$ by Theorem 3.2.3. Moreover, space $\mathbb{S}(K_p) \subset L^r(K_p)$ is dense in $L^r(K_p)$ by Theorem 3.1.4, $1 \leqslant r < +\infty$, thus

$$\left\| T_{\langle\cdot\rangle^\alpha}\varphi_k(\cdot) - \varphi^{\langle\alpha\rangle}(\cdot) \right\|_{L^r(K_p)} \to 0, \quad k \to +\infty.$$

Hence, $\varphi \in \mathbb{S}(K_p)$ has any order L^r-strong p-type derivatives, and $D^{\langle\alpha\rangle}\varphi(x) = \varphi^{\langle\alpha\rangle}(x)$ by routine argument in some real analysis courses.

The sense of this Theorem. Regard $T_{\langle\cdot\rangle^\alpha}$ and $T_{\langle\cdot\rangle^{-\alpha}}$ ($\alpha \geqslant 0$) as "derivative operator" and "integral operator", respectively, then they are closed as operations in $\mathbb{S}(K_p)$, i.e., a test function $\varphi \in \mathbb{S}(K_p)$ has any order point-wise and L^r-strong derivatives, and they are all still in $\mathbb{S}(K_p)$; also, $\varphi \in \mathbb{S}(K_p)$ has any order point-wise and L^r-strong integrals, they are still in $\mathbb{S}(K_p)$; moreover, for any $\alpha \in [0, +\infty)$, $x \in K_p$,

$$\varphi^{\langle\alpha\rangle}(x) = D^{\langle\alpha\rangle}\varphi(x), \quad \varphi_{\langle\alpha\rangle}(x) = I_{\langle\alpha\rangle}\varphi(x).$$

That is, the p-type derivative operation and p-type integral operation are closed operations in the test function space $\mathbb{S}(K_p)$. Thus, we may regard

$\mathbb{S}(K_p)$ as a Schwartz type space on a local field which plays role as that of the Schwartz space $\mathbb{S}(\mathbb{R}^n)$ in $\mathbb{R}^n$.

Remark 1. Theorem 3.3.1 holds for the foundational function space $\mathbb{B}(K_p)$, that is, for $\psi \in \mathbb{B}(K_p)$ and $\alpha \geqslant 0$, the following relations hold

$$D^{\langle \alpha \rangle} \psi(x) = \psi^{\langle \alpha \rangle}(x) \in \mathbb{B}(K_p), x \in K_p$$

and

$$I_{\langle \alpha \rangle} \psi(x) = \psi_{\langle \alpha \rangle}(x) \in \mathbb{B}(K_p), x \in K_p.$$

Remark 2. By Theorem 3.3.1 and by $\overline{\mathbb{S}(K_p)} \subset L^1(K_p) \subset \mathbb{S}^*(K_p)$, we only need to study properties of point-wise p-type derivatives and integrals of $\varphi \in \mathbb{S}(K_p)$.

Theorem 3.3.2 *If $\varphi \in \mathbb{S}(K_p)$, then for any $\alpha \in [0, +\infty)$, the Fourier transformation formula holds:*

$$\left[\varphi^{\langle \alpha \rangle}(\cdot) \right]^{\wedge}(\xi) = \langle \xi \rangle^{\alpha} \varphi^{\wedge}(\xi), \quad \xi \in \Gamma_p. \tag{3.3.8}$$

Proof. Since

$$\varphi^{\langle \alpha \rangle}(x) = \int_{\Gamma_p} \left\{ \int_{K_p} \langle \xi \rangle^{\alpha} \varphi(t) \overline{\chi}_{\xi}(t-x) \, dt \right\} d\xi$$

$$= \int_{\Gamma_p} \langle \xi \rangle^{\alpha} \left\{ \int_{K_p} \varphi(t) \overline{\chi}_{\xi}(t) \overline{\chi}_{\xi}(-x) \, dt \right\} d\xi$$

$$= \int_{\Gamma_p} \langle \xi \rangle^{\alpha} \chi_x(\xi) \left\{ \int_{K_p} \varphi(t) \overline{\chi}_{\xi}(t) dt \right\} d\xi$$

$$= \int_{\Gamma_p} \langle \xi \rangle^{\alpha} \varphi^{\wedge}(\xi) \chi_x(\xi) d\xi = \left(\langle \cdot \rangle^{\alpha} \varphi^{\wedge}(\cdot) \right)^{\vee}(x),$$

then the uniqueness of Fourier transformations implies (3.3.8), thus the proof is complete.

The sense of this Theorem. The formula (3.3.8) shows that the Fourier transformation of the p-type derivative $\varphi^{\langle \alpha \rangle}(x)$ of $\varphi \in \mathbb{S}(K_p)$ satisfies a similar formula in the case of $\mathbb{R}^n$.

Theorem 3.3.3 *If $\varphi \in \mathbb{S}(K_p)$, then for $\alpha \geqslant 0$, and $x \in K_p$*

$$\left(\varphi^{\langle \alpha \rangle}(\cdot) \right)_{\langle \alpha \rangle}(x) = \varphi(x) = \left(\varphi_{\langle \alpha \rangle}(\cdot) \right)^{\langle \alpha \rangle}(x). \tag{3.3.9}$$

Proof. By definitions of p-type derivatives and p-type integrals, The uniqueness of Fourier transformations, and Theorem 3.3.2.

The sense of this Theorem. The formula (3.3.9) reveals that the p-type derivatives and the p-type integrals are inverse operations mutually, this is an essential and important property of all kind calculus.

Theorem 3.3.2 and Theorem 3.3.3 hold for functions $\psi \in \mathbb{B}(K_p)$.

3.3.3 p-type derivatives and p-type integrals of $T \in \mathbb{S}^*(K_p)$

Definition 3.3.3 (p-type derivative of a distribution) *Let $\alpha > 0$. If for a distribution $T \in \mathbb{S}^*(K_p)$ on a local field K_p, there exists a distribution $S \in \mathbb{S}^*(K_p)$, such that*

$$\langle S, \varphi \rangle = \langle T, \varphi^{\langle \alpha \rangle} \rangle, \quad \forall \varphi \in \mathbb{S}(K_p),$$

then the distribution S is said to be an α-order p-type derivative of distribution T, denoted by $S = T^{\langle \alpha \rangle}$. Thus, an α-order p-type derivative S of $T \in \mathbb{S}^*(K_p)$ is a distribution $T^{\langle \alpha \rangle} \in \mathbb{S}^*(K_p)$ satisfying

$$\langle T^{\langle \alpha \rangle}, \varphi \rangle = \langle T, \varphi^{\langle \alpha \rangle} \rangle, \quad \forall \varphi \in \mathbb{S}(K_p). \tag{3.3.10}$$

If there exists a distribution $Q \in \mathbb{S}^*(K_p)$, such that for $\alpha > 0$

$$\langle Q, \varphi \rangle = \langle T, \varphi_{\langle \alpha \rangle} \rangle, \quad \forall \varphi \in \mathbb{S}(K_p),$$

then the distribution Q is said to be an α-order p-type integral of distribution T, denoted by $Q = T_{\langle \alpha \rangle}$. That is, the α-order p-type derivative Q of $T \in \mathbb{S}^*(K_p)$ is a distribution $T_{\langle \alpha \rangle} \in \mathbb{S}^*(K_p)$ satisfying

$$\langle T_{\langle \alpha \rangle}, \varphi \rangle = \langle T, \varphi_{\langle \alpha \rangle} \rangle, \quad \forall \varphi \in \mathbb{S}(K_p). \tag{3.3.11}$$

We have an important theorem.

Theorem 3.3.4 *If $T \in \mathbb{S}^*(K_p)$, then for any $\alpha \geqslant 0$ holds*

$$\left(T^{\langle \alpha \rangle} \right)_{\langle \alpha \rangle} = T = \left(T_{\langle \alpha \rangle} \right)^{\langle \alpha \rangle} \tag{3.3.12}$$

in distribution sense.

Proof. For a distribution $T \in \mathbb{S}^*(K_p)$, take $\varphi \in \mathbb{S}(K_p)$, by Definition and Theorem 3.3.3,

$$\left\langle \left(T^{\langle \alpha \rangle} \right)_{\langle \alpha \rangle}, \varphi \right\rangle = \left\langle T^{\langle \alpha \rangle}, \varphi_{\langle \alpha \rangle} \right\rangle = \left\langle T, (\varphi_{\langle \alpha \rangle})^{\langle \alpha \rangle} \right\rangle = \left\langle T, \left(\varphi^{\langle \alpha \rangle} \right)_{\langle \alpha \rangle} \right\rangle$$
$$= \left\langle T_{\langle \alpha \rangle}, \varphi^{\langle \alpha \rangle} \right\rangle = \left\langle \left(T_{\langle \alpha \rangle} \right)^{\langle \alpha \rangle}, \varphi \right\rangle.$$

The formula (3.3.12) is proved.

The sense of this Theorem. The p-type derivatives and p-type integrals in the distribution sense are inverse operations mutually. The essential and important property of the p-type calculus in the Theorem 3.3.3 holds for distributions.

Theorem 3.3.5 For $T \in \mathbb{S}^*(K_p)$ and any $\alpha > 0$, it holds in distribution sense
$$\left[T^{\langle\alpha\rangle}\right]^\wedge = \langle\xi\rangle^\alpha T^\wedge, \xi \in \Gamma_p. \tag{3.3.13}$$

Proof. For a distribution $T \in \mathbb{S}^*(K_p)$, take any $\varphi \in \mathbb{S}(K_p)$, then its α-order p-type derivative $T^{\langle\alpha\rangle}$ satisfies
$$\left\langle T^{\langle\alpha\rangle}, \varphi \right\rangle = \left\langle T, \varphi^{\langle\alpha\rangle} \right\rangle.$$

Since $\varphi \in \mathbb{S}(K_p) \xleftrightarrow{\text{one-one, iso.}} \varphi^\wedge \in \mathbb{S}(\Gamma_p)$, and $(\varphi^\wedge)^\wedge = \tilde{\varphi}$, then
$$\left\langle \left(T^{\langle\alpha\rangle}\right)^\wedge, \varphi^\wedge \right\rangle = \left\langle T^{\langle\alpha\rangle}, (\varphi^\wedge)^\wedge \right\rangle = \left\langle T^{\langle\alpha\rangle}, \tilde{\varphi} \right\rangle = \left\langle T, (\tilde{\varphi})^{\langle\alpha\rangle} \right\rangle.$$

Moreover,
$$(\tilde{\varphi}(x))^{\langle\alpha\rangle} = \int_{\Gamma_p} \langle\xi\rangle^\alpha \left\{ \int_{K_p} \tilde{\varphi}(t)\overline{\chi}_\xi(t-x)\,dt \right\} d\xi$$
$$= \int_{\Gamma_p} \langle\xi\rangle^\alpha \left\{ \int_{K_p} \tilde{\varphi}(t)\overline{\chi}_\xi(t)\,dt \right\} \chi_x(\xi)\,d\xi$$
$$= \int_{\Gamma_p} \langle\xi\rangle^\alpha \varphi^\wedge(-\xi)\chi_x(\xi)\,d\xi = [\langle\cdot\rangle^\alpha \varphi^\wedge(\cdot)]^\vee(-x),$$

thus by Theorem 3.1.23 (v), it follows that
$$\left\langle \left(T^{\langle\alpha\rangle}\right)^\wedge, \varphi^\wedge \right\rangle = \left\langle T, (\tilde{\varphi})^{\langle\alpha\rangle} \right\rangle = \left\langle T, \left[(\langle\cdot\rangle^\alpha \varphi^\wedge(\cdot))^\vee\right]^\sim \right\rangle$$
$$= \langle T^\wedge, \langle\cdot\rangle^\alpha \varphi^\wedge \rangle = \langle \langle\cdot\rangle^\alpha T^\wedge, \varphi^\wedge \rangle,$$

hence, $\left(T^{\langle\alpha\rangle}\right)^\wedge = \langle\xi\rangle^\alpha T^\wedge$ holds in distribution sense. The proof is complete.

The sense of this Theorem. (3.3.13) is the Fourier transformation formula of $T^{\langle\alpha\rangle}$ of p-type derivative of distribution $T \in \mathbb{S}^*(K_p)$.

The following is an important example.

Example 3.3.1 Find α-order p-type derivative of a character $\chi_\xi \in \Gamma_p$, $\alpha > 0$.

Solution. A character $\chi_\xi(x)$ dose not belong to $\mathbb{S}(K_p)$, and it is not compactly supported, so that it has to be regarded as a distribution
$$\chi_\xi \in L_{\text{loc}}(K_p) \subset \mathbb{S}^*(K_p).$$

We find its Fourier transformation first, then by Theorem 3.3.4, we evaluate $(\chi_\xi(\cdot))^{\langle\alpha\rangle}(x)$.

For any $\varphi \in \mathbb{S}(K_p)$, it has for $\xi \in \Gamma_p$

$$\langle (\chi_\xi)^\wedge, \varphi^\wedge \rangle = \langle \chi_\xi, (\varphi^\wedge)^\wedge \rangle = \langle \chi_\xi, \tilde{\varphi} \rangle = \int_{K_p} \chi_\xi(x)\varphi(-x)\,dx$$

$$= \int_{K_p} \varphi(-x)\overline{\chi}_\xi(-x)\,dx = \int_{K_p} \varphi(x)\overline{\chi}_\xi(x)\,dx$$

$$= \varphi^\wedge(\xi) = \langle \delta_\xi, \varphi^\wedge \rangle,$$

so that $(\chi_\xi)^\wedge = \delta_\xi$, this implies $\chi_\xi = (\delta_\xi)^\vee$ (the definition of δ_ξ is in Example 3.1.10).

By Theorem 3.3.5, we have

$$\left[(\chi_\xi)^{\langle\alpha\rangle}\right]^\wedge = \langle \xi \rangle^\alpha (\chi_\xi)^\wedge = \langle \xi \rangle^\alpha \delta_\xi,$$

thus

$$(\chi_\xi)^{\langle\alpha\rangle} = [\langle \cdot \rangle^\alpha \delta_\xi]^\vee \qquad (3.3.14)$$

in distribution sense.

Then, we prove $[\langle \cdot \rangle^\alpha \delta_\xi]^\vee = \langle \xi \rangle^\alpha [\delta_\xi]^\vee = \langle \xi \rangle^\alpha \chi_\xi$.

In fact, take any $\varphi \in \mathbb{S}(K_p)$, we deduce

$$\left\langle [\langle \cdot \rangle^\alpha \delta_\xi]^\vee, \varphi \right\rangle = \langle \langle \cdot \rangle^\alpha \delta_\xi, \varphi^\vee \rangle = \langle \delta_\xi, \langle \cdot \rangle^\alpha \varphi^\vee \rangle = \langle \xi \rangle^\alpha \varphi^\vee(\xi)$$

$$= \langle \xi \rangle^\alpha \langle \delta_\xi, \varphi^\vee \rangle = \langle \xi \rangle^\alpha \langle \delta_\xi^\vee, \varphi \rangle = \langle \langle \xi \rangle^\alpha \delta_\xi^\vee, \varphi \rangle.$$

Substitute in (3.3.14), we have for $\xi \in \Gamma_k$

$$(\chi_\xi)^{\langle\alpha\rangle} = \langle \xi \rangle^\alpha \chi_\xi. \qquad (3.3.15)$$

The sense of this Example. The formula (3.3.15) shows that a character χ_ξ with p-type derivative $(\chi_\xi)^{\langle\alpha\rangle}$ is an eigen-function of corresponding eigen-equation $y^{\langle\alpha\rangle}(x) = \langle \xi \rangle^\alpha y(x)$, and $\lambda = \langle \xi \rangle, \xi \in \Gamma_p$, is an eigne-value. This is the other essential property of p-type calculus.

3.3.4 *Background of establishing for p-type calculus*

1. New problems — "logical derivatives"

The sequence of functions in the following figs called Radamacher Function System, appears in many scientific fields. But for any function in Radamacher function System, there is no classical derivative at the discontinuous points, and the derivatives are equal to zero at the continuous points. Thus, their classical derivatives have no real meaning.

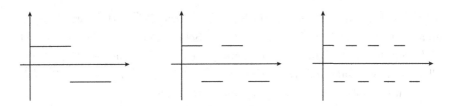

However, as a signal, a Rasamacher function has velocity, certainly, so to find its rate of change as a motion is a natural and important problem for scientists[7],[21].

In 1969, Mathematician J. E. Gibbs with M. J. Millard published paper[18] titled "Walsh functions as solution of a logical differential equations" and the other paper[19], they used the point of view of theory of Galois fields to define so called "logical derivative" on a finite dyadic group $G_0 = \{0, 1\}$, and gave a solution of a logical differential equation. In 1971, Gibbs with B. Ireland gave some generalizations, and published the paper titled "Some generalizations of the Logical Derivatives"[20]. Thus, scientists have called this kind of derivatives the "**logical derivative**" or the "**Gibbs derivative**".

In 1973, Mathematician P. L. Butzer with his students studied the Gibbs derivatives, published the paper titled "Walsh–Fourier series and the concept of a derivative"[10], and gave the definition:

Let $G_0 = \{x = (x_0, x_1, \cdots, x_k, \cdots) : x_k \in \{0, 1\}, k \in \mathbb{P}\}$ be **the dyadic compact Abelian group** with $x \oplus y = (x_k + y_k(\text{mod} z))$, and $f : G_0 \to \mathbb{C}$ a Haar measurable complex function on G_0. Suppose that

$$e_j = \left(x_k^j\right), \quad x_k^j = \begin{cases} 1, & k = j, \\ 0, & k \neq j, \end{cases} \quad j, k \in \mathbb{N}.$$

If for any $x \in G_0$, the series

$$\frac{1}{2} \sum_{j=0}^{+\infty} 2^j \{f(x \oplus e_{j+1}) - f(x)\} \qquad (3.3.16)$$

converges, then the sum is said to be a **point-wise logical derivative** of $f(x)$ at x, denoted by $f^{(1)}(x)$. If the series in (3.3.16) converges in $L^r(G_0)$-sense, $1 \leqslant r < +\infty$, then the sum is said to be an L^r-**strong logical derivative** of $f(x)$, denoted by $D^{(1)} f(x)$.

Higher order logical derivatives are defined inductively.

Later, Serbian mathematician R. S. Stankovic[69],[70], Hungarian mathematicians G. Gát[16],[17], J. Pál[50−52], F. Schipp and W. R. Wade, as well as P. Simon[66],[67], American mathematician C. W. Onneweer[42−49], and Chinese mathematicians W.X. Zheng, W.Y. Su, F.X. Ren[7],[62],[90],[98],[118], have studied lots of interesting and magnetic topics.

Remark 1. Why do not we use the difference $\dfrac{f(x \oplus \Delta x) - f(x)}{\Delta x}$, and take limit when $\Delta x \to 0$ as a derivative of $f(x)$, but use the sum in (3.3.16)? Because:

① When $\Delta x \to 0$, the limit of difference $\dfrac{f(x \oplus \Delta x) - f(x)}{\Delta x}$ is zero for those functions similar to the Radamaharer ones, so it has no meaning to describe the rate of change of a signal.

② (3.3.16) can be rewritten as
$$\sum_{j=0}^{+\infty} \frac{1}{2^{-j+1}} \{f(x \oplus e_{j+1}) - f(x)\},$$
note that, in the dyadic compact Abelian group G_0, the dyadic difference of $(x \oplus e_{j+1})$ and x is $e_{j+1} = 2^{-j-1}$ in the dyadic sense; and $f(x \oplus e_{j+1}) - f(x)$ is the increment of functions, so that the sum in (3.3.16) is **the total rate of change**, (except for a strictly positive factor of proportionality), it is true that the series in the (3.3.16) can describe the total rate of change of a function on G_0. Since G_0 is totally disconnected, thus the rate of change at one point does not make sense for most scientific problems but a global rate of change is essential.

③ For the p-adic locally compact Abelian group
$$G = \{x = (x_{-s}, x_{-s+1}, \cdots, x_{-1}, x_0, x_1, \cdots, x_k, \cdots):$$
$$x_k \in \{0, 1, \cdots, p-1\}, k = -s, -s+1, \cdots, s \in \mathbb{Z}\},$$
Chinese mathematicians defined logical derivatives and integrals on G_0. See (3.3.17) later.

Remark 2. Some important properties of derivatives are:

① As an operation, "derivative" must have **inverse operation "integral"**, so that when one tries to establish a new "calculus", he must have to define the inverse operation of the new derivative, "integral". It is true that mathematicians gave the definition of the logical integral:

Let $W^{\wedge}(k) = \begin{cases} k^{-1}, & k \subset \mathbb{N}, \\ 0, & k = 0, \end{cases}$ then they call the form

$$I_{\langle 1 \rangle} f(x) = W * f(x) = \int_{G_0} f(t) W(t \ominus x) \, dt$$

an "L^r-**strong logical integral**" of $f(x)$, where the convolution $W * f(x)$ is in dyadic sense, and the sign "$\ominus$" is the inverse operation of $\oplus$ on G_0. We have[119],[122]

$$I_{\langle 1 \rangle} \left(D^{\langle 1 \rangle} f \right) = f, \quad D^{\langle 1 \rangle} \left(I_{\langle 1 \rangle} f \right) = f$$

under certain conditions.

② The new derivative has **Fourier transformation formula** under certain conditions

$$\left(D^{\langle 1 \rangle} f \right)^{\wedge} (k) = k f^{\wedge}(k), \quad k \in \mathbb{P}.$$

③ The new derivative has an essential property: there are **eigen-equation** $y^{\langle 1 \rangle}(x) = \lambda y(x)$, **eigen-functions** $\{w_k(x) : k \in \mathbb{P}\}$, $x \in G_0$, and **eigen-values** $\lambda = k$, $k \in \mathbb{P}$.

The property ③ is a much deeper one of derivatives, its sense is: if some one tries to establish a new calculus, he has to consider a corresponding differential equation called an eigen-equation, such that the new derivative of solution of eigen-equation satisfies the eigen-equation as an eigen-function. Eigen-equation is a corner stone in physics for describing motion in nature. Moreover, an eigen-function is a character of the corresponding character group, for example, $y_t(x) = e^{itx}$ is a character of usual addition group of $\mathbb{R}$, and e^{itx} is an eigen-function of the eigen-equation $\dfrac{dy}{dx} = \lambda y$ with eigen-value $\lambda = it$ in the classical case; Walsh function $w_k(x) = \chi_k(x) = e^{\frac{2\pi}{p} i k \odot x}$, $p \geqslant 2$ a prime, is a character of p-aidc compact group, also an eigen-function of eigen-equation $y^{\langle 1 \rangle} = \lambda y$.

④ The new derivative satisfies the **direct and inverse approximation theorems** and the **equivalent theorem** in the construction theory of function, for example, the equivalent theorem on the space $X(G_0)$:

The following 4 statements are equivalent for $\alpha > 0$ and $s \in \mathbb{P}$:

(i) $f^{\langle s \rangle} \in \text{Lip}(X(G_0), \alpha)$;

(ii) $\omega \left(X(G_0), f^{\langle s \rangle}, \delta \right) = O(\delta^\alpha)$, $\delta \to 0$;

(iii) $E_{2^n}(X(G_0), f) = O\left(2^{-n(\alpha+s)}\right)$, $n \to +\infty$;

(iv) $\|f(\cdot) - S_{2^n}(f, \cdot)\|_{X(G_0)} = O\left(2^{-n(\alpha+s)}\right)$, $n \to +\infty$,

where $\text{Lip}, \omega, E_{2^n}, S_{2^n}$ are the Lipschitz class, continuous modulus, best approximation, 2^n-part sum of the Fourier series of $f \in X(G_0)$, respectively, on the function space

$$X(G_0) = \begin{cases} C(G_0), \\ L^r(G_0), 1 \leqslant r < +\infty. \end{cases}$$

This ④ is the other essential and deep property of a derivative. Since in "construction theory of function" (or "approximation theory"), the direct and inverse theorems (that is, Jackson and Bernstein Theorems) show the very deep relationship between a function and its derivative: the more faster the best approximation of a function is to zero, the more higher smoothness of a function is; conversely, the more higher the smoothness of a function is, then the more faster the best approximation of function tends to zero. The *Construction Theory of Function*, is a main mathematical branch, it describes some essential and deep and constructive properties of functions, for example, the direct and inverse theorems, the equivalent theorems. So that, if some one tries to establish a new kind of calculus, he has to ask the new concept whether has the approximation properties as above statements. In fact, the Gibbs derivatives is a kind of derivatives in the global point of view, it satisfies the above deep propertis.

2. p-adic analysis — study of p-adic Walsh system

In 1976–1983, W.X. Zheng, W.Y. Su and F.X. Ren obtained lots of results about p-adic Walsh system (see [62], [76]∼[78], [118]∼[122], [124]∼[126]).

Let $p \geqslant 2$ be a prime, and

$$G = \{x = (x_{-s}, x_{-s+1}, \cdots, x_{-1}, x_0, x_1, \cdots) : x_j \in \{0, 1, \cdots, p-1\},$$
$$j \geqslant -s, \ s \in \mathbb{P}\}.$$

It is well-known that G is a locally compact, totally disconnected, non-trivial topological group with the mod p operation, and it is isomorphic to the addition group of a local field, $G \leftrightarrow K^+$.

Let $f : G \to \mathbb{C}$ be a Haar measurable complex function on G, then if someone try to define a rate of change of f, then he must consider:

(i) A rate of change has a global sense on the domain $\mathfrak{D}_f$ of f, that is, the rate of change at $x \in \mathfrak{D}_f$ is affected by all points in $\mathfrak{D}_f$, but is not just determined by those points in a neighborhood of x.

(ii) A rate of change at $x \in \mathfrak{D}_f$ has to be a summation which terms are certain linear combinations of some ratio of differences of f at points in $\mathfrak{D}_f$ with differences of variables.

(iii) Principles for establishing a rate of change for functions on group G_0 have to be ①, ②, ③, ④ in Remark 2.

Chinese mathematicians generalized the definition (3.3.16) to a locally compact group:

For a Haar measurable complex function $f : G \to \mathbb{C}$ on p-aidc locally compact group G, $\forall N \in \mathbb{N}$, take a sum

$$\sum_{k=-N}^{+N} p^k \left\{ \sum_{j=0}^{p-1} A_j f\left(x \oplus jp^{-k-1}\right) \right\} \tag{3.3.17}$$

with $A_0 = \dfrac{p-1}{2}$, $A_k = \dfrac{\omega^k}{1-\omega^k}$, $k = 1, 2, \cdots, p-1$, if the limit of the sum in (3.3.17) is finite at $x \in G$ as $N \to +\infty$, then the limit, denoted by $f^{\langle 1 \rangle}(x)$, is said to be a **point-wise logical derivative** of $f(x)$ at x.

$\sum_{j=0}^{p-1} A_j f\left(x \oplus jp^{-k-1}\right)$ is certain liner combinations of function values, and $p^k = \dfrac{1}{p} \dfrac{1}{p^{-k-1}}$ can be regarded as $\left(x \oplus jp^{-k-1}\right) - (x \oplus (j-1)p^{-k-1})$, so that

$$p^k \left\{ \sum_{j=0}^{p-1} A_j f\left(x \oplus jp^{-k-1}\right) \right\} = \frac{1}{p} \frac{\sum_{j=0}^{p-1} A_j f\left(x \oplus jp^{-k-1}\right)}{p^{-k-1}},$$

then take summation for k from $-\infty$ to $+\infty$, thus we have a global rate of change at x.

If the sum in (3.3.17) has $L^r(G)$ $(1 \leqslant r \leqslant +\infty)$ sense limit, then the limit is said to be an **L^r-strong logical derivative**, denoted by $D^{\langle 1 \rangle} f(x)$. Higher order logical derivatives are defined by induction.

"Point-wise logical integral" or "$L^r(G)$-strong logical integral" are defined as a point-wise limit or L^r-strong limit of the convolution

$$V_n * f(x) = \int_G V_n\left(t \ominus x\right) f(t) dt$$

as $n \to +\infty$, respectively, where $V_n^\wedge(t) = \begin{cases} t^{-1}, & t \in [p^{-n}, +\infty), \\ 0, & t \in [0, p^{-n}). \end{cases}$

Later, they published many papers for studying the topics, continuously, and proved lots of interesting properties of logical derivatives and logical integrals on p-adic locally compact group G, and on its character group — Walsh system[82],[122],[126].

In 1981 to 1989, Z.L. He defined logical derivatives and logical integrals on an $\mathfrak{a}$-adic group, and proved certain properties with $\mathfrak{a} = (p_{-s}, p_{-s+1}, \cdots, p_{-1}, p_0, p_1, \cdots)$; $p_j \geqslant 2$ is a prime, $j \geqslant s, s \in \mathbb{P}$ (see [25]~[29]).

In 1985, W.X. Zheng defined logical derivatives and logical integrals on a local field K[128], and proved many properties. In 1990, W.X. Zheng, W.Y. Su, H.K. Jiang completed the definitions in [128] (see [117]).

In 1992, W.Y. Su defined p-type derivatives and p-type integrals on a local field K by virtue of so called pseudo-differential operators; generalized the concept of derivatives and integrals to the distribution spaces; proved many essential and deep properties; and pointed out a foundation principle for establishing new calculus[71],[73] to introduce the p-type calculus on local fields.

In 1993, H.K. Jiang defined logical derivatives and logical integrals on a locally compact $\mathfrak{a}$-adic group, and proved many properties of them[32].

The harmonic analysis over local fields starts a quite new area, which contains interesting and important, deep and new properties (see [31], [74], [79]–[81], [83]–[89], [91]–[97], [105], [107], [112]–[116], [126]–[130], [139]–[148]).

3. A glace of properties of p-type calculus on local fields

The definitions of derivatives introduced by Gibbs, Butzer, Onneweer, Zheng$\cdots$, are in the series forms. However, Su introduced p-type derivatives and p-type integrals by virtue of pseudo-differential operators, this kind calculus has some deep properties, such as, closed property, continuous property, and complies with certain principles on $\mathbb{S}(K_p)$, so one can generalize the p-type calculus to the distribution space $\mathbb{S}^*(K_p)$, keeping properties as those on $\mathbb{S}(K_p)$.

Let us emphasis some basic essential and deep properties of p-type calculus (also of Newton calculus) so that we may get a foundation principle for establishing other new calculus.

In the p-type calculus, we have the following:

(1) p-type derivative and p-type integral as operators are inverse operators of each other.

For any $\alpha > 0$, we have
$$\left(\varphi^{\langle\alpha\rangle}(\cdot)\right)_{\langle\alpha\rangle}(x) = \varphi(x) = \left(\varphi_{\langle\alpha\rangle}(\cdot)\right)^{\langle\alpha\rangle}(x), \quad \forall \varphi \in \mathbb{S}(K_p), \quad x \in K_p$$
and
$$\left(T^{\langle\alpha\rangle}\right)_{\langle\alpha\rangle} = T = \left(T_{\langle\alpha\rangle}\right)^{\langle\alpha\rangle}, \quad \forall T \in \mathbb{S}^*(K_p).$$

This is the first essential property for new calculus.

(2) There is a relationship between Fourier transformations $\varphi^\wedge$ of $\varphi \in \mathbb{S}(K_p)$ with $\left(\varphi^{\langle\alpha\rangle}\right)^\wedge$ of $\varphi^{\langle\alpha\rangle}$. Also, that for $T^\wedge$ of $T \in \mathbb{S}^*(K_p)$ with $\left(T^{\langle\alpha\rangle}\right)^\wedge$ of $T^{\langle\alpha\rangle}$.

For any $\alpha > 0$, we have
$$\left[\phi^{\langle\alpha\rangle}(\cdot)\right]^\wedge(\xi) = \langle\xi\rangle^\alpha \phi^\wedge(\xi), \quad \xi \in \Gamma_p, \quad \forall \varphi \in \mathbb{S}(K_p)$$
and
$$\left[T^{\langle\alpha\rangle}\right]^\wedge = \langle\xi\rangle^\alpha T^\wedge, \quad \forall T \in \mathbb{S}^*(K_p).$$

This is the second essential property for new calculus, specially in the signal analysis area and in some application area.

(3) Approximation equivalent theorem in function spaces $X(K_p)$.

For $\alpha > 0$ and $s \geqslant 0$, the following statements are equivalent:
(i) $f^{\langle s \rangle} \in \text{Lip}\,(X(K_p), \alpha)$;
(ii) $\omega\left(X(K_p), f^{\langle s \rangle}, \delta\right) = O\left(\delta^\alpha\right)$, $\delta \to 0$;
(iii) $E_{p^n}\left(X(K_p), f\right) = O\left(p^{-n(\alpha+s)}\right)$, $n \to +\infty$;

where $\text{Lip}, \omega, E_{p^n}$ are the Lipschitz class, continuous modulus, best approximation on the space $X(K_p) = \begin{cases} C(K_p), \\ L^r(K_p), 1 \leqslant r < +\infty, \end{cases}$ respectively.

This is the third essential and deep properties for new calculus in the point of view of the construction theory of function as well as approximation theory.

(4) The eigen-equation, eigen-function and eigen-value of p-type calculus are corresponding to the character equation, character function and character value of local fields.

For the p-type calculus, $y^{\langle 1 \rangle} = \lambda y$ is the eigen-equation (as in Newton calculus, it is $y' = \lambda y$); then we determine in Example 3.3.1 that $y = \chi_\xi(x)$ is an eigen-function, i.e., equation $(\chi_\xi)^{\langle\alpha\rangle}(x) = \langle\xi\rangle^\alpha \chi_\xi(x)$ in (3.3.15) at $\alpha = 1$ becomes $(\chi_\xi)^{\langle 1 \rangle}(x) = \langle\xi\rangle \chi_\xi(x)$ which is the eigen-equation corresponding to the p-type calculus. Moreover, $\lambda = \langle\xi\rangle$ is an eigen-value.

As we see, $\chi_\xi(x)$ in (3.3.15) is a character in character group $\Gamma_p = \{\chi_\xi : \xi \in K_p\}$ of local field K_p, so we may call the equation $(\chi_\xi)^{\langle 1 \rangle}(x) = \langle \xi \rangle \chi_\xi(x)$ as a character equation of K_p and $\lambda = \langle \xi \rangle$ is a character value of K_p.

This is the forth essential and deep property for certain calculus, specially in the point of view of group theory; also in that of the physics science in theory and applications.

3.4 Operator and construction theory of function on Local fields

The operator theory and construction theory of function (the approximation theory) on a local field K_p are important parts as those on the Euclidian space.

3.4.1 Operators on a local field K_p

1. Type of operators

Definition 3.4.1 (s-(r,s) type, w-$(1,1)$ type) *A linear operator $T: L^r(K_p) \to L^s(K_p)$ is said to be an s-(r,s) type on a local field K_p, $1 \leqslant r, s \leqslant +\infty$, if there exists a constant $c > 0$, such that*

$$\|Tf\|_{L^s(K_p)} \leqslant c\|f\|_{L^r(K_p)}, \quad \forall f \in L^r(K_p). \tag{3.4.1}$$

Also, say that the space $L^r(K_p)$ is embedded continuously into $L^s(K_p)$ by operator T.

If for any given $\lambda > 0$, there exists a constant $c > 0$, such that

$$|\{x \in K_p : Tf(x) > \lambda\}| \leqslant c\frac{\|f\|_{L^1(K_p)}}{\lambda}, \tag{3.4.2}$$

*then the operator T is said to be a **w-$(1,1)$ type** on local field K_p.*

2. H–L maximum operator

In the Section 3.1, the Hardy–Littlewood (H–L) maximum operator M for $g \in L_{\text{loc}}(K_p)$ is defined as (Definition 3.1.10)

$$Mg(x) = \sup_{k \in \mathbb{Z}} \frac{1}{p^{-k}} \int_{x+B^k} |g(z)|\,dz = \sup_{\substack{x \in S \\ S:\,\text{ball in } K_p}} \frac{1}{|S|} \int_S |g(z)|\,dz.$$

And Theorem 3.1.14 shows that M is a w-$(1,1)$ type operator: if $f \in L^1(K_p)$, then for $\forall \lambda > 0$,

$$|\{x \in K_p : Mf(x) > \lambda\}| \leqslant \frac{c}{\lambda} \|f\|_{L^1(K_p)}. \tag{3.4.3}$$

We now prove that, H–L maximum operator M is an s-(r,r) type operator.

Theorem 3.4.1 *The H–L maximum operator M is an s-(r,r), $r \geqslant 1$, type operator on K_p: there exists a constant $c > 0$, such that $M : L^r(K_p) \to L^r(K_p)$ satisfies*

$$\|Mf\|_{L^r(K_p)} \leqslant c \|f\|_{L^r(K_p)}, \quad \forall f \in L^r(K_p). \tag{3.4.4}$$

Proof. To prove the inequality (3.4.4), take $\lambda > 0$, and denote

$$E_\lambda = \{x \in K_p : Mf(x) > \lambda\}.$$

For $x \in E_\lambda$, there exists a ball $S \subset K_p$, such that $x \in S$, and

$$\frac{1}{|S|} \int_S |f(t)| \, dt > \lambda \tag{3.4.5}$$

(if not, it would be $Mf(x) \leqslant \lambda$). Denote all balls in $S \subset K_p$ satisfied (3.4.5) by $\{S_x\}_{x \in E_\lambda}$, then

$$\bigcup_{x \in E_\lambda} S_x \supset E_\lambda.$$

On the other hand, by Wiener covering Lemma 3.1.13, there exists disjoint finite balls

$$\{S_j\}_{j=1}^N \subset \bigcup_{x \in E_\lambda} S_x,$$

such that for η, $0 < \eta < 1$, holds

$$\eta |E_\lambda| < \sum_{j=1}^N |S_j| < \frac{1}{\lambda} \sum_{j=1}^N \int_{S_j} |f(t)| \, dt = \frac{1}{\lambda} \int_{\bigcup_{j=1}^N S_j} |f(t)| \, dt,$$

$\left(\text{by (3.4.5)}, |S| < \frac{1}{\lambda} \int_S |f(t)| \, dt\right)$. Thus,

$$|E_\lambda| = |\{x \in K_p : Mf(x) > \lambda\}| < \frac{1}{\eta \lambda} \int_{\bigcup_{j=1}^N S_j} |f(t)| \, dt \equiv \frac{c}{\lambda} \int_{\bigcup_{j=1}^N S_j} |f(t)| \, dt.$$

$$\tag{3.4.6}$$

Since it has $|S_j| < \dfrac{1}{\lambda} \int_{S_j} |f(t)|\, dt$ on each S_j, i.e., $Mf(x) > \lambda$ for $x \in S_j$, this shows that $\bigcup_{j=1}^{N} S_j \subset E_\lambda$ holds, then (3.4.6) becomes

$$|E_\lambda| \leqslant \frac{c}{\lambda} \int_{E_\lambda} |f(t)|\, dt, \tag{3.4.7}$$

this is

$$|\{x \in K_p : Mf(x) > \lambda\}| \leqslant \frac{c}{\lambda} \int_{\{x \in K_p : Mf(x) > \lambda\}} |f(t)|\, dt, \tag{3.4.8}$$

the inequality in (3.4.8) is a very useful inequality.

Further, by

$$K_p = \left\{x \in K_p : |f(x)| \leqslant \frac{\lambda}{2}\right\} \cup \left\{x \in K_p : |f(x)| > \frac{\lambda}{2}\right\} \equiv G_1 \cup G_2,$$

it follows the decomposition

$$f(x) = f_1(x) + f_2(x),$$

where

$$f_1(x) = \begin{cases} f(x), & x \in G_1, \\ 0, & x \in G_2, \end{cases} \quad f_2(x) = \begin{cases} 0, & x \in G_1, \\ f(x), & x \in G_2. \end{cases}$$

Clearly, $Mf_1(x) \leqslant \dfrac{\lambda}{2}$, thus

$$\{x \in K_p : Mf(x) > \lambda\} \subset \left\{x \in K_p : Mf_2(x) > \frac{\lambda}{2}\right\}.$$

By (3.4.8), for $f_2(x)$, it has

$$|\{x \in K_p : Mf(x) > \lambda\}| \leqslant \left|\left\{x \in K_p : Mf_2(x) > \frac{\lambda}{2}\right\}\right|$$

$$\leqslant \frac{2c}{\lambda} \int_{G_2} |f_2(x)|\, dx = \frac{2c}{\lambda} \int_{G_2} |f(x)|\, dx,$$

so that we have the other inequality

$$|\{x \in K_p : Mf(x) > \lambda\}| \leqslant \frac{c'}{\lambda} \int_{\{x \in K_p : |f(x)| > \frac{\lambda}{2}\}} |f(x)|\, dx. \tag{3.4.9}$$

To prove the s-(r,r) type of M, using an integral equality in harmonic analysis[65], we have

$$\|Mf\|_{L^r(K_p)}^r = r \int_0^{+\infty} \lambda^{r-1} |\{x \in K_p : Mf(x) > \lambda\}|\, d\lambda$$

$$\leqslant r \int_0^{+\infty} \lambda^{r-1} \frac{c}{\lambda} \int_{\{x \in K_p : |f(x)| > \frac{\lambda}{2}\}} |f(x)| \, dx d\lambda$$

$$= cr \int_0^{+\infty} \lambda^{r-2} \int_{\{x \in K_p : |f(x)| > \frac{\lambda}{2}\}} |f(x)| \, dx d\lambda$$

$$= cr \int_{K_p} |f(x)| \left\{ \int_0^{2|f|} \lambda^{r-2} d\lambda \right\} dx$$

$$= cr \int_{K_p} |f(x)| \frac{1}{r-1} (2|f(x)|)^{r-1} dx$$

$$= c' \frac{r}{r-1} \int_{K_p} |f(x)|^r \, dx = c' r' \, \|f\|^r_{L^r(K_p)},$$

then we get (3.4.4). The proof is complete.

About the operator theory on a local field K_p, in the point of view of the abstract harmonic analysis, $(K_p, \mathbb{F}, d\mu)$ is a measure space, we may study various topics. Compared with that of on $\mathbb{R}^n$, there are very ripe results about operator theory on local fields, it is quite young, and lots of new topics are worth to study, such as, singular integral operators, multiplier theory on K_p, and so on.

3.4.2 Construction theory of function on a local field K_p

1. Approximation theory on compact group $D \subset K_p$ in K_p

Similar to the classical case of Fourier series on $\mathbb{R}$, we may introduce the normal complete orthogonal system $\{\chi_n(x)\}_{n=0}^{+\infty}$ on compact group $D = \{x \in K_p : |x| \leqslant 1\}$, then establish the construction theory of function, including the theory of Fourier series and Dirichlet kernel; $(c, 1)$ summation and Fejér kernel, varies approximation theorems; and some approximation identity kernels.

(1) **Orthonormal complete system on $D \subset K_p$.**

Let $\chi \in \Gamma_p$ be a basic character of local field K_p, it is trivial($\chi|_D \equiv 1$) on D but non-trivial ($\chi|_{B^{-1}} \neq 1$) on B^{-1}. Denoted by $\chi|_D$ the restriction of χ on D. Then the relationship $\chi_u \in \Gamma_p \leftrightarrow u \in K_p$ between the character group Γ_p with K_p holds, and

$$\chi_u(x) = \chi(ux), \quad x \in K_p. \tag{3.4.10}$$

Theorem 3.4.2 *For $\chi_u, u \in K_p$ defined in (3.4.10), we have*
(i) $\chi_u|_D \in \Gamma_G$;
(ii) $\chi_u|_D = \chi_v|_D \Leftrightarrow u - v \in D$.

Proof. (i) holds since D is a subgroup of K_p.
To prove (ii), we deduce

$$\chi_u|_D = \chi_v|_D \Leftrightarrow \chi_u|_D(x) = \chi_v|_D(x), \forall x \in D$$

$$\Leftrightarrow 1 = [\chi_u|_D(x)][(\chi_v|_D)(x)]^{-1} = \left\{[\chi_u|_D]\left[(\chi_v|_D)^{-1}\right]\right\}(x)$$

$$= \left[(\chi_u)(\chi_v)^{-1}\right]\Big|_D(x) = [(\chi_u)(\chi_{-v})]|_D(x) = \chi_{u-v}|_D(x)$$

$$\Leftrightarrow u - v \in D.$$

By this theorem, we assert that: if $\{u(n)\}_{n=0}^{+\infty}$ is a complete set of all coset representatives of D in the additive group K_p^+ of K_p, then $\{\chi_{u(n)}\}_{n=0}^{+\infty}$ is a complete family of characters on D.

Theorem 3.4.3 *Let $\{u(n)\}_{n=0}^{+\infty}$ be a complete family of coset representatives of compact group D in K_p, then $\{\chi_n \equiv \chi_{u(n)}\}_{n=0}^{+\infty}$ is a complete family of characters, and $\{\chi_n\}_{n=0}^{+\infty}$ can be an **orthonormal complete system** on D.*

The following is a natural order of $\{u(n)\}_{n=0}^{+\infty}$.
Let $\{B^{-k}/D : k \in \mathbb{N}\}$ be a complete family of coset representatives of D in K_p:

$$B^{-1}/D = \{0 \cdot \beta^{-1} + D, 1 \cdot \beta^{-1} + D, \cdots, (p-1) \cdot \beta^{-1} + D\};$$

$$B^{-2}/D = \{0 \cdot \beta^{-2} + 0 \cdot \beta^{-1} + D, 0 \cdot \beta^{-2} + 1 \cdot \beta^{-1} + D, \cdots,$$

$$0 \cdot \beta^{-2} + (p-1) \cdot \beta^{-1} + D, 1 \cdot \beta^{-2} + 0 \cdot \beta^{-1} + D,$$

$$1 \cdot \beta^{-2} + 1 \cdot \beta^{-1} + D, \cdots, 1 \cdot \beta^{-2} + (p-1) \cdot \beta^{-1} + D,$$

$$\cdots$$

$$(p-1) \cdot \beta^{-2} + 0 \cdot \beta^{-1} + D, (p-1) \cdot \beta^{-2} + 1 \cdot \beta^{-1} + D, \cdots,$$

$$(p-1) \cdot \beta^{-2} + (p-1) \cdot \beta^{-1} + D\};$$

$$\cdots$$

Clearly, $\{B^{-k}/D : k \in \mathbb{N}\}$ is a countable set. Denote by

$$u(0) = 0 \cdot \beta^{-1}, \quad u(1) = 1 \cdot \beta^{-1}, \cdots, u(p-1) = (p-1) \cdot \beta^{-1};$$

$$u(p) = 1 \cdot \beta^{-2} + 0 \cdot \beta^{-1}, \quad u(p+1) = 1 \cdot \beta^{-2} + 1 \cdot \beta^{-1}, \cdots,$$

$$u(2p-1) = 1 \cdot \beta^{-2} + (p-1) \cdot \beta^{-1};$$

$$\cdots$$

Thus we have a set $\{u(n)\}_{n=0}^{+\infty}$, and the corresponding set $\{\chi_n \equiv \chi_{u(n)}\}_{n=0}^{+\infty}$.

By Pontryagin dual theorem, Γ_D is a discrete group, it is a countable set, thus[100]

$$\Gamma_D \leftrightarrow \{\chi_n \equiv \chi_{u(n)}\}_{n=0}^{+\infty}.$$

Since

$$\int_D \chi_{u(n)}(x)\overline{\chi}_{u(n)}(x)dx = \int_D |\chi_{u(n)}(x)|^2 dx = \int_D 1 \cdot dx = 1$$

and

$$\int_D \chi_{u(n)}(x)\overline{\chi}_{u(m)}(x)dx = \int_D \chi_{u(n)-u(m)}(x)dx = 0, m \neq n,$$

we get the integral formula:

$$\int_D \chi_{u(n)}(x)\overline{\chi}_{u(m)}(x)dx = \begin{cases} 1, & m = n, \\ 0, & m \neq n, \end{cases}$$

so that $\{\chi_n \equiv \chi_{u(n)}\}_{n=0}^{+\infty}$ is a normally complete orthogonal system on D.

(2) **Fourier series theory on** $D \subset K_p$.

Definition 3.4.2 (Fourier series on D) Let $\{\chi_n\}_{n=0}^{+\infty}$ be a normally complete orthogonal system on compact subgroup D in K_p, for $f \in L^1(D)$, we define

Fourier coefficients:

$$c_n = \int_D f(x)\overline{\chi}_n(x)dx = \int_D f(x)\overline{\chi}_{u(n)}(x)dx, \quad n = 0, 1, 2, \cdots. \quad (3.4.11)$$

Fourier series:

$$\sum_{k=0}^{+\infty} c_n \chi_{u(n)}(x). \quad (3.4.12)$$

Dirichlet kernel:

$$D_n(x) = \begin{cases} \sum_{k=0}^{n-1} \chi_k(x), & n \geqslant 1, \\ 0, & n = 0. \end{cases} \quad (3.4.13)$$

The partial sum of Fourier series:

$$S_n(f, x) = \sum_{k=0}^{n-1} c_k \chi_k(x).$$

The convolution representation of partial sum:

$$S_n(f, x) = \int_D f(x - t) D_n(t) dt = \int_D f(t) D_n(x - t) dt. \quad (3.4.14)$$

Compared with the classical case of Fourier series, we have

Theorem 3.4.4 *Dirichlet kernel has the following properties*

(i) $\int_{|x|\leq 1} D_n(x)dx = 1, \forall n \geq 1$;

(ii) $|D_n(x)| \leq n, \forall n \geq 0, x \in D$;

(iii) $D_n(x) = D_{mp^k+t}(x) = D_{p^k}(x)D_m(\beta^{-k}x) + \chi_m(\beta^{-k}x)D_t(x)$, for $\forall n = mp^k + t$, with $m \in \mathbb{P}, k \in \mathbb{P}, 0 \leq t < p^k$, $x \in D$;

(iv) $|D_n(x)| \leq \dfrac{p}{|x|}$, $x \neq 0, x \in D$;

(v) $D_{p^k}(x) = \prod_{\mu=0}^{k-1} D_p(\beta^{-\mu}x), k \geq 1, x \in D$.

Proof. (i) By the integral $\int_{|x|\leq 1} \chi_n(x)dx = \begin{cases} 1, & n=1, \\ 0, & n>1, \end{cases}$ we may prove (i).

(ii) By $|D_n(x)| = \left|\sum_{k=0}^{n-1} \chi_k(x)\right| \leq n$, we may have (ii).

(iii) By $\forall m \in \mathbb{P}, k \in \mathbb{P}, 0 \leq t < p^k$, it follows for $x \in D$,

$$D_{m \cdot p^k + t}(x) = \sum_{\mu=0}^{m-1}\sum_{\nu=0}^{p^k-1} \chi_{\mu \cdot p^k + \nu}(x) + \sum_{\nu=0}^{t-1} \chi_{m \cdot p^k + \nu}(x)$$

$$= \sum_{\mu=0}^{m-1} \chi_{\mu \cdot p^k}(x) \sum_{\nu=0}^{p^k-1} \chi_\nu(x) + \chi_{m \cdot p^k}(x) \sum_{\nu=0}^{t-1} \chi_\nu(x)$$

$$= D_m(\beta^{-k}x) D_{p^k}(x) + \chi_m(\beta^{-k}x) D_t(x).$$

This is (iii).

(iv) For any $x \in D$, let $|x| = p^{-k+1}, k \geq 1$, and $n = m \cdot p^k + t, 0 \leq t < p^k$. Thus, for $|x| = p^{-k+1}$ with $x = x_{k-1}\beta^{k-1} + x_k\beta^k + \cdots$, and $x_j \in GF(p)$, $j \geq k-1$, we have

$$\chi_n(x) = \chi_{u(n)}(x) = \chi(u(n)x), \quad 0 \leq n < p^k.$$

So that

$$|x| = p^{-k+1} \Rightarrow D_{p^k}(x) = \sum_{j=0}^{p^k-1} \chi_j(x) = 0,$$

then by (iii), we have

$$D_{m \cdot p^k + t}(x) = D_{p^k}(x)D_m(\beta^{-k}x) + \chi_m(\beta^{-k}x)D_t(x) = \chi_m(\beta^{-k}x)D_t(x).$$

Then by (ii), $|D_t(x)| \leqslant t$, it follows for $x \neq 0, x \in D$,
$$|D_n(x)| = |D_{m \cdot p^k + t}(x)| = |\chi_m(\beta^{-k}x) D_t(x)| = |D_t(x)| \leqslant t < p^k = p \cdot p^{k-1} = \frac{p}{|x|}.$$
This is (iv).

(v) When $k = 1$, $D_{p^1}(x) = \prod_{\mu=0}^{1-1} D_p(\beta^{-\mu}x)$ is an identity. For $k > 1$, we have equality $D_{p^{k+1}}(x) = D_p(\beta^{-k}x) D_{p^k}(x)$, $k \in \mathbb{N}$. In fact,
$$D_{p^{k+1}}(x) = D_{p \cdot p^k + 0}(x) = D_{p^k}(x) D_p(\beta^{-k}x) + \chi_p(\beta^{-k}x) D_0(x)$$
$$= D_{p^k}(x) D_p(\beta^{-k}x) + 0 = D_p(\beta^{-k}x) D_{p^k}(x).$$

Then by (iii) it follows for $x \in D$,
$$D_{m \cdot p^k + t}(x) = D_{p^k}(x) D_m(\beta^{-k}x) + \chi_m(\beta^{-k}x) D_t(x).$$
Take $m = p$, $t = 0$, (v) is obtained.

Theorem 3.4.5 *Let $f : D \to \mathbb{C}$ be a Haar measurable function on D. We have*

(i) *If $f \in L^1(D)$, and c_n is in (3.4.11), then $\lim_{n \to +\infty} c_n = 0$;*

(*Riemann–Legesgue lemma*)

(ii) *If $f \in L^1(D)$, and $c_n = 0, \forall n \in \mathbb{P}$, then $f = 0$;* (*uniqueness theorem*)

(iii) *If $f \in L^1(D)$, then $S_n f(x) = \int_D f(t) D_n(x - t) dt = D_n * f(x)$;*

(*partial sum of Fourier series*)

and
$$\lim_{n \to +\infty} S_{p^n} f(x) = f(x), \quad a.e.. \qquad (3.4.15)$$

Moreover, (3.4.15) holds at the continuous points of $f(x)$, and if $f(x)$ is continuous on D, then (3.4.15) holds uniformly on D;

(iv) *If $f \in L^2(D)$, then $\int_D |f|^2 dx = \sum_{n=0}^{+\infty} |c_n|^2$;* (*Parseval formula*)

(v) *If $f \in L^r(D), 1 \leqslant r < +\infty$, then*
$$\lim_{n \to +\infty} \|S_{p^n} f(\cdot) - f(\cdot)\|_{L^r(D)} = 0. \qquad (3.4.16)$$

For proofs please refer to [100].

Theorem 3.4.6 *Let $f : D \to \mathbb{C}$ be a Haar measurable function on D. We have*

(i) *If $f \in L^1(D)$, and*

$$\int_D \frac{|f(x_0 - z) - f(x_0)|}{|z|} dz < +\infty, \tag{3.4.17}$$

then

$$\lim_{n \to +\infty} S_n f(x_0) = f(x_0). \quad \text{(Dini convergence theorem)}$$

(ii) If $f \in L^1(D)$, and f is constant on each $x_0 + B^k$ for some $k \in \mathbb{Z}$, then

$$\lim_{n \to +\infty} S_n f(x_0) = f(x_0).$$

(iii) If $f \in L^1(D)$, and f is a radial function, i.e., $f(x) = \alpha_k$ for $|x| = p^{-k}$, $k \geqslant 0$. Then for $x \neq 0$, we have

$$\lim_{n \to +\infty} S_n f(x) = f(x).$$

Proof. To prove (i), for any $\varepsilon > 0$, by assumption (3.4.17), there exists $k \geqslant 0$, such that

$$\int_{B^k} \frac{|f(x_0 - z) - f(x_0)|}{|z|} dz < \frac{\varepsilon}{2p}.$$

Let

$$g_\varepsilon(z) = \{f(x_0 - z) - f(x_0)\}(1 - \Phi_{B^k}(z)),$$

by Theorem 3.4.4 (iv), we have

$$|S_n f(x_0) - f(x_0)| = \left| \int_D \{f(x_0 - z) - f(x_0)\} D_n(z) dz \right|$$

$$\leqslant p \int_{B^k} \frac{|f(x_0 - z) - f(x_0)|}{|z|} dz$$

$$+ \left| \int_D g_\varepsilon(z) D_n(z) dz \right|. \tag{3.4.18}$$

In virtue of selection of $k \geqslant 0$, the first term in (3.4.18) is $\leqslant \frac{\varepsilon}{2}$; and the support of function g_ε in the second term in (3,4,18) is in the complementary set $K_p \backslash B^k$, thus, if $n = m \cdot p^k + t$, then by the Theorem 3.4.4 (iii), we have

$$D_n(z) = \sum_{\mu=0}^{t-1} \chi_{m \cdot p^k + \mu}(z),$$

with $|z| > p^{-k}$, $0 \leqslant t < p^k$. So that the second term in (3.4.18) becomes

$$\int_{K_p} g_\varepsilon(z) D_n(z) dz = \int_{K_p} g_\varepsilon(z) \sum_{\mu=0}^{t-1} \chi_{m \cdot p^k + \mu}(z) dz = \sum_{\mu=0}^{t-1} \int_{K_p} g_\varepsilon(z) \chi_{m \cdot p^k + \mu}(z) dz$$

$$= \sum_{\mu=0}^{t-1} \overline{\int_{K_p} \overline{g_\varepsilon(z)} \overline{\chi}_{m \cdot p^k + \mu}(z) dz} = \sum_{\mu=0}^{t-1} \overline{c}_{m \cdot p^k + \mu},$$

where $c_{m \cdot p^k + \mu}$ are the Fourier coefficients of $g_\varepsilon(z)$, and the sum $\sum_{\mu=0}^{t-1} \overline{c}_{m \cdot p^k + \mu}$ has at most t terms with $0 \leqslant t < p^k$. On the other hand, Riemann–Lebesgue lemma implies

$$c_{mp^k + \mu} = o(1), \quad n = mp^k + \mu \to \infty,$$

thus

$$\left| \int_{K_p} g_\varepsilon(z) D_n(z) dz \right| \leqslant \sum_{\mu=0}^{t-1} \left| c_{m \cdot p^k + \mu} \right| \to 0.$$

We conclude that when $n \to +\infty$, the last two terms in (3.4.18) tends to zero, this implies (i).

The conchusions (ii) and (iii) hold as corollaries of on (i). The proof is complete.

(3) **$(c, 1)$-summation method** $D \subset K_p$.

There are also Fejér kernel and Fejér operator on D.

Let

$$c_k = \int_D f(x) \overline{\chi}_k(x) dx = \int_D f(x) \overline{\chi}_{u(k)}(x) dx, \quad k = 0, 1, \cdots,$$

be the Fourier coefficients of $f(x)$, and $S_n = \sum_{k=0}^{n-1} c_k \chi_k(x)$ the partial sum of Fourier series of $f(x)$. Let

$$\sigma_n = \frac{S_1 + S_2 + \cdots + S_n}{n}, \qquad (3.4.19)$$

$\{\sigma_n\}_{n=0}^{+\infty}$ is said to be a **$(c, 1)$-mean** of $f(x)$. And

$$K_n(x) = \begin{cases} \dfrac{1}{n} \sum_{k=1}^{n} D_k(x), & n \geqslant 1, \\ 0, & n = 0, \end{cases} \qquad (3.4.20)$$

is called **Fejér kernel**. Equivalently,

$$K_n(x) = \begin{cases} \sum_{k=0}^{n-1} \left(1 - \dfrac{k}{n}\right) \chi_k(x), & n \geqslant 1, \\ 0, & n = 0. \end{cases}$$

The **Fejér integral operator** is defined as
$$\sigma_n f(x) = \int_{K_p} f(x-t) K_n(t) dt = \int_{K_p} f(t) K_n(x-t) dt. \qquad (3.4.21)$$

Theorem 3.4.7 *Fejer kernel has the following properties*

(i) $\int_D K_n(x) dx = 1$, $\forall n \geqslant 1$;

(ii) $|K_n(x)| \leqslant \dfrac{n+1}{2}$, $\forall n \geqslant 1, x \in D$;

(iii) $|K_n(x)| \leqslant \dfrac{p}{|x|}$, $\forall n \geqslant 1$, $x \neq 0, x \in D$;

(iv) $\forall n = mp^k + t, m \geqslant 0, k \geqslant 0, 0 \leqslant t < p^k$, then
$$nK_n(x) = p^k D_{p^k}(x) m K_m(\beta^{-k}x) + t D_{p^k}(x) D_m(\beta^{-k}x)$$
$$+ D_m(\beta^{-k}x) p^k K_{p^k}(x) + \chi_m(\beta^{-k}x) \cdot tK_t(x), x \in D;$$

(v) $\forall p^l \leqslant n < p^{l+1}$, $|x| = p^{-k+1}$, $1 \leqslant k \leqslant l-1$, *there exists a constant* $A > 0$, *independent of* n *and* k, *such that* $\int_{|x|=p^{-k+1}} |K_n(x)| dx \leqslant A p^{\frac{k-l}{2}}$;

(vi) $\int_D |K_n(x)| dx \leqslant B$, *constant* B *is independent of* n;

(vii) $\int_{p^{-k} \leqslant |x| \leqslant 1} |K_n(x)| dx = o(1)$, $n \to +\infty$, $\forall k \geqslant 0$.

Proof. For (i), we can obtain by
$$\int_D K_n(x) dx = \frac{1}{n} \sum_{k=1}^n \int_D D_k(x) dx = \frac{1}{n} \sum_{k=1}^n \int_D 1 dx = 1.$$

For (ii), since $|D_n(x)| \leqslant n$, it follows
$$|K_n(x)| \leqslant \frac{1}{n} \sum_{k=1}^n |D_k(x)| \leqslant \frac{1}{n} \sum_{k=1}^n k = \frac{1}{n} \frac{n(n+1)}{2} = \frac{n+1}{2}.$$

For (iii), since $|D_n(x)| \leqslant \dfrac{p}{|x|}$ implies $|K_n(x)| \leqslant \dfrac{p}{|x|}, x \neq 0, x \in D$.

For (iv), by Theorem 3.4.4 (iii), $\forall m \in \mathbb{P}, k \in \mathbb{P}, 0 \leqslant t < p^k$,
$$D_{m \cdot p^k + t}(x) = D_{p^k}(x) D_m(\beta^{-k}x) + \chi_m(\beta^{-k}x) D_t(x),$$
then for $x \in D$
$$nK_n(x) = \sum_{\mu=0}^{m-1} \sum_{\nu=1}^{p^k} D_{\mu p^k + \nu}(x) + \sum_{\nu=1}^t D_{mp^k + \nu}(x),$$

by computing it we get (iv).

For (v), by assumption, $p^l \leqslant n < p^{l+1}$, $|x| = p^{-k+1}$, $1 \leqslant k \leqslant l-1$, thus,

$$1 \leqslant k \leqslant l-1 \Rightarrow l > k \geqslant 1;$$
$$p^l \leqslant n < p^{l+1} \Rightarrow p^{l-k} \leqslant np^{-k} < p^{l-k+1};$$
$$n = mp^k + t, 0 \leqslant t < p^k \Rightarrow p^{l-k-1} < m < p^{l-k+1}.$$

On the other hand, since $|x| = p^{-k+1} > p^{-1}(k > 2 \Rightarrow -k < -2 \Rightarrow p^{-k+1} > p^{-2+1} = p^{-1})$, so that $D_{p^k}(x) = 0$. Then by (iv), we have

$$nK_n(x) = p^k D_{p^k}(x) m K_m\left(\beta^{-k}x\right) + t D_{p^k}(x) D_m\left(\beta^{-k}x\right)$$
$$+ D_m\left(\beta^{-k}x\right) p^k K_{p^k}(x) + \chi_m\left(\beta^{-k}x\right) \cdot tK_t(x)$$
$$= D_m\left(\beta^{-k}x\right) p^k K_{p^k}(x) + \chi_m\left(\beta^{-k}x\right) \cdot tK_t(x)$$
$$= n\left\{\frac{1}{n}D_m\left(\beta^{-k}x\right) p^k K_{p^k}(x) + \frac{1}{n}\chi_m\left(\beta^{-k}x\right) \cdot tK_t(x)\right\}$$
$$\equiv nI_1^n(x) + nI_2^n(x).$$

Estimate $I_2^n(x)$: since $|x| = p^{-k+1}$ implies $t^2 < \left(p^k\right)^2 = \dfrac{p^2}{|x|^2}$, thus

$$|I_2^n(x)| \leqslant \frac{t}{n}|K_t(x)| \leqslant \frac{t^2}{n} < \frac{p^2}{n|x|^2}.$$

Then

$$\int_{|x|=p^{-k+1}} |I_2^n(x)|\, dx \leqslant \frac{A}{n}\int_{|x|=p^{-k+1}} |x|^{-2} dx = \frac{A}{n} p^{2k-1}\int_{|x|=p^{-k+1}} dx$$
$$= \frac{A}{n} p^{2k-1} p^{-k+1}\left(1 - p^{-1}\right) < \frac{A}{n} p^k \leqslant Ap^{k-l},$$

where constant $A > 0$ is independent of n and k.

Estimate $I_1^n(x)$: since $|I_1^n(x)| \leqslant \dfrac{p^{2k}}{n}\left|D_m\left(\beta^{-k}x\right)\right|$, thus

$$\int_{|x|=p^{-k+1}} |I_1^n(x)|\, dx \leqslant \frac{p^{2k}}{n}\int_{|x|=p^{-k+1}} \left|D_m\left(\beta^{-k}x\right)\right| dx$$
$$= \frac{p^{2k}p^{-k}}{n}\int_{|x|=1} \left|D_m\left(\beta^{-1}x\right)\right| dx$$
$$\leqslant \frac{p^k}{n}\int_{|x|\leqslant 1} \left|D_m\left(\beta^{-1}x\right)\right| dx$$
$$\leqslant \frac{p^k}{n}\left\{\int_{|x|\leqslant 1} \left|D_m\left(\beta^{-1}x\right)\right|^2 dx\right\}^{\frac{1}{2}}$$

$$\leqslant \frac{p^k \sqrt{m}}{n} = \frac{mp^k}{n} m^{-\frac{1}{2}} \leqslant A p^{\frac{k-l}{2}}.$$

Note that $l > k$, and combining the above two estimating results, we have (v).

For (vi), $\forall n$, we may assume that $p^l \leqslant n < p^{l+1}$. Then, $\forall x$, it has

$$|K_n(x)| \leqslant n < p^{l+1} \quad \text{and} \quad \int_{|x|<p^{-l+1}} |K_n(x)|\, dx \leqslant p^{l+1} p^{-l+1} = p^2.$$

Then by (v),

$$\int_{|x|\geqslant p^{-l+2}} |K_n(x)|\, dx = \sum_{k=1}^{l-1} \int_{|x|=p^{-k+1}} |K_n(x)|\, dx$$

$$\leqslant A \sum_{k=1}^{l-1} p^{\frac{k-l}{2}} = A \sum_{j=1}^{l-1} p^{\frac{-j}{2}} \leqslant A \left(1 - p^{-\frac{1}{2}}\right)^{-1} \leqslant A,$$

Take $B = p^2 + A$, (vi) is proved.

Finally, for (vii), since $p^l \leqslant n < p^{l+1}$ implies $l = l(n) \to +\infty$ as $n \to +\infty$, thus

$$\int_{p^{-k}\leqslant |x|\leqslant 1} |K_n(x)|\, dx = \sum_{j=1}^{k+1} \int_{|x|=p^{-j+1}} |K_n(x)|\, dx \leqslant A \sum_{j=1}^{k+1} p^{\frac{j-l}{2}}$$

$$\leqslant A p^{\frac{k-l}{2}} = o(1), \quad n \to +\infty,$$

this is (vii). The proof is complete.

Theorem 3.4.8 Fejér integral $\sigma_n f(x) = \int_D f(x-t) K_n(t)\, dt$ has the properties

(i) If $f \in L^r(D), 1 \leqslant r < +\infty$, then $\lim\limits_{n \to +\infty} \|\sigma_n f(\cdot) - f(\cdot)\|_{L^r(D)} = 0$.

(ii) If $f(x)$ is continuous on D, then $\lim\limits_{n \to +\infty} \sigma_n f(x) = f(x)$ is uniformly on D.

(iii) If $f(x)$ is bounded on D, and $f(x)$ is continuous at $x_0 \in D$, then

$$\lim_{n \to +\infty} \sigma_n f(x_0) = f(x_0).$$

(iv) If $f(x)$ is bounded oscillating function on D, then

$$\lim_{n \to +\infty} \sigma_n f(x) = f(x)$$

holds at all continuous points of $f(x)$.

(v) If $f(x)$ is bounded variation continuous function on D, then

$$\lim_{n\to+\infty} S_n f(x) = f(x)$$

holds uniformly on D, where a **bounded variation function** on D is defined as: for any $k \geq 0$, let the system $\{S_\mu^k\}_{\mu=1}^{p^k}$ be the set of all cosets of B^k in D, i.e., $\{S_\mu^k\}_{\mu=1}^{p^k} = \{D/B^k\}$. Then $f(x)$ is said to be a bounded variation function on D, if

$$V^* f \equiv \sup_{k\geq 0} \sum_{\mu=1}^{p^k} \sup_{x,y\in S_\mu^k} |f(x) - f(y)| < +\infty;$$

a **total variation** of $f(x)$, denoted by Vf, defined as

$$Vf \equiv \limsup_{k\geq 0} \sum_{\mu=1}^{p^k} \sup_{x,y\in S_\mu^k} |f(x) - f(y)| < +\infty.$$

Proof. By

$$\sigma_n f(x) - f(x) = \int_D \{f(x-t) - f(x)\} K_n(t) dt,$$

and by the Minkovski inequality, we have

$$\|\sigma_n f(\cdot) - f(\cdot)\|_{L^r(D)} \leq \int_D \|f(\cdot - t) - f(\cdot)\|_{L^r(D)} |K_n(t)| dt.$$

Since

$$f \in L^r(D), 1 \leq r < +\infty \Rightarrow \|f(\cdot - t) - f(\cdot)\|_{L^r(D)} = o(1), \quad |t| \to 0;$$

$$f \in L^r(D), 1 \leq r < +\infty \Rightarrow \|f(\cdot - t) - f(\cdot)\|_{L^r(D)} \leq 2\|f\|_{L^r(D)}.$$

For any $\varepsilon > 0$, choose $k = k(\varepsilon) > 0$, such that for $|t| < p^{-k}$, we have

$$\|f(\cdot - t) - f(\cdot)\|_{L^r(D)} \leq \frac{\varepsilon}{2A},$$

with $A > 0$ to be the constant in the Theorem 3.4.7 (v). Thus,

$$\|\sigma_n f(\cdot) - f(\cdot)\|_{L^r(D)}$$
$$\leq \frac{\varepsilon}{2A} \int_{|t|\leq p^{-k}} |K_n(t)| dt + 2\|f\|_{L^r(D)} \int_{p^{-k}<|t|\leq 1} |K_n(t)| dt$$
$$\leq \frac{\varepsilon}{2A} \int_D |K_n(t)| dt + 2\|f\|_{L^r(D)} \cdot o(1) \leq \frac{\varepsilon}{2} + o(1),$$

for the chosen $k = k(\varepsilon) > 0$ above, by the Theorem 3.4.7 (vii), as $n \to +\infty$, we have

$$\lim_{n\to+\infty} \|\sigma_n f(\cdot) - f(\cdot)\|_{L^r(D)} = 0,$$

so, (i) holds.

If $f(x)$ is continuous on D, then by compactness of D, for each $\varepsilon > 0$, there exists $k = k(\varepsilon) > 0$, such that for $|t| < p^{-k}$, one has
$$|f(x-t) - f(x)| < \frac{\varepsilon}{A}, \quad \text{uniformly on } x \in D,$$
then
$$|\sigma_n f(x) - f(x)| \leqslant \int_D |f(x-t) - f(x)| |K_n(t)| \, dt < \frac{\varepsilon}{A} \int_D |K_n(t)| \, dt < \varepsilon,$$
thus, (ii) holds.

For (iii), similar to the method used in that of (ii).

For (iv), by virtue of the following two facts:

① If $f(x)$ is a bounded variation function on D, then $f(x)$ is continuous on $D \backslash E$, where E is a null measure set in D, i.e., $\mu E = 0$†.

② If $f(x)$ is a bounded variation function on D, then the Fourier coefficients of $f(x)$ satisfy $|nc_n| \leqslant pV^* f$, $n \geqslant 1$; and $\limsup_n |nc_n| \leqslant pV^* f$; especially, we have $c_n = O\left(\dfrac{1}{n}\right)$.††

Then by $c_n = O\left(\dfrac{1}{n}\right)$, and (iii), we get (iv).

Finally, to prove (v), we need the following two facts about the Fourier series:

③ If $S_n f \to Sf$, then $\sigma_n f \to \sigma f$.

④ If $\sigma_n f \to \sigma f$, and $c_n = O\left(\dfrac{1}{n}\right)$, then $S_n f \to \sigma f$.

Then by the assumption of (v), $f(x)$ is a bounded variation function on D, so that
$$c_n = O\left(\frac{1}{n}\right).$$
Since $f(x)$ is continuous on D, thus (iv) gives $\sigma_n f \to \sigma f$ on D, thus, (v) holds.

Remark 1. The proof of †.

In fact, let
$$\mathrm{OSC} f(x) = \limsup_{k \geqslant 0} \sup_{y \in B^k} |f(x+y) - f(x)|,$$
then
$$f(x) \text{ is continuous at } x \in D \text{ if and only if } \mathrm{OSC} f(x) = 0.$$
Thus, let

$$E_0 = \{x \in D : \mathrm{OCS}f(x) > 1\},$$
$$E_k = \{x \in D : 2^{-k} < \mathrm{OCS}f(x) < 2^{-k+1}\}, \quad k = 1, 2, \cdots.$$

It is easy to see that $E_k, k = 0, 1, 2, \cdots$ are finite sets (for example, $\forall F \subset E_0$, we have
$$\sum_{x \in F} \mathrm{OCS}f(x) \leqslant V^*f.$$
Thus, card $F < V^*f$, and E_0 is finite). Then, the set of all discontinuous points of $f(x)$, $E = \bigcup_{k \geqslant 0} E_k$ is a countable set.

Remark 2. The proof of ††.

In fact, let $n > 0$, $|u(n)| = p^k$, $k \geqslant 1$, and $p^{k-1} \leqslant n < p^k$. Then, take a sequence $\{a_j^k\}$, such that it satisfies $\int_{S_j^{k-1}} \overline{\chi}_n(x)dx = 0$ on $S_j^k = a_j^k + B^k$. So we have
$$c_n = \int_D f(x)\overline{\chi}_n(x)dx = \sum_{j=1}^{p^{k-1}} \int_{S_j^{k-1}} f(x)\overline{\chi}_n(x)dx$$
$$= \sum_{j=1}^{p^{k-1}} \int_{S_j^{k-1}} \{f(x) - f(a_j^{k-1})\}\overline{\chi}_n(x)dx,$$
and
$$|c_n| \leqslant \sum_{j=1}^{p^{k-1}} \sup_{x,y \in S_j^{k-1}} |f(x) - f(a_j^{k-1})| \cdot p^{-k+1} \leqslant p\frac{1}{n}V^*f.$$

We summarize that:
$$f(x) \text{ is a bounded variation function on } D \Rightarrow c_n = O\left(\frac{1}{n}\right);$$
$$f \in L^1(D) \Rightarrow c_n = o\left(\frac{1}{n}\right).$$

(4) Approximation theory on $D \subset K_p$.

Now we turn to study approximation theory on D, including the continuous modulus, Lipschitz class, the best approximation of function; moreover, show the equivalent approximation theorem on compact group D.

Definition 3.4.3 (Continuous modulus, Lip class, the best approximation on D) Let

$$f \in X(D) = \begin{cases} C(D), \\ L^r(D), & 1 \leqslant r < +\infty. \end{cases}$$

For $\delta > 0$,

$$\omega(X(D), f, \delta) = \sup_{h \in D, |h| \leqslant \delta} \|f(\cdot + h) - f(\cdot)\|_{X(D)}$$

is said to be a **continuous modulus** of $f(x)$ in the space $X(D)$.

For $\alpha > 0$,

$$\text{Lip}(X(D), \alpha) = \{f \in X(D) : \|f(\cdot + h) - f(\cdot)\|_{X(D)} = O(|h|^\alpha), h \in D\}$$

is said to be a **Lipschitz class in the space** $X(D)$.

Then

$$E_n(X(D), f) = \inf_{p_n \in \mathbb{P}_n} \|f - p_n\|_{X(D)}$$

is said to be **the best approximation** of $f(x)$ by n-order polynomials of characters $p_n \in \mathbb{P}_n$ in $X(D)$, where

$$\mathbb{P}_n = \{p_k(x) : 0 \leqslant k \leqslant n, x \in D\}$$

with $n \in \{0, 1, 2, \cdots\}$ is a non-negative integer, and $p_k(x)$ is a polynomial of characters

$$p_k(x) = a_k \chi_k(x) + a_{k-1} \chi_{k-1}(x) + \cdots + a_1 \chi_1(x) + a_0,$$

coefficients $a_j \in \mathbb{C}$, $0 \leqslant j \leqslant k$, for $k \in \{0, 1, 2, \cdots\}$ with $k \leqslant n$.

These concepts in local fields are similar to those of in $\mathbb{R}$, they are very fundamental, very important and very essential in the abstract harmonic analysis over local fields.

① Basic properties of continuous modulus $\omega(X(D), f, \delta)$

(a) $\omega(X(D), f, \delta)$ is a monotone increasing function of $\delta \geqslant 0$;

(b) $\lim_{\delta \to +0} \omega(X(D), f, \delta) = 0$;

(c) $\omega(X(D), f, \delta) = o(\delta), \delta \to +0$ implies $f = c$, a.e.;

(d) $f \in L^r(D)$, $1 \leqslant r < +\infty$, implies $\omega(L^r(D), f, p^{-k}) = o(1), k \to +\infty$;

$f \in C(D)$ implies $\omega(L^\infty(D), f, p^{-k}) = o(1), k \to +\infty$;

(e) $f \in L^1(D)$ implies $|c_n| \leqslant \omega(L^1(D), f, p|u(n)|^{-1})$, $n \to +\infty$.

Proof. We only prove (e): take $n > 0$, then $\exists h \in D$, satisfies

$|hu(n)| = p$, such that $|\chi_n(h) - 1| \geq 1$, and holds

$$\chi_n(h) c_n = \int_D f(x+h) \overline{\chi}_n(x) dx.$$

Thus,

$$(\chi_n(h) - 1) c_n = \int_D \{f(x+h) - f(x)\} \overline{\chi}_n(x) dx,$$

that is

$$|c_n| \leq \frac{1}{|\chi_n(h) - 1|} \int_D |\{f(x+h) - f(x)\}| |\overline{\chi}_n(x)| dx$$

$$\leq \frac{1}{|\chi_n(h) - 1|} \int_D |\{f(x+h) - f(x)\}| dx \leq \omega\left(L^1(D), f, |h|\right)$$

$$= \omega\left(L^1(D), f, p|u(n)|^{-1}\right).$$

② Basic properties of Lipschitz class $\operatorname{Lip}(X(D), \alpha)$

(a) $f \in \operatorname{Lip}(X(D), \alpha)$ implies $\exists M > 0$, such that $\omega(X(D), f, \delta) \leq M\delta^\alpha$;

(b) $f \in \operatorname{Lip}(X(D), \alpha)$, $g \in L^1(D)$ implies $f * g \in \operatorname{Lip}(X(D), \alpha)$;

(c) $f \in \operatorname{Lip}(X(D), \alpha)$ implies $c_n = O(n^{-\alpha})$, $n \to +\infty$;

$f \in \operatorname{Lip}(X(D), \alpha), \alpha > \dfrac{1}{2}$ implies $\displaystyle\sum_{n=0}^{+\infty} |c_n| < +\infty$.

Proof. We only prove (c):

The first result of (c) comes from (a). The second result is proved as follows. Since

$$\sum_{j=0}^{+\infty} |c_j| = |c_0| + \sum_{j=1}^{+\infty} \sum_{|u(n)|=p^k} |c_n| \leq |c_0| + \sum_{j=1}^{+\infty} p^{\frac{k}{2}} \left\{ \sum_{|u(n)|=p^k} |c_n|^2 \right\}^{\frac{1}{2}}.$$

Fix k, $k \geq 1$, take $h \neq 0$, then

$$\omega^2(L^2(D), f, |h|) \geq \int_D |f(x+h) - f(x)|^2 dx = \sum_{j=0}^{+\infty} |\chi_j(h) - 1|^2 |c_j|^2$$

$$\geq \sum_{|u(n)|=p^k} |\chi_n(h) - 1|^2 |c_n|^2. \tag{3.4.22}$$

Now we assume that $|u(n)| = p^k$, then $\exists h \in D, |h| = p^{-k+1}$, such that

$$|\chi_n(h) - 1| > 1,$$

this is because: at least one of the values of $\chi_n(h)$ has negative real part on the $p-1$ cosets of B^k in $\{B^{k-1}\backslash B^k\}$. Take summation on the $p-1$ cosets for (3.4.22), we have

$$(p-1)\omega^2\left(L^2(D), f, p^{-k+1}\right) \geqslant \sum_{|u(n)|=p^k}^{+\infty} |c_n|^2.$$

So that

$$\sum_{k=0}^{+\infty} |c_n| \leqslant |c_0| + A \sum_{k=1}^{+\infty} p^{\frac{k}{2}} \omega\left(L^2(D), f, p^{-k+1}\right). \quad (3.4.23)$$

Then we suppose that $f \in \text{Lip}(X(D), \alpha)$, $\alpha > \dfrac{1}{2}$, thus

$$\omega\left(L^2(D), f, p^{-k+1}\right) \leqslant \omega\left(L^\infty(D), f, p^{-k+1}\right) = O\left(p^{-k\alpha}\right).$$

Finally, by substituting (3.4.23), we get

$$\sum |c_n| \leqslant |c_0| + A \sum_{k=1}^{+\infty} p^{\frac{k}{2}} p^{-k\alpha} = |c_0| + A \sum_{k=1}^{+\infty} p^{k(\frac{1}{2}-\alpha)} < +\infty.$$

The proof is complete.

Theorem 3.4.9 *On the compact group $D \subset K_p$ of local field K_p, the following statements are equivalent: for $s \geqslant 0$,*

(i) $f^{\langle s \rangle} \in \text{Lip}(X(D), \alpha)$, $\alpha > 0$;

(ii) $\omega\left(X(D), f^{\langle s \rangle}, \delta\right) = O(\delta^\alpha)$, $\delta \to 0$;

(iii) $E_{p^n}(X(D), f) = O\left(p^{-n(\alpha+s)}\right)$, $n \to +\infty$;

(iv) $\|f(\cdot) - S_{p^n}(X(D), f, \cdot)\|_{X(D)} = O\left(p^{-n(\alpha+s)}\right)$, $n \to +\infty$,

where Lip, ω, E_{p^n}, S_{p^n} *are the Lipschitz class, continuous modulus, the best approximation, the p^n-partial sum of Fourier series on space* $X(D) = \begin{cases} C(D), \\ L^r(D), \ 1 \leqslant r < +\infty, \end{cases}$ *respectively.*

The special case of this theorem for $p=2$ is listed in the Section 3.3.4, Remark 2 ④.

For the approximation identity kernels and approximation operators on Walsh system, and on the compact group D in a local field K_p are listed:

③ Abel–Poisson type kernel[77].

$$\lambda_r(x) = \frac{1}{1-r} \prod_{k=1}^{+\infty} \frac{1 - r^{p^{k-1}}}{1 - \omega^{x_k} r^{p^{k-1}}}, \quad \text{with } \omega = e^{\frac{2\pi i}{p}}.$$

Theorem 3.4.10 *For $\alpha > 0$, $s \in \mathbb{P}$, if $D^{\langle s \rangle} f \in \mathrm{Lip}\,(X(D), \alpha)$, then as $r \to 1$,*

$$\|\lambda_r * f(\cdot) - f(\cdot)\|_{X(D)} = \begin{cases} O\left((1-r)^{s+\alpha}\right), & 0 < \alpha < 1, \\ O\left((1-r)^s \ln \dfrac{1}{1-r}\right), & \alpha = 1, \\ O((1-r)^s), & \alpha > 1. \end{cases}$$

Conversely, if $f \in X(D)$ satisfies

$$\|\lambda_r * f(\cdot) - f(\cdot)\|_{X(D)} = O\left((1-r)^{s+\alpha}\right), \quad r \to 1,$$

then

$$D^{\langle s \rangle} f \in \mathrm{Lip}\,(X(D), \alpha).$$

④ de la Vallée–Poussin type kernel[31].

$$V_n(x) = \frac{1}{c_n} \left\{ \frac{1}{2} \chi_n^{\frac{1}{2}}(x) + \frac{1}{2} \overline{\chi_n^{\frac{1}{2}}}(x) \right\}^{\varphi(k)}, \quad x \in D,$$

with $\varphi(k) = 2p^{bk}$, $b > 0$, χ_n is a character of D.

The approximation theorem holds for $V_n(x)$.

⑤ Typical means kernel[125],[126].

$$K_{n,\lambda}(x) = \sum_{k=0}^{n-1} \left(1 - \left(\frac{k}{n}\right)^\lambda \right) \overline{w}_k(x), \quad \lambda > 0, \quad x \in D,$$

$w_k(x)$ are in the Walsh system. The approximation theorem holds for $K_{n,\lambda}(x)$.

2. Approximation theory on the locally compact group K_p^+ in K_p

Definition 3.4.4 (Continuous modulus, Lip class, the best approximation on K_p) *Denote $X(K_p)$ the set of complex valued Haar measurable functions $f: K_p \to \mathbb{C}$ on K_p that*

$$X(K_p) = \begin{cases} C(K_p), \\ L^r(K_p), 1 \leqslant r < +\infty, \end{cases}$$

with norms

$$\|f\|_{X(K_p)} = \begin{cases} \|f\|_{C(K_p)} = \sup_{x \in K_p} |f(x)|, \\ \|f\|_{L^r(K_p)} = \left\{ \displaystyle\int_{K_p} |f(x)|^r dx \right\}^{\frac{1}{r}}, 1 \leqslant r < +\infty. \end{cases}$$

For $\delta > 0$,
$$\omega(X(K_p), f, \delta) = \sup_{h \in K_p, |h| \leqslant \delta} \|f(\cdot + h) - f(\cdot)\|_{X(K_p)} \qquad (3.4.24)$$
is said to be a **continuous modulus** of $f(x)$ in $X(K_p)$.

For $\alpha > 0$,
$$\text{Lip}(X(K_p), \alpha) = \left\{ f \in X(K_p) : \|f(\cdot + h) - f(\cdot)\|_{X(K_p)} = O(|h|^\alpha), h \in K_p \right\} \qquad (3.4.25)$$
is said to be a **Lipschitz calss** in $X(K_p)$.

Then
$$E_n(X(K_p), f) = \inf_{\varphi \in \mathbb{S}_n(K_p)} \|f - \varphi\|_{X(K_p)} \qquad (3.4.26)$$
is said to be a the **best approximation** of $f(x)$ by the element in $\mathbb{S}_n(K_p)$, where
$$\mathbb{S}_n(K_p) = \{\varphi \in \mathbb{S}(K_p) : \text{index}\varphi = (k, l), k \leqslant n, l \in \mathbb{Z}\}$$
is the set of all $\varphi \in \mathbb{S}(K_p)$ with index pair $(k, l) \in \mathbb{Z} \times \mathbb{Z}$ (Theorem 3.1.7) and $k \leqslant n$ for fixed $n \in \mathbb{Z}$.

Theorem 3.4.11 *The following statements are equivalent each other for $\alpha > 0$ and $s \geqslant 0$ on K_p,*

(i) $f^{\langle s \rangle} \in \text{Lip}(X(K_p), \alpha)$;

(ii) $\omega\left(X(K_p), f^{\langle s \rangle}, \delta\right) = O(\delta^\alpha), \delta \to 0$;

(iii) $E_{p^n}(X(K_p), f) = O\left(p^{-n(\alpha+s)}\right), n \to +\infty$.

This Theorem has been listed in the Section 3.3.4, Subsection 3 (3).

We recall the results of approximation identity kernels and approximation operators on $\mathbb{R}^1$ before showing the results of those on K_p.

(1) Approximation identity kernels and approximation operators on $\mathbb{R}^1$

Definition 3.4.5 (Approximation identity kernel on $\mathbb{R}$) *Let a function set*
$$\{\kappa(x, \rho) : x \in \mathbb{R}\}$$
*with a positive parameter $\rho \in J \subseteq [0, +\infty)$ satisfy: $\kappa(x, \rho) \in L^1(\mathbb{R})$, $\forall \rho > 0$; and a normalized condition $\int_{\mathbb{R}} \kappa(u, \rho) du = \sqrt{2\pi}$ for $\rho > 0$. Then it is said to be a **kernel** on $\mathbb{R}$.*

A kernel $\{\kappa(x,\rho) : x \in \mathbb{R}\}$ is said to be **a real**, or **bounded**, or **continuous**, or **absolutely continuous kernel**, respectively, if $\kappa(x,\rho)$ is real, or bounded, or continuous, or absolutely continuous for each $\rho > 0$; and it is said to be **even**, if $\kappa(x,\rho) = \kappa(-x,\rho)$; or **positive**, if $\kappa(x,\rho) \geqslant 0$ for each $\rho > 0$, respectively.

Furthermore, $\{\kappa(x,\rho) : x \in \mathbb{R}\}_{\rho \in J \subseteq \mathbb{R}}$ is said to be an **approximation identity kernel**, if $\exists M > 0$, such that $\|\kappa(\cdot,\rho)\|_1 \leqslant M$, $\forall \rho > 0$, and

$$\lim_{\rho \to +\infty} \int_{|u| \leqslant \delta} |\kappa(u,\rho)|\, du = 0, \text{ for } \delta > 0.$$

Correspondingly, the integral

$$I(f,x,\rho) = \frac{1}{\sqrt{2\pi}} \int_{\mathbb{R}} f(x-u)\kappa(x,\rho)\, du, \quad \forall \rho > 0$$

is said to be the **approximation identity integral generated by kernel** $\{\kappa(u,\rho)\}_{\rho>0}$, or the **singular integral with kernel** $\{\kappa(x,\rho)\}_{\rho>0}$.

An approximation identity integral is a convolution type operator, sometimes, as a form

$$I(f,x,\rho) = \frac{\rho}{\sqrt{2\pi}} \int_{\mathbb{R}} f(x-u)\kappa(x\rho)\, du, \quad \forall \rho > 0.$$

(A) Gauss–Weierstrass singular integral operator

Let $W(x) = 2^{-1/2} e^{-x^2/4}$, take $\rho = t^{-1/2}, t \to 0^+$. Then

$$W(f,x,t) = \frac{\rho}{\sqrt{2\pi}} \int_{-\infty}^{+\infty} f(x-u) W(\rho u)\, du = \frac{1}{\sqrt{4\pi t}} \int_{-\infty}^{+\infty} f(x-u) e^{-u^2/4t}\, du$$

is the Gauss–Weierstrass singular integral operator. It has the following properties:

Let $X \equiv X(\mathbb{R})$, we have

① $\|W(f,\cdot,t)\|_X \leqslant \|f\|_X$, $t > 0$;

② $\lim\limits_{t \to 0^+} W(f,x,t) = f(x)$, a.e. $x \in \mathbb{R}$ for $f \in L^r(\mathbb{R}), 1 \leqslant r < \infty$;

③ $\|W(f,\cdot,t) - f(\cdot)\|_X = O\left(\omega^*\left(X,f,t^{1/2}\right)\right)$, $t \to 0^+$;

④ $0 < \alpha < 2$, then
$$f \in \mathrm{Lip}^*\alpha \Leftrightarrow \|W(f,\cdot,t) - f(\cdot)\|_X = O\left(t^{\alpha/2}\right), \quad t \to 0^+;$$

⑤ $\alpha = 2$, then
$$f \in \mathrm{Lip}^*2 \Rightarrow \|W(f,\cdot,t) - f(\cdot)\|_X = O(t), \quad t \to 0^+,$$

where $\omega^*(X,f,t)$ and $\mathrm{Lip}^*\alpha$ are the second order continuous modulus and second order Lipschitz class.

(B) Fejér singular integral operator

Let $F(x) = \dfrac{1}{\sqrt{2\pi}} \left[\dfrac{\sin(x/2)}{x/2} \right]^2$, take $\rho \to +\infty$. Then

$$\sigma(f, x, t) = \dfrac{\rho}{\sqrt{2\pi}} \int_{-\infty}^{+\infty} f(x-u) F(\rho u)\, du$$

is the Fejér integral operator. It has the following properties:

① $\|\sigma(f, \cdot, t)\|_X \leqslant \|f\|_X$, $t > 0$;

② $\lim\limits_{t \to 0^+} \sigma(f, x, t) = f(x)$, a.e. $x \in \mathbb{R}$, $\forall f \in L^r(\mathbb{R})$, $1 \leqslant r < \infty$;

③ $\|\sigma(f, \cdot, \rho) - f(\cdot)\|_X = O(\omega^*(X, f, \rho^{-1}))$, $\rho \to +\infty$;

④ $0 < \alpha < 1$, then

$$f \in \text{Lip } \alpha \Leftrightarrow \|\sigma(f, \cdot, \rho) - f(\cdot)\|_X = O(\rho^{-\alpha}), \quad \rho \to +\infty;$$

⑤ $\alpha = 1$, then

$$f \in \text{Lip}^* 1 \Rightarrow \|\sigma(f, \cdot, \rho) - f(\cdot)\|_X = O(\rho^{-1}), \quad \rho \to +\infty.$$

(C) Cauchy–Poisson singular integral operator

Let $p(x) = \sqrt{\dfrac{2}{\pi}} \dfrac{1}{1+x^2}$, take $\rho = y^{-1}, y \to 0^+$. Then

$$P(f, x, y) = \dfrac{y}{\pi} \int_{-\infty}^{+\infty} \dfrac{f(x-u)}{y^2 + u^2}\, du$$

is the Cauchy–Poisson singular integral operator. It has the following properties:

① $\|P(f, \cdot, y)\|_X \leqslant \|f\|_X$, $y > 0$;

② $\lim\limits_{y \to 0^+} P(f, x, y) = f(x)$, a.e. $x \in \mathbb{R}$ for $f \in L^r(\mathbb{R})$, $1 \leqslant r < \infty$;

③ $\|P(f, \cdot, y) - f(\cdot)\|_X = O(\omega^*(X, f, y))$, $y \to 0^+$;

④ $0 < \alpha < 1$, then

$$f \in \text{Lip } \alpha \Leftrightarrow \|P(f, \cdot, y) - f(\cdot)\|_X = O(y^\alpha), \quad y \to 0^+,$$

⑤ $\alpha = 1$, then

$$f \in \text{Lip}^* 1 \Rightarrow \|P(f, \cdot, y) - f(\cdot)\|_X = O(y), \quad y \to 0^+.$$

(2) Approximation identity kernels and approximation operators on K_p

We use a local field K_p as an underlying space, with non-Archimedean valued norm $|x|$, $x \in K_p$.

(A) Poisson type kernel

$$R_y(x) = p^{-s}|y|^{-1}\Phi_s\left(xy^{-1}\right)\left\{1+\sum_{k=1}^{m}c_k\chi_{xy^{-1}}\left(\alpha_k\beta^{-s-1}\right)\right\}, \quad x \in K_p,$$

where $y \in K_p$ is a parameter, $s \in \mathbb{Z}$, $\alpha_k \in K_p$, $c_k \in \mathbb{C}$, $k = 1, 2, \cdots, m$; $m \in \{1, 2, \cdots, p-1\}$.

Theorem 3.4.12[83]   It holds on the K_p^+

(i) For any $m \geqslant 0$, the p-type derivative $D^{\langle m \rangle}R_y(x) = R_y^{\langle m \rangle}(x) \in L^1(K_p)$ exists, and

$$\left\|R_y^{\langle m \rangle}\right\|_{L^1(K_p)} = O\left(|y|^{-m}\right), \quad |y| \to 0.$$

(ii) For $1 \leqslant r < +\infty$, any $m \geqslant 0$ and $\alpha > 0$, then

$$D^{\langle m \rangle}f \in \text{Lip}\left(L^r(K_p), \alpha\right) \quad \text{implies} \quad \left\|R_y^*f(\cdot) - f(\cdot)\right\|_{L^r(K_p)}$$
$$= O\left(|y|^{m+\alpha}\right), \quad |y| \to 0;$$

(iii) For $1 \leqslant r < +\infty$, any $m \geqslant 0$ and $\alpha > 0$, if $f \in L^r(K_p)$ satisfies

$$\left\|R_y^*f(\cdot) - f(\cdot)\right\|_{L^r(K_p)} = O\left(|y|^{m+\alpha}\right), \quad |y| \to 0,$$

then $D^{\langle m \rangle}f \in L^r(K_p)$, and $\omega\left(L^r(K_p), D^{\langle m \rangle}f, \delta\right) = \begin{cases} O(\delta^\alpha), & 0 < \alpha < 1, \\ O(\delta|\ln\delta|), & \alpha = 1. \end{cases}$

(B) de la Vallee–Poussin type kernel

$$V_y(x) = \frac{1}{c_n}\Phi_0(yx)\left\{\frac{1}{2}\chi_{y^{-1}}^{\frac{1}{2}}(x) + \frac{1}{2}\overline{\chi}_{y^{-1}}^{\frac{-1}{2}}(x)\right\}^{\varphi(y)}, \quad x, y \in K_p,$$

where $\varphi(y) = 2|y|^{-b}$, $b > 4$. It has any order p-type derivatives and approximation theorem holds (see [31], [32]).

(C) Radial approximation identity kernel

If $\omega_0 \in L^1(\Gamma_p)$ satisfies

① $\omega_0 \in L^1(\Gamma_p)$ is radial, that is, $\omega_0(\xi) = \omega_0(|\xi|)$, $\xi \in \Gamma_p$;

② $\exists w \in L^1(K_p)$, such that $w^\wedge(\xi) = \omega_0(\xi)$;

③ $\exists \alpha > 0$, such that $\lim_{|\xi| \to 0}\frac{\omega_0(\xi) - 1}{|\xi|^\alpha} = c \neq 0$;

④ $\frac{\omega_0(\xi) - 1}{|\xi|^\alpha} = \mu^\wedge(\xi)$, $\alpha > 0$, with $\mu \in L^1(K_p)$, and $\|\mu\|_{L^1(K_p)} = 1$.

Then, ω_0 is said to be a radial approximation identity kernel.

ω_0 holds the direct and inverse approximation theorems[129]–[131].

Remark. We list the approximation identity kernels and approximation operators on $\mathbb{R}$ in the following (3) and (4) for comparing with those on K_p[9].

(3) **Kernels and operators on $C([0, 2\pi])$ or $L^1([0, 2\pi])$**

Dirichlet kernel

$$D_n(x) = 1 + 2\sum_{k=1}^{n} \cos kx = \begin{cases} \dfrac{\sin \dfrac{2n+1}{2}x}{\sin \dfrac{x}{2}}, & x \neq 2n\pi, \\ 2n+1, & x = 2n\pi. \end{cases}$$

Dirichlet integral operator

$$S_n(f, x) = f * D_n(x) = \frac{1}{2\pi} \int_{-\pi}^{\pi} f(x-u) D_n(u) \, du.$$

Fejér kernel

$$F_n(x) = \frac{1}{n+1} \sum_{k=0}^{n} D_k(x) = \begin{cases} \dfrac{1}{n+1} \left(\dfrac{\sin \dfrac{(n+1)x}{2}}{\sin \dfrac{x}{2}} \right)^2, & x \neq 2n\pi, \\ n+1, & x = 2n\pi. \end{cases}$$

Fejér integral operator

$$\sigma_n(f, x) = f * F_n(x) = \frac{1}{2\pi} \int_{-\pi}^{\pi} f(x-u) F_n(u) \, du.$$

Abel–Poisson kernel

$$p_r(x) = 1 + 2 \sum_{k=1}^{+\infty} r^k \cos kx = \frac{1-r^2}{1 - 2r\cos x + r^2}.$$

Abel–Poisson integral operator

$$P_r(f, x) = f * p_r(x) = \frac{1}{2\pi} \int_{-\pi}^{\pi} f(x-u) p_r(u) \, du.$$

Rogosinski kernel

$$b_n(x) = \frac{1}{2} + \left[D_n\left(x + \frac{\pi}{2n+1}\right) + D_n\left(x - \frac{\pi}{2n-1}\right) \right].$$

Rogosinski integral operator

$$B_n(f, x) = f * b_n(x) = \frac{1}{2\pi} \int_{-\pi}^{\pi} f(x-u) b_n(u) \, du.$$

Jackson kernel

$$j_n(x) = \frac{3}{n(2n^2+1)} \left(\frac{\sin \frac{nx}{2}}{\sin \frac{x}{2}} \right)^2.$$

Jackson integral operator

$$J_n(f,x) = f * j_n(x) = \frac{1}{2\pi} \int_{-\pi}^{\pi} f(x-u) j_n(u) \, du.$$

Weierstrass kernel

$$w_t(x) = \sum_{k=-\infty}^{+\infty} e^{-tk^2} e^{ikx}.$$

Weierstrass integral operator

$$W_t(f,x) = f * w_t(x) = \frac{1}{2\pi} \int_{-\pi}^{\pi} f(x-u) w_t(u) \, du.$$

de la Vallée–Poussin kernel

$$v_n(x) = \frac{(n!)^2}{(2n)!} \left(2\cos \frac{x}{2} \right)^{2n}.$$

de la Vallée–Poussin integral operator

$$V_n(f,x) = f * v_n(x) = \frac{1}{2\pi} \int_{-\pi}^{\pi} f(x-u) v_n(u) \, du.$$

(iv) **Kernels and operators on $C(\mathbb{R})$ or $L^1(\mathbb{R})$**

Fejér kernel

$$F(x) = \frac{1}{\sqrt{2\pi}} \left(\frac{\sin \frac{x}{2}}{\frac{x}{2}} \right)^2.$$

Fejer integral operator

$$\sigma(f,x,\rho) = \frac{1}{\sqrt{2\pi}} \int_{\mathbb{R}} f(x-u) \rho F(\rho u) \, du, \quad \rho \to +\infty.$$

Gauss–Weierstrass kernel

$$W(x) = \frac{1}{\sqrt{2}} e^{\frac{-x^2}{4}}.$$

Gauss–Weierstrass integral operator

$$W(f,x,t) = \frac{1}{\sqrt{4\pi t}} \int_{-\pi}^{\pi} f(x-u) e^{\frac{-u^2}{4t}} \, du, \quad t \to 0.$$

Cauchy–Poisson kernel

$$p(x) = \sqrt{\frac{2}{\pi}} \frac{1}{1+x^2}.$$

Cauchy–Poisson integral operator

$$P(f,x,y) = \frac{y}{\pi} \int_{\mathbb{R}} f(x-u) \frac{1}{y^2+u^2} du, \quad y \to 0.$$

Jackson–de la Vallée–Poussin kernel

$$n(x) = \frac{3}{\sqrt{8\pi}} \left(\sin \frac{x}{2}\right)^4 \cdot \left(\frac{x}{2}\right)^{-4}.$$

Jackson–de la Vallée–Poussin integral operator

$$N(f,x,\rho) = \frac{12}{\pi \rho^3} \int_{\mathbb{R}} f(x-u) \frac{\sin^4 \frac{\rho u}{2}}{u^4} du, \quad \rho \to +\infty.$$

Picard kernel

$$c(x) = \sqrt{\frac{\pi}{2}} e^{-|x|}.$$

Picard integral operator

$$C(f,x,\rho) = \frac{\rho}{2} \int_{\mathbb{R}} f(x-u) e^{-\rho|u|} du, \quad \rho \to +\infty.$$

Chinese mathematicians have completed a lot of excellent work in the harmonic analysis and approximation on local fields. The proceedings of workshop dedicated to the memory of J.E. Gibbs held in 2007, Nis, Serbia, lists the remarkable contributions of the past 40 years of Chinese mathematicians[98].

Exercises

1. Summarize the properties of approximation identity kernels and operators on $\mathbb{R}^n$ and K_p.
2. Construct new approximation kernels and operators on local field K_p.
3. Consider a new frame of construction theory of function on K_p^*.

Chapter 4

Function Spaces on Local Fields

The function spaces on a local field K_p, such as the continuous function space $C(K_p)$, the space of continuous functions with zero limit as $|x| \to +\infty$, $C_0(K_p)$, the space of r-power integrable functions $L^r \equiv L^r(K_p), 1 \leqslant r \leqslant +\infty$, the test function class $\mathbb{S}(K_p)$, the distribution space $\mathbb{S}^*(K_p)$, and so on, have introduced in the Chapter 3. In this chapter, we will introduce the B-type space, F-type space, and Hölder type space, Lebesgue type space, Sobolev type space. See [42], [47]~[49], [73], [84], [96], [105], [139], [140].

4.1 B-type spaces and F-type spaces on local fields

4.1.1 B-type spaces, F-type spaces

Let $f : K_p \to \mathbb{C}$ be a complex Haar measurable function on K_p. Let $\mathbb{S}(K_p)$ and $\mathbb{S}^*(K_p)$ be the test function class and distribution space on a local field K_p, respectively. Correspondingly, $\mathbb{S}(\Gamma_p)$ and $\mathbb{S}^*(\Gamma_p)$ are the test function class and distribution space on the character group Γ_p. Since $K_p \xleftrightarrow{\text{iso.}} \Gamma_p$, we may regard $\mathbb{S}(K_p)$ and $\mathbb{S}(\Gamma_p)$ are identical. Moreover, so are $\mathbb{S}^*(K_p)$ and $\mathbb{S}^*(\Gamma_p)$.

Definition 4.1.1 (Two norms) *Take a function sequence in $\mathbb{S}(K_p)$* $\varphi \equiv \{\varphi_j(x)\}_{j=0}^{+\infty} \subset \mathbb{S}(K_p)$ *satifying*
(i) supp $\varphi_j \subset K_p$ *is a compact set, $j \in \mathbb{P}$;*
(ii) ① supp $\varphi_0 \subset \{x \in K_p : |x| < p\}$,

$$\text{supp } \varphi_j \subset \{x \in K_p : p^{j-1} < |x| < p^{j+1}\}, \quad j \in \mathbb{N};$$

② $\left|\varphi_j^{(s)}(x)\right| \leqslant c_s p^{-j+js}, s \in (0, +\infty), \quad j \in \mathbb{P}, \ x \in K_p,$

where $\varphi_j^{\langle s \rangle}(x)$ is the s-order point-wise p-type derivative of $\varphi_j(x)$;

(iii) $\sum_{j=0}^{+\infty} \varphi_j(x) = 1$, $x \in K_p$.

Denoted by
$$\mathbb{A}(K_p) = \left\{ \varphi \equiv \{\varphi_j\}_{j \in \mathbb{P}} \subset \mathbb{S}(K_p) : \varphi_j \text{ with (i)}\sim\text{(iii)} \right\} \quad (4.1.1)$$
for this function class.

Similarly, we define
$$\mathbb{A}(\Gamma_p) = \left\{ \psi \equiv \{\psi_j\}_{j \in \mathbb{P}} \subset \mathbb{S}(\Gamma_p) : \psi_j \text{ with (i)}'\sim\text{(iii)}' \right\}, \quad (4.1.2)$$
where (i)$'\sim$(iii)$'$ are: for $\psi = \{\psi_j(\xi)\}_{j=0}^{+\infty} \subset \mathbb{S}(\Gamma_p)$,

(i)$'$ supp $\psi_j \subset \Gamma_p$ is a compact set, $j \in \mathbb{P}$;

(ii)$'$ ① supp $\psi_0 \subset \{\xi \in \Gamma_p : |\xi| < p\}$,

supp $\psi_j \subset \{\xi \in \Gamma_p : p^{j-1} < |\xi| < p^{j+1}\}$, $j \in \mathbb{N}$,

② $|T_{\langle s \rangle} \psi_j^\vee(x)| \leqslant c_s p^{-j+js}$, $s \in (0, +\infty)$, $j \in \mathbb{P}$, $x \in K_p$,

where $T_{\langle s \rangle} \psi_j^\vee(x)$ is the s-order point-wise p-type derivative of $\psi_j^\vee(x)$;

(iii)$'$ $\sum_{j=0}^{+\infty} \psi_j(\xi) = 1$, $\xi \in \Gamma_p$.

For a given sequence $\{a_j(x)\}_{j=0}^{+\infty} \subset \mathbb{S}(K_p)$, we define two kinds of norms

$$\|a_j\|_{l_t(L^r(K_p))} = \left\{ \sum_{j=0}^{+\infty} \left[\|a_j(\cdot)\|_{L^r(K_p)} \right]^t \right\}^{\frac{1}{t}} = \left\{ \sum_{j=0}^{+\infty} \left[\int_{K_p} |a_j(x)|^r \, dx \right]^{\frac{t}{r}} \right\}^{\frac{1}{t}}, \quad (4.1.3)$$

$$\|a_j\|_{L^r(l_t(K_p))} = \left\| \left\{ \sum_{j=0}^{+\infty} |a_j(\cdot)|^t \right\}^{\frac{1}{t}} \right\|_{L^r(K_p)} = \left(\int_{K_p} \left\{ \sum_{j=0}^{+\infty} |a_j(x)|^t \right\}^{\frac{r}{t}} dx \right)^{\frac{1}{r}}, \quad (4.1.4)$$

with $0 < r, t \leqslant +\infty$; for $r, t = +\infty$, take a modification as usual (essential supremum).

Similarly, for given $\{a_j(\xi)\}_{j=0}^{+\infty} \subset \mathbb{S}(\Gamma_p)$, the norms $\|a_j\|_{l_t(L^r(\Gamma_p))}$, $\|a_j\|_{L^r(l_t(\Gamma_p))}$ can be defined.

We may prove that[101] for $0 < r, t \leqslant +\infty$, the above norms are "**semi-norms**", and for $1 \leqslant r, t \leqslant +\infty$, they are norms.

On a local field K_p, there is a **non-homogeneous unit decomposition**

$$1 = \Phi_{B^0}(x) + \sum_{j=1}^{+\infty} \Phi_{B^{-j}\setminus B^{-j+1}}(x), \quad x \in K_p, \qquad (4.1.5)$$

where Φ_A is the characteristic function $\Phi_A(x) = \begin{cases} 1, & x \in A \\ 0, & x \notin A \end{cases}$ of a set $A \subset K_p$.

Also, a **homogeneous unit decomposition**

$$1 = \sum_{j=-\infty}^{+\infty} \Phi_{B^{-j}\setminus B^{-j+1}}(x), \quad x \in K_p. \qquad (4.1.6)$$

On the character group Γ_p of K_p, **non-homogeneous unit decomposition** is

$$1 = \Phi_{\Gamma^0}(\xi) + \sum_{j=1}^{+\infty} \Phi_{\Gamma^j\setminus\Gamma^{j-1}}(\xi), \quad \xi \in \Gamma_p. \qquad (4.1.7)$$

Also a **homogeneous unit decomposition** is

$$1 = \sum_{j=-\infty}^{+\infty} \Phi_{\Gamma^j\setminus\Gamma^{j-1}}(\xi), \quad \xi \in \Gamma_p. \qquad (4.1.8)$$

Definition 4.1.2 (Littlewood–Paley decomposition of a distribution) *Let $f \in \mathbb{S}^*(K_p)$ be a distribution on the test function space $\mathbb{S}(K_p)$. The Littlewood–Paley decomposition of f is defined as*

$$f = \sum_{j=0}^{+\infty} f_j, \qquad (4.1.9)$$

if f_j satisfies

(i) supp $f_0^\wedge \subset \Gamma^0$;

(ii) supp $f_j^\wedge \subset \Gamma^j\setminus\Gamma^{j-1}, j = 1, 2, \cdots,$

where $f_j^\wedge, j \in \mathbb{P} = \{0,1,2,\cdots\}$ is the Fourier transformation of f_j in distribution sense.

We have a property[93],[96] : $\forall f \in \mathbb{S}^*(K_p)$ has a Littlewood–Paley decomposition $f = \sum_{j=0}^{+\infty} f_j$, where

$$\begin{cases} f_0(x) = f * p^0 \Phi_{B^0}(x) \equiv f * \varphi_0(x), \\ f_j(x) = f * \left(p^j \Phi_{B^j} - p^{j-1}\Phi_{B^{j-1}}\right)(x) \equiv f * \varphi_j(x), \end{cases} \quad x \in K_p, \; j = 1,2,3,\cdots,$$

and
$$[p^j \Phi_{B^j}(\cdot)]^\wedge (\xi) = \Phi_{\Gamma^j}(\xi), \quad \xi \in \Gamma_p, \quad j = 0, 1, 2, \cdots.$$

Definition 4.1.3 (B-type space and F-type space) *Take function sequence* $\{\psi_j(\xi)\}_{j=0}^{+\infty} \subset \mathbb{A}(\Gamma_p)$ *as*
$$\psi_0(\xi) = \Phi_{\Gamma^0}(\xi), \quad \psi_j(\xi) = \Phi_{\Gamma^j \backslash \Gamma^{j-1}}(\xi), \quad j \in \mathbb{N},$$
denoted by
$$\{\psi_j(\xi)\}_{j=0}^{+\infty} = \{\Phi_{\Gamma^0}, \Phi_{\Gamma^j \backslash \Gamma^{j-1}}\}_{j=1}^{+\infty}. \tag{4.1.10}$$
The set $\{\Phi_{\Gamma^0}, \Phi_{\Gamma^j \backslash \Gamma^{j-1}}\}_{j=1}^{+\infty}$ *has the following properties:*

① supp $\Phi_{\Gamma^0} \subset \Gamma$, supp $\Phi_{\Gamma^j \backslash \Gamma^{j-1}} \subset \Gamma^j \backslash \Gamma^{j-1}$, *all supports are compact;*

② $\Phi_{\Gamma^0}(\xi) + \sum_{j=1}^{+\infty} \Phi_{\Gamma^j \backslash \Gamma^{j-1}}(\xi) = 1, \xi \in \Gamma_p;$

③ $\left|T_{(\cdot)^s} \Phi_{\Gamma^0}^{\vee}(x)\right| \leqslant c_s p^{-0+0 \cdot s}, \left|T_{(\cdot)^s} \Phi_{\Gamma^j \backslash \Gamma^{j-1}}^{\vee}(x)\right| \leqslant c_s p^{-j+j \cdot s}, s \geqslant 0,$

$j \in \mathbb{N}, x \in K_p$. *Then* $\{\psi_j(\xi)\}_{j=0}^{+\infty} = \{\Phi_{\Gamma^0}, \Phi_{\Gamma^j \backslash \Gamma^{j-1}}\}_{j=1}^{+\infty} \subset \mathbb{A}(\Gamma_p).$

We define

(i) *For* $0 < r \leqslant +\infty, 0 < t \leqslant +\infty, s \in \mathbb{R}$, *the set*
$$B_{rt}^s(K_p) = \left\{f \in \mathbb{S}^*(K_p) : \|f\|_{B_{rt}^s(K_p)} < +\infty\right\}$$
*is said to be a **B-type space** on local field* K_p, *where* $\{\psi_j(\xi)\}_{j=0}^{+\infty} \subset \mathbb{A}(\Gamma_p)$ *is in (4.1.10), and*
$$\|f\|_{B_{rt}^s(K_p)} = \left\|p^{sj} (\psi_j f^\wedge)^\vee (\cdot)\right\|_{l_t(L^r(K_p))}$$
$$= \left\{\sum_{j=0}^{+\infty} \left\|p^{sj} (\psi_j(\cdot) f^\wedge(\cdot))^\vee (\cdot)\right\|_{L^r(K_p)}^t\right\}^{\frac{1}{t}}$$
$$= \left\{\sum_{j=0}^{+\infty} p^{sjt} \left\{\int_{K_p} \left|(\psi_j(\cdot) f^\wedge(\cdot))^\vee (x)\right|^r dx\right\}^{\frac{t}{r}}\right\}^{\frac{1}{t}},$$

if $r, t = +\infty$, *with usual modification;*

B-type space, sometimes, called Besov type space, or Lipschitz type space.

(ii) *For* $0 < r < +\infty, 0 < t \leqslant +\infty, s \in \mathbb{R}$, *the set*
$$F_{rt}^s(K_p) = \left\{f \in \mathbb{S}^*(K_p) : \|f\|_{F_{rt}^s(K_p)} < +\infty\right\}$$
*is said to be an **F-type space** on local field* K_p, *where* $\{\psi_j(\xi)\}_{j=0}^{+\infty} \subset \mathbb{A}(\Gamma_p)$ *is in (4.1.10), and*

$$\|f\|_{F^s_{rt}(K_p)} = \left\|p^{sj}\left(\psi_j f^\wedge\right)^\vee(\cdot)\right\|_{L^r(l_t(K_p))}$$

$$= \left\|\left\{\sum_{j=0}^{+\infty}\left|p^{sj}\left(\psi_j(\cdot)f^\wedge(\cdot)\right)^\vee(\cdot)\right|^t\right\}^{\frac{1}{t}}\right\|_{L^r(K_p)}$$

$$= \left(\int_{K_p}\left\{\sum_{j=0}^{+\infty}p^{sjt}\left|\left(\psi_j(\cdot)f^\wedge(\cdot)\right)^\vee(x)\right|^t\right\}^{\frac{r}{t}}dx\right)^{\frac{1}{r}}.$$

If $r = +\infty$, $0 < t < +\infty$, $-\infty < s < +\infty$, then

$$F^s_{\infty t}(K_p) = \left\{f \in \mathbb{S}^*(K_p): \exists\,\{\psi_j\} \subset \mathbb{A}(\Gamma_p), \{f_j\} \subset L^\infty(K_p), \text{ s.t.}\right.$$

$$\left. f = \sum_{j=0}^{+\infty}\left(\psi_j f_j^\wedge\right)^\vee, \|f\|_{F^s_{\infty t}(K_p)} < +\infty\right\},$$

with

$$\|f\|_{F^s_{\infty t}(K_p)} = \left\|p^{sj}f_j\right\|_{L^\infty(l_t(K_p))} = \left\|\left(\sum_{j=0}^{+\infty}p^{sjt}|f_j(\cdot)|^t\right)^{\frac{1}{t}}\right\|_{L^\infty(K_p)}.$$

Remark 1. We define the non-homogeneous B-type and F-type spaces above. Similarly, the homogeneous B-type and F-type spaces can be defined. That is, we just need to take

$$\{\psi_j(\xi)\}_{j=-\infty}^{+\infty} = \{\Phi_{\Gamma^j\setminus\Gamma^{j-1}}\}_{j=-\infty}^{+\infty}. \tag{4.1.11}$$

Correspondingly, we just need to take the homogeneous decompositions

$$K_p = \bigcup_{j=-\infty}^{+\infty} B^{-j}\setminus B^{-j+1} \text{ and } \Gamma_p = \bigcup_{j=-\infty}^{+\infty} \Gamma^j\setminus\Gamma^{j-1}.$$

Remark 2. We may prove that the norms are independent of selections of $\{\psi_j(\xi)\}_{j=0}^{+\infty}$ or $\{\psi_j(\xi)\}_{j=-\infty}^{+\infty}$. For the case in $\mathbb{R}^n$, we refer to Triebel[101]: the norms of spaces $B^s_{rt}(\mathbb{R}^n)$ and $F^s_{rt}(\mathbb{R}^n)$ determined by any sequences $\psi = \{\psi_j(\xi)\}_j$ are equivalent. For the case in K_p, the same assertions holds.

Theorem 4.1.1 (i) If $0 < r, t \leqslant +\infty$, $-\infty < s < +\infty$, then $B^s_{rt}(K_p)$ is a quasi-Banach space; if $1 \leqslant r, t \leqslant +\infty$, then it is a Banach space, and

$$\mathbb{S}(K_p) \subset B^s_{rt}(K_p) \subset \mathbb{S}^*(K_p).$$

(ii) If $0 < r < +\infty$, $0 < t \leqslant +\infty$, $-\infty < s < +\infty$, then $F_{rt}^s(K_p)$ is a quasi-Banach space; if $1 \leqslant r < +\infty$, $1 \leqslant t \leqslant +\infty$, then it is a Banach space, and
$$\mathbb{S}(K_p) \subset F_{rt}^s(K_p) \subset \mathbb{S}^*(K_p).$$

(iii) The quasi-norms of spaces $B_{rt}^s(K_p)$ and $F_{rt}^s(K_p)$ are independent of selections of sequence $\psi = \{\psi_j(\xi)\}_{j=0}^{+\infty} \subset \mathbb{A}(\Gamma_p)$, that is, two quasi-norms $\left\| p^{sj}\left(\psi_j^{(1)} f^\wedge\right)^\vee (\cdot) \right\|_{l_t(L^r(K_p))}$ and $\left\| p^{sj}\left(\psi_j^{(2)} f^\wedge\right)^\vee (\cdot) \right\|_{l_t(L^r(K_p))}$ of $B_{rt}^s(K_p)$ depending on $\psi^{(1)} = \left\{\psi_j^{(1)}(\xi)\right\}_{j=0}^{+\infty}$ and $\psi^{(2)} = \left\{\psi_j^{(2)}(\xi)\right\}_{j=0}^{+\infty}$, respectively, are equivalent to each other. So are those of $F_{rt}^s(K_p)$.

Theorem 4.1.2 The spaces $B_{rt}^s(K_p)$ and $F_{rt}^s(K_p)$, $0 < r, t \leqslant +\infty$, hold

(i) For $0 < t_0 \leqslant t_1 \leqslant +\infty$, $-\infty < s < +\infty$,

① If $0 < r \leqslant +\infty$, then $B_{rt_0}^s(K_p) \subset B_{rt_1}^s(K_p)$;

② If $0 < r < +\infty$, then $F_{rt_0}^s(K_p) \subset F_{rt_1}^s(K_p)$.

(ii) For $0 < t_0 \leqslant +\infty$, $0 < t_1 \leqslant +\infty$, $-\infty < s < +\infty$, and for $\varepsilon > 0$,

③ If $0 < r \leqslant +\infty$, then $B_{rt_0}^{s+\varepsilon}(K_p) \subset B_{rt_1}^s(K_p)$.

④ If $0 < r < +\infty$, then $F_{rt_0}^{s+\varepsilon}(K_p) \subset F_{rt_1}^s(K_p)$.

(iii) For $0 < t \leqslant +\infty$, $0 < r < +\infty$, $-\infty < s < +\infty$, then

⑤ $B_{r\min(r,t)}^s(K_p) \subset F_{rt}^s(K_p) \subset B_{r\max(r,t)}^s(K_p)$.

Theorem 4.1.3 The spaces $B_{rt}^s(K_p)$ and $F_{rt}^s(K_p)$, $0 < r, t \leqslant +\infty$, hold

(i) If $1 \leqslant r < +\infty$, $0 < t < +\infty$, $-\infty < s < +\infty$, then
$$\{B_{rt}^s(K_p)\}^* = B_{r't'}^{-s}(K_p);$$

(ii) If $1 \leqslant r < +\infty$, $1 < t < +\infty$, $-\infty < s < +\infty$, then
$$\{F_{rt}^s(K_p)\}^* = F_{r't'}^{-s}(K_p),$$

where r', t' are the conjugate numbers of r, t, respectively:
$$r' = \begin{cases} \dfrac{r}{r-1}, & 1 \leqslant r < +\infty, \\ +\infty, & 0 < r < 1, \end{cases}$$

$$t' = \begin{cases} \dfrac{t}{t-1}, & 1 \leqslant t < +\infty, \\ +\infty, & 0 < t < 1. \end{cases}$$

(iii) If $0 < r < 1$, $0 < t < +\infty$, $-\infty < s < +\infty$, then
$$\{B_{rt}^s(K_p)\}^* = B_{r't'}^{-s+\left(\frac{1}{r}-1\right)}(K_p) = B_{\infty t'}^{-s+\left(\frac{1}{r}-1\right)}(K_p).$$

(iv) If $0 < r < 1$, $0 < t < +\infty$, $-\infty < s < +\infty$, then
$$\{F_{rt}^s(K_p)\}^* = B_{r'\infty}^{-s+\left(\frac{1}{r}-1\right)}(K_p) = B_{\infty\infty}^{-s+\left(\frac{1}{r}-1\right)}(K_p).$$

(v) If $0 < r$, $t \leqslant +\infty$, $s > \sigma_r = \left(\frac{1}{r}-1\right)_+$, the positive part of $\frac{1}{r}-1$, then
$$B_{rt}^s(K_p) \subset L_{\text{loc}}^1(K_p).$$

The proofs of the above theorems are routine, and we refer to Triebel[101].

4.1.2 Special cases of B-type spaces and F-type spaces

The certain useful spaces are special cases of B-type spaces and F-type spaces. We list the special cases in that of underlying space $\mathbb{R}^n$.

If $s > 0$, $s \notin \mathbb{N}$, then $B_{\infty\infty}^s(\mathbb{R}^n) = C^s(\mathbb{R}^n)$ — Hölder space.

If $1 < r < +\infty$, $-\infty < s < +\infty$, then $F_{r2}^s(\mathbb{R}^n) = H_r^s(\mathbb{R}^n)$ — Bessel potential space.

If $1 \leqslant r < +\infty$, $1 \leqslant t < +\infty$, $s > 0$, then $B_{rt}^s(\mathbb{R}^n) = \Lambda_{rt}^s(\mathbb{R}^n)$ — Besov space.

If $0 < r < +\infty$, then $F_{r2}^0(\mathbb{R}^n) = h_r(\mathbb{R}^n)$ — locally non-homogeneous Hardy space.

If $r = +\infty$, then $F_{\infty 2}^0(\mathbb{R}^n) = \text{bmo}(\mathbb{R}^n)$ — bmo space.

Denote by $\dot{B}_{rt}^s(\mathbb{R}^n)$ and $\dot{F}_{rt}^s(\mathbb{R}^n)$ homogeneous B-type and F-type space, respectively.

If $0 < r < +\infty$, then $\dot{F}_{r2}^0(\mathbb{R}^n) = H_r(\mathbb{R}^n)$ — locally homogeneous Hardy space.

If $r = +\infty$, then $\dot{F}_{\infty 2}^0(\mathbb{R}^n) = \text{BMO}(\mathbb{R}^n)$ — BMO space.

However, "function space theory" on local fields is quite young, compared with that of on Euclidean spaces. There are lots of open problems to be studied, such as, Bessel potential space, Riesz fractional order integrals, Hardy space, BMO space on local fields and so on.

4.1.3 Hölder type spaces on local fields

The Hölder space, Lebesgue space, Sobolev space on $\mathbb{R}^n$ play very important roles in partial differential equations. On local fields, we will consider how to define the Hölder type space, Lebesgue type space and Sobolev type space, and consider what roles they will play in harmonic analysis over local fields. Moreover, we try to establish a partial differential equation theory on local fields, and expect to apply the function space theory to this PDE theory.

Definition 4.1.4 (Hölder type space) We define Hölder type space $C^\sigma(K_p)$ for $\sigma \in (-\infty, +\infty)$ on local field K_p,

(i) If $\sigma = 0$, then $C^0(K_p) = C(K_p)$, where $C(K_p)$ is the bounded continuous function space on K_p.

(ii) If $\sigma \in (0, +\infty)$, then $C^\sigma(K_p)$ is defined as the set of distributions for $f \in \mathbb{S}^*(K_p)$ satisfying:

(a) f has the Littlewood–Paley decomposition (4.1.9), $f = \sum_{j=0}^{+\infty} f_j$;

(b) f_j satisfies $\|f_j\|_{L^\infty(K_p)} \leqslant c p^{-j\sigma}$, $j \in \mathbb{P}$.

For (i) and (ii), let

$$\|f\|_{L^\infty(K_p)} = \sup_{j \in \mathbb{P}} \left\{ p^{j\sigma} \|f_j\|_{L^\infty(K_p)} \right\}, \qquad (4.1.12)$$

then $C^\sigma(K_p)$ becomes a Banach space under the norm (4.1.12).

(iii) If $\sigma \in (-\infty, 0)$, then $C^\sigma(K_p) = B^\sigma_{\infty\infty}(K_p)$, where $B^\sigma_{\infty\infty}(K_p)$ is the special case of the B-type space, that is, $B^s_{rt}(K_p)$ with $s = \sigma$, $r = t = \infty$.

Hölder type spaces have the following important properties[73].

Theorem 4.1.4 For $0 \leqslant \sigma_1 < \sigma_2 < +\infty$, the inclusion relationship for Hölder type spaces $C^{\sigma_2}(K_p) \subset C^{\sigma_1}(K_p)$ holds.

Theorem 4.1.5 For Hölder type space $C^\sigma(K_p), \sigma \in \mathbb{R}$, it follows that

$$\sigma = s \in \mathbb{R} \Rightarrow C^s(K_p) = B^s_{\infty\infty}(K_p).$$

Moreover, the inclusion relationship $\mathbb{S}(K_p) \subset C^\sigma(K_p) \subset \mathbb{S}^*(K_p)$ holds.

Proof. Only need to prove that the theorem holds for $\sigma = s > 0$, since theorem holds for $\sigma = s \leqslant 0$ by definition. Take $\{\psi_j(\xi)\}_{j=0}^{+\infty} \subset \mathbb{A}(\Gamma_p)$ as ∎

$$\psi_0(\xi) = \Phi_{\Gamma^0}(\xi), \quad \psi_j(\xi) = \Phi_{\Gamma^j \setminus \Gamma^{j-1}}(\xi), \quad j \in \mathbb{N},$$

then for $f \in B^s_{\infty\infty}(K_p)$, if $s > 0$,
$$B^s_{\infty\infty}(K_p) = \left\{ f \in \mathbb{S}^*(K_p) : \|f\|_{B^s_{\infty\infty}(K_p)} < +\infty \right\},$$
where
$$\|f\|_{B^s_{\infty\infty}(K_p)} = \left\| p^{sj} [\psi_j f^\wedge]^\vee (\cdot) \right\|_{l_\infty(L^\infty(K_p))}$$
$$= \sup_j \left\{ \sup_x \left| p^{sj} [\psi_j(\cdot) f^\wedge(\cdot)]^\vee (x) \right| \right\}.$$
On the other hand, for $f \in C^s(K_p)$, $s > 0$,
$$\|f\|_{C^s(K_p)} = \sup_j \left\{ p^{sj} \|f_j\|_{L^\infty(K_p)} \right\} < +\infty.$$
Then
$$p^{sj} [\psi_j(\cdot) f^\wedge(\cdot)]^\vee (x) = p^{sj} f_j(x),$$
thus
$$\|f\|_{B^s_{\infty\infty}}(K_p) = \left\| p^{sj} [\psi_j f^\wedge]^\vee (\cdot) \right\|_{l_\infty(L^\infty(K_p))}$$
$$= \sup_j \left\{ \sup_x \left| p^{sj} [\psi_j(\cdot) f^\wedge(\cdot)]^\vee (x) \right| \right\}$$
$$= \sup_j \left\{ \sup_x \left| p^{sj} f_j(x) \right| \right\} = \sup_j \left\{ p^{sj} \|f_j\|_{L^\infty(K_p)} \right\}$$
$$= \|f\|_{C^s(K_p)},$$
we conclude that
$$f \in C^s(K_p) \Leftrightarrow f \in B^s_{\infty,\infty}(K_p), \quad s > 0.$$
The inclusion relationship $\mathbb{S}(K_p) \subset C^\sigma(K_p) \subset \mathbb{S}^*(K_p)$ is clear. The proof is complete.

The above Theorem 4.1.5 shows that $C^\sigma(K_p) = B^\sigma_{\infty\infty}(K_p)$ holds for $\sigma \in [-\infty, +\infty)$, thus the Definition 4.1.4 of Hölder type space is well defined for $\sigma \in \mathbb{R}$.

Theorem 4.1.6 *Hölder type space $C^\sigma(K_p)$ has the following properties for $\sigma \in [0, +\infty)$:*

(i) *If $f \in C^\sigma(K_p)$, then $\forall \lambda \in [0, \sigma]$, the p-type derivative $f^{\langle \lambda \rangle} \equiv T_{\langle \cdot \rangle^\lambda} f$ of f exists, and $f^{\langle \lambda \rangle} \in C^{\sigma-\lambda}(K_p)$.*

(ii) *If $T_{\langle \cdot \rangle^\sigma} f \in C(K_p)$, then $\forall \lambda \in [0, \sigma]$, the p-type derivative $f^{\langle \lambda \rangle}$ of f exists, and $f^{\langle \lambda \rangle} = T_{\langle \cdot \rangle^\lambda} f \in C^{\sigma-\lambda}(K_p)$.*

Proof. Since $\mathbb{S}(K_p) \subset B_{\infty\infty}^\sigma(K_p) = C^\sigma(K_p)$, $\forall \sigma \in [0, +\infty)$, and $\mathbb{S}(K_p)$ is dense in $\mathbb{S}^*(K_p)$, so we only need to prove the theorem for the case that $f \in \mathbb{S}^*(K_p)$ is a function.

For (i), by assumption $f \in C^\sigma(K_p)$, $\sigma \in [0, +\infty)$, then f has Littlewood–Paley decomposition $f = \sum_{j=0}^{+\infty} f_j$.

We evaluate: $\forall \lambda \in [0, \sigma]$,

$$T_{\langle \cdot \rangle^\lambda} f(x) = \int_{\Gamma_p} \int_{K_p} \langle \xi \rangle^\lambda f(t) \overline{\chi_\xi}(t-x) \, dt d\xi$$

$$= \int_{\Gamma_p} \langle \xi \rangle^\lambda \chi_x(\xi) \left\{ \int_{K_p} f(t) \overline{\chi_\xi}(t) dt \right\} d\xi$$

$$= \int_{\Gamma_p} \langle \xi \rangle^\lambda f^\wedge(\xi) \chi_x(\xi) d\xi = \int_{\Gamma_p} \langle \xi \rangle^\lambda \sum_{j=0}^{+\infty} f_j^\wedge(\xi) \chi_x(\xi) d\xi$$

$$= \int_{\Gamma^0} \langle \xi \rangle^\lambda f_0^\wedge(\xi) \chi_x(\xi) d\xi + \sum_{j=1}^{+\infty} \int_{\Gamma^j \setminus \Gamma^{j-1}} \langle \xi \rangle^\lambda f_j^\wedge(\xi) \chi_x(\xi) d\xi$$

$$= \int_{\Gamma^0} f_0^\wedge(\xi) \chi_x(\xi) d\xi + \sum_{j=1}^{+\infty} \int_{\Gamma^j \setminus \Gamma^{j-1}} p^{j\lambda} f_j^\wedge(\xi) \chi_x(\xi) d\xi$$

$$= \int_{\Gamma^0} f_0^\wedge(\xi) \chi_x(\xi) d\xi + \sum_{j=1}^{+\infty} p^{j\lambda} \int_{\Gamma^j \setminus \Gamma^{j-1}} f_j^\wedge(\xi) \chi_x(\xi) d\xi$$

$$= \int_\Gamma f_0^\wedge(\xi) \chi_x(\xi) d\xi + \sum_{j=1}^{+\infty} p^{j\lambda} \int_\Gamma f_j^\wedge(\xi) \chi_x(\xi) d\xi$$

$$= f_0(x) + \sum_{j=1}^{+\infty} p^{j\lambda} f_j(x) = \sum_{j=0}^{+\infty} p^{j\lambda} f_j(x).$$

Then
$$T_{\langle \cdot \rangle^\lambda} f(x) = \sum_{j=0}^{+\infty} p^{j\lambda} f_j(x), \quad \forall \lambda \in [0, \sigma].$$

Hence, for $f \in C^\sigma(K)$, we have
$$\|T_{\langle \cdot \rangle^\sigma} f(\cdot)\|_{L^\infty(K)} = \sup_j \left\{ \|p^{j\sigma} f_j(\cdot)\|_{L^\infty(K)} \right\} \leqslant \sup_j \left\{ p^{j\sigma} \cdot c p^{-j\sigma} \right\} \leqslant c.$$

This implies that
$$f^{\langle \lambda \rangle}(x) = T_{\langle \cdot \rangle^\lambda} f(x) = \sum_{j=0}^{+\infty} p^{j\sigma} f_j(x)$$

exists.

Next, we prove $f^{\langle\lambda\rangle}(x) \in C^{\sigma-\lambda}(K_p)$, $\lambda \in [0,\sigma]$.

To prove $f^{\langle\sigma\rangle}(x) = T_{\langle\cdot\rangle^\sigma}f(x) \in C(K_p)$, take $h \in K_p$ with $p^{-j_0-1} \leqslant |h| \leqslant p^{-j_0}$, then

$$T_{\langle\cdot\rangle^\sigma}f(x+h) - T_{\langle\cdot\rangle^\sigma}f(x)$$
$$= \sum_{j=0}^{+\infty} T_{\langle\cdot\rangle^\sigma}[f_j(x+h) - f_j(x)]$$
$$= \int_{\Gamma_p}\int_{K_p} f_0(t)\overline{\chi}_{x+h-t}(\xi)dtd\xi + \sum_{j=1}^{+\infty}\int_{\Gamma_p}\int_{K_p} \langle\xi\rangle^\sigma f_j(t)\overline{\chi}_{x+h-t}(\xi)dtd\xi$$
$$- \int_{\Gamma_p}\int_{K_p} f_0(t)\overline{\chi}_{x-t}(\xi)dtd\xi - \sum_{j=1}^{+\infty}\int_{\Gamma_p}\int_{K_p} \langle\xi\rangle^\sigma f_j(t)\overline{\chi}_{x-t}(\xi)dtd\xi$$
$$= \int_{\Gamma_p}\int_{K_p} f_0(t)\left[\overline{\chi}_\xi(x+h-t) - \overline{\chi}_\xi(x-t)\right]dtd\xi$$
$$+ \sum_{j=1}^{+\infty}\int_{\Gamma_p}\int_{K_p} \langle\xi\rangle^\sigma f_j(t)\left[\overline{\chi}_\xi(x+h-t) - \overline{\chi}_\xi(x-t)\right]dtd\xi.$$

By the continuity of characters, $\forall \varepsilon > 0$, $\exists \delta > 0$, such that for $|h| < \delta$ holds

$$\left|\overline{\chi}_\xi(x+h-t) - \overline{\chi}_\xi(x-t)\right| < \varepsilon.$$

For the non-Archimedean norm $|x| \in \{p^k : k \in \mathbb{Z}\}$, if $p^{-j_0-1} \leqslant |h| \leqslant p^{-j_0}$, with $|h|$ small enough, then $x+h-t \in B^k$ and $x-t \in B^k$. However, the totally disconnected property of K_p implies that: $\overline{\chi}_\xi(x+h-t) - \overline{\chi}_\xi(x-t) = 0$, thus $T_{\langle\cdot\rangle^\sigma}f(x+h) - T_{\langle\cdot\rangle^\sigma}f(x) = 0$. Hence it follows that: $\forall \varepsilon > 0$, $\exists \delta > 0$, such that for $|h| < \delta$,

$$\left|T_{\langle\cdot\rangle^\sigma}f(x+h) - T_{\langle\cdot\rangle^\sigma}f(x)\right| < \varepsilon,$$

This implies that

$$T_{\langle\cdot\rangle^\sigma}f \in C(K_p).$$

Similarly, $T_{\langle\cdot\rangle^\lambda}f(x) = \sum_{j=0}^{+\infty} p^{j\lambda}f_j(x)$, $\forall \lambda \in [0,\sigma)$. Then,

$$\left\|T_{\langle\cdot\rangle^\lambda}f\right\|_{L^\infty(K_p)} = \sup_j \left\{p^{j\lambda}\|f_j\|_{L^\infty(K_p)}\right\}$$

$$\leqslant \sup_j \{p^{j\lambda} c p^{-j\sigma}\} \leqslant c \sup_j \{p^{-j(\sigma-\lambda)}\}.$$

Thus, it follows that
$$\left\|T_{\langle\cdot\rangle^\lambda} f\right\|_{L^\infty(K_p)} \leqslant c p^{-j(\sigma-\lambda)},$$

That is $T_{\langle\cdot\rangle^\lambda} f \in C^{\sigma-\lambda}(K_p)$, $0 \leqslant \lambda < \sigma$.

Combining the case of $\lambda = \sigma$, it follows that
$$T_{\langle\cdot\rangle^\lambda} f \in C^{\sigma-\lambda}(K_p), \quad 0 \leqslant \lambda < \sigma,$$

thus, (i) is proved.

To prove (ii), for $f \in \mathbb{S}^*(K_p)$, if $T_{\langle\cdot\rangle^\sigma} f \in C(K_p)$, take the Littlewood–Paley decomposition of f,
$$f = f * p^0 \Phi_{B^0}(x) + \sum_{j=0}^{+\infty} f * \left\{p^j \Phi_{B^j}(x) - p^{j-1}\Phi_{B^{j-1}}(x)\right\} \equiv u_0(x) + \sum_{j=1}^{+\infty} u_j(x),$$

with
$$u_0(x) = f * p^0 \Phi_{B^0}(x), \quad u_j(x) = f * \left\{p^j \Phi_{B^j}(x) - p^{j-1}\Phi_{B^{j-1}}(x)\right\}, \quad j \in \mathbb{N}.$$

Since the Fourier transformation of $p^j \Phi_{B^j}(x)$ is $[p^j \Phi_{B^j}(\cdot)]^\wedge(\xi) = \Phi_{\Gamma^j}(\xi)$, $j \in \mathbb{P} = \{0\} \cup \mathbb{N}$, so
$$\operatorname{supp} u_0^\wedge \subset \Gamma^0, \quad \operatorname{supp} u_j^\wedge \subset \Gamma^j \setminus \Gamma^{j-1}, \quad j \in \mathbb{N}.$$

Then, substitute the Littlewood–Paley decomposition of f into $T_{\langle\cdot\rangle^\lambda} f(x)$, $0 \leqslant \lambda \leqslant \sigma$, we get

$$T_{\langle\cdot\rangle^\lambda} f(x) = \int_{\Gamma_p} \left\{\int_{K_p} \langle\xi\rangle^\lambda f(t) \overline{\chi}_\xi(t-x)\, dt\right\} d\xi = \int_{\Gamma_p} \langle\xi\rangle^\lambda f^\wedge(\xi) \chi_x(\xi) d\xi$$

$$= \int_{\Gamma_p} \langle\xi\rangle^\lambda \left\{u_0(\cdot) + \sum_{j=1}^{+\infty} u_j(\cdot)\right\}^\wedge (\xi) \chi_x(\xi) d\xi$$

$$= \int_{\Gamma_p} \langle\xi\rangle^\lambda u_0^\wedge(\xi) \chi_x(\xi) d\xi + \sum_{j=1}^{+\infty} \int_{\Gamma^j \setminus \Gamma^{j-1}} \langle\xi\rangle^\lambda u_j^\wedge(\xi) \chi_x(\xi) d\xi$$

$$= \int_{\Gamma^0} f^\wedge(\xi) \chi_x(\xi) d\xi + \sum_{j=1}^{+\infty} \int_{\Gamma^j \setminus \Gamma^{j-1}} p^{j\lambda} f^\wedge(\xi) \chi_x(\xi) d\xi$$

$$= \sum_{j=0}^{+\infty} p^{j\lambda} u_j(x).$$

By $T_{(\cdot)^\sigma} f \in C(K_p)$, the above series is convergent at any $x \in K_p$, so $p^{j\lambda} u_j(x) = o(1)$, thus, $\|u_j\|_{L^\infty(K_p)} \leqslant cp^{-j\lambda}$, $j \in \mathbb{P}$. Therefore $f \in C^\lambda(K_p)$ for $0 \leqslant \lambda \leqslant \sigma$.

Take $\lambda = \sigma$, we get $f \in C^\sigma(K_p)$, and by (i) in this theorem, we have
$$T_{(\cdot)^\lambda} f \in C^{\sigma-\lambda}(K_p), \quad \lambda \in [0, \sigma].$$
(ii) is proved.

The proof of Theorem 4.1.6 is complete.

Remark. The above theorems describe essential and important properties of Hölder type spaces. Theorem 4.1.4 says: the spaces with higher p-type smoothness are contained in those with lower p-type smoothness. And Theorem 4.1.6 shows: the Hölder type space $C^\sigma(K_p)$ is the nice and suitable space in which the functions with p-type differentiability live. Thus, by virtue of p-type calculus to describe the smoothness of those functions defined on local fields will be appropriate.

4.1.4 Lebesgue type spaces and Sobolev type spaces

1. Lebesgue type spaces

Lebesgue type space $L_r^s(K_p)$, $s \in (-\infty, +\infty)$, $1 \leqslant r \leqslant +\infty$, is defined as
$$L_r^s(K_p) = \left\{ f \in \mathbb{S}^*(K_p) : \|f\|_{L_r^s(K_p)} = \left\|(\langle \cdot \rangle^s f^\wedge(\cdot))^\vee(\cdot)\right\|_{L^r(K_p)} < +\infty \right\}.$$

2. Sobolev type spaces

Sobolev type $W^s(K_p)$, $s \in [0, +\infty)$, is defined as
$$W^s(K_p) = \left\{ f \in \mathbb{S}^*(K_p) : \|f\|_{W^s(K_p)} = \left\|(\langle \cdot \rangle^s f^\wedge(\cdot))^\vee(\cdot)\right\|_{L^2(K_p)} < +\infty \right\},$$
it is a special Lebesgue type space $L_2^s(K_p)$, $s \in [0, +\infty)$.

For the well-defined function spaces above, we study the bounded properties of the pseudo-differential operators T_α.

Theorem 4.1.7 *Let $m \in \mathbb{R}, \rho > 1, \delta \geqslant 0$; or $m + 3(1 - \rho) < 0$.*
$$T_\sigma f(x) = \int_{\Gamma_p} \left\{ \int_{K_p} \sigma(x, \xi) f(t) \overline{\chi}_\xi(t - x) \, dt \right\} d\xi$$
is the pseudo-differential operator with symbol $\sigma \in S_{\rho,\delta}^m(K_p)$. We have

(i) *If $\alpha > m$, then $T_\sigma : C^\alpha(K_p) \to C^{\alpha-m}(K_p)$ is bounded, and holds*

$$\|T_\sigma u\|_{C^{\alpha-m}(K_p)} \leqslant c \|u\|_{C^\alpha(K_p)}.$$

(ii) If $s > m$, then $T_\sigma : W^s(K_p) \to W^{s-m}(K_p)$ is bounded, and holds

$$\|T_\sigma u\|_{W^{s-m}(K_p)} \leqslant c \|u\|_{W^s(K_p)}.$$

(iii) If $\alpha > \dfrac{1}{2}$, then $W^\alpha(K_p) \subset C^{\alpha-\frac{1}{2}}(K_p)$.

Proof. (i) Prove $T_\sigma : C^\alpha(K_p) \to C^{\alpha-m}(K_p)$ and $\|T_\sigma u\|_{C^{\alpha-m}(K_p)} \leqslant c\|u\|_{C^\alpha(K_p)}$.

Take the L-P decomposition of $u \in C^\alpha(K_p)$:

$$u(x) = u * p^0 \Phi_{B^0}(x) + \sum_{j=1}^{+\infty} u * \left\{ p^j \Phi_{B^j}(x) - p^{j-1} \Phi_{B^{j-1}}(x) \right\}$$

$$\equiv u_0(x) + \sum_{j=1}^{+\infty} u_j(x),$$

with

$$u_0(x) = f * p^0 \Phi_{B^0}(x),$$
$$u_j(x) = f * \left\{ p^j \Phi_{B^j}(x) - p^{j-1} \Phi_{B^{j-1}}(x) \right\}, \quad j \in \mathbb{N}.$$

By virtue of the decomposition theorem of symbol $\sigma(x, \xi) \in S_{\rho\sigma}^m(K_p)$, for $m < 0, \rho \geqslant 1$, or $m \leqslant 0, \rho > 1$; or $m + 3(1-\rho) < 0$, the series

$$\sigma(x, \xi) = \sum_{k,j=0}^{+\infty} \omega_{kj}(x) \psi_{kj}(\xi)$$

converges absolutely and uniformly, where

$$\omega_{kj}(x) = \begin{cases} \displaystyle\int_{\Gamma_p} \sigma(x, \xi) \Phi_{\Gamma^0}(\xi) \overline{\chi}_{v(k)}(\xi) d\xi, & j = 0, \\ \displaystyle\int_{\Gamma_p} \sigma(x, \eta) \Phi_{\Gamma^0 \backslash \Gamma^{-1}}(\xi) \overline{\chi}_{v(k)}(\xi) d\xi, & j > 0, \end{cases} \quad (4.1.13)$$

with $|\eta| = p^j |\xi|$, and

$$\psi_{kj}(\xi) = \begin{cases} \Phi_{\Gamma^0}(\xi) \chi_{v(k)}(\xi), & j = 0, \\ \Phi_{\Gamma^j \backslash \Gamma^{j-1}}(\xi) \chi_{v(k)}(\theta) = \Phi_{\Gamma^0 \backslash \Gamma^{-1}}(\theta) \chi_{v(k)}(\theta), & j > 0, \end{cases} \quad (4.1.14)$$

with $|\theta| = p^{-j} |\xi|$; $\{v(k)\}_{k=0}^{+\infty}$ is the complete set of coset of $D \subset K_p$ in K_p, and

$$\{v(k)\}_{k=0}^{+\infty} \leftrightarrow \{\chi_{v(k)}\}_{k=0}^{+\infty},$$

the character set $\{\chi_{v(k)}\}_{k=0}^{+\infty}$ is the complete orthonormal basis of compact subgroup D, and

$$|\omega_{kj}(x)| \leqslant c_\gamma p^{j(m+(1-\rho)\gamma)} |v(k)|^{-\gamma}. \tag{4.1.15}$$

Then, combining (4.1.13)~(4.1.15), definition of $C^\alpha(K_p)$, and supp $\psi_{kj} \cap$ supp $u_l^\wedge = \varnothing$, $j \neq l$, we have

$$T_\sigma u(x) = \int_{\Gamma_p} \sigma(x,\xi) u^\wedge(\xi) \chi_x(\xi) d\xi = \int_{\Gamma_p} \sigma(x,\xi) \sum_{l=0}^{+\infty} u_l^\wedge(\xi) \chi_x(\xi) d\xi$$

$$= \sum_{l=0}^{+\infty} \int_{\Gamma_p} \sum_{k,j=0}^{+\infty} \omega_{kj}(x) \psi_{kj}(\xi) u_l^\wedge(\xi) \chi_x(\xi) d\xi$$

$$= \sum_{l=0}^{+\infty} \sum_{k,j=0}^{+\infty} \int_{\Gamma_p} \omega_{kj}(x) \psi_{kj}(\xi) u_l^\wedge(\xi) \chi_x(\xi) d\xi \equiv \sum_{l=0}^{+\infty} \sum_{k=0}^{+\infty} I_{kl}.$$

On the other hand, by $T_\sigma u(x) = \sum_{l=0}^{+\infty} T_\sigma u_l(x)$,

$$|T_\sigma u_l(x)| \equiv \left| \sum_{k=0}^{+\infty} I_{kl} \right| = \left| \sum_{k=0}^{+\infty} \int_{\Gamma_p} \omega_{kl}(x) \varphi_{kl}(\xi) u_l^\wedge(\xi) \chi_x(\xi) d\xi \right|$$

$$= \left| \sum_{k=0}^{+\infty} \int_{\Gamma_p} \omega_{kl}(x) \varphi_{kl}(\xi) |\xi|^{-\alpha} |\xi|^\alpha u_l^\wedge(\xi) \Phi_{\Gamma^l \setminus \Gamma^{l-1}}(\xi) \chi_x(\xi) d\xi \right|$$

$$\leqslant \left| \sum_{k=0}^{+\infty} p^{-l\alpha} |\omega_{kl}(x)| \int_{\Gamma_p} |\xi|^\alpha u_l^\wedge(\xi) \chi_{v(k)}(\theta) \chi_x(\xi) d\xi \right|$$

$$\leqslant c p^{-l\alpha} \sum_{k=0}^{+\infty} |\omega_{kl}(x)| \|u_l\|_{C^\alpha(K_p)}$$

$$\leqslant c p^{-l(\alpha-m)} \sum_{k=0}^{+\infty} p^{l(1-\rho)\gamma} |v(k)|^{-\gamma} \|u_l\|_{C^\alpha(K_p)},$$

where $|\theta| = p^{-l} |\xi|$. Thus, by $\alpha - m > 0$, $\rho > 1$, it follows that

$$|T_\sigma u_l(x)| \leqslant c p^{-l(\alpha-m)} \|u\|_{C^\alpha(K_p)}.$$

Moreover,

$$(T_\sigma u_l(\cdot))^\wedge(\xi) = \sum_{k=0}^{+\infty} I_{kl}^\wedge(\xi) = \sum_{k=0}^{+\infty} \int_{\Gamma_p} \omega_{kl}^\wedge(\xi - \eta) u_l^\wedge(\eta) \chi_{v(k)}(\theta) d\eta$$

$$= \sum_{k=0}^{+\infty} \int_{\Gamma_p} \omega_{kl}^{\wedge}(\eta) u_l^{\wedge}(\xi-\eta) \Phi_{\Gamma^l \setminus \Gamma^{l-1}}(\xi-\eta) \chi_{v(k)}(\theta') d\eta,$$

where $|\theta'| = p^{-l}|\xi - \eta|$.

Since $(I_{kl})^{\wedge}(\xi) \subset \Gamma_l$, by [93], we have

$$T_\sigma u \in C^{s-m}(K_p) \quad \text{and} \quad \|T_\sigma u\|_{C^{\alpha-m}(K_p)} \leqslant c \|u\|_{C^\alpha(K_p)}.$$

(i) is proved.

(ii) Prove $T_\sigma : W^s(K_p) \to W^{s-m}(K_p)$ and $\|T_\sigma u\|_{W^{s-m}(K_p)} \leqslant c \|u\|_{W^s(K_p)}$.

By

$$\|T_\sigma u_l\|_{L^2(K_p)}$$

$$= \left\{ \int_{K_p} |T_\sigma u_l(x)|^2 dx \right\}^{\frac{1}{2}}$$

$$= \left\{ \int_{K_p} \left| \sum_{k=0}^{+\infty} \int_{\Gamma_p} \omega_{kl}(x) \Phi_{\Gamma^l \setminus \Gamma^{l-1}}(\xi) \chi_{v(k)}(\theta) u_l^{\wedge}(\xi) \chi_x(\xi) d\xi \right|^2 dx \right\}^{\frac{1}{2}}$$

$$\leqslant \sum_{k=0}^{+\infty} \left\{ \int_{K_p} \left| \omega_{kl}(x) \int_{\Gamma_p} u_l^{\wedge}(\xi) \chi_{v(k)}(\theta) \chi_x(\xi) d\xi \right|^2 dx \right\}^{\frac{1}{2}}$$

$$\leqslant \sum_{k=0}^{+\infty} \|\omega_{kl}\|_{L^\infty(K_p)} \|u_l\|_{L^2(K_p)}$$

$$\leqslant \sum_{k=0}^{+\infty} c_\gamma p^{l(m+(1-\rho)\gamma)} |v(k)|^{-\gamma} \|u_l\|_{L^2(K_p)}$$

$$\leqslant c_\gamma p^{l(m+(1-\rho)\gamma)} c_l p^{-ls} \|u\|_{W^s(K_p)} \leqslant c_l p^{-l(s-m)} \|u\|_{W^s(K_p)}.$$

And by [93], we have

$$T_\sigma u \in W^{s-m}(K_p) \quad \text{and} \quad \|T_\sigma u\|_{W^{s-m}(K_p)} \leqslant c \|u\|_{W^s(K_p)}.$$

Thus (ii) is proved.

(iii) Prove $W^\alpha(K_p) \subset C^{\alpha-\frac{1}{2}}(K_p)$.

Take $u \in W^\alpha(K_p)$, and its L–P decomposition

$$u = \sum_{j=0}^{+\infty} u_j,$$

where
$$(u_j)^\wedge(\xi) = \psi_j(\xi) u^\wedge(\xi), \quad j \in \mathbb{P},$$

with $\psi_j(\xi) = \begin{cases} \Phi_{\Gamma^0}(\xi), & j=0, \\ \Phi_{\Gamma^j \setminus \Gamma^{j-1}}(\xi), & j>0, \end{cases}$ and $\operatorname{supp} u_j^\wedge(\xi) \subset \begin{cases} \Gamma^0, & j=0, \\ \Gamma^j \setminus \Gamma^{j-1}, & j>0. \end{cases}$

Moreover, $\|u_j\|_{L^2(K_p)} \leqslant c_j p^{-j\alpha}$, with $\left\{\sum_{j=0}^{+\infty} |c_j|^2\right\}^{\frac{1}{2}} < +\infty.$

Since on local field K_p, it holds $\psi_j(\xi) = \psi_1(\eta)$, $j \in \mathbb{P}$, with $|\xi| = p^j = p^{j-1}|\eta|$, then

$$u_j(x) = p^{j-1}(\psi_1)^\vee * u_j(x) = p^{j-1}\int_{K_p} (\psi_1)^\vee(y) u_j(x-z)\, dz, \quad |y| = p^{-j+1}|z|.$$

And by $\left\|(\psi_1)^\vee\right\|_{L^2(K_p)} = (p^{j-1})^{-\frac{1}{2}}$, it follows that

$$\|u_j\|_{L^\infty(K_p)} \leqslant p^{j-1} \|u_j\|_{L^2(K_p)} \left\|(\psi_1)^\vee\right\|_{L^2(K_p)}$$
$$\leqslant c_j \cdot p^{j-1} \cdot (p^{j-1})^{-\frac{1}{2}} p^{-j\alpha} \|u\|_{W^\alpha(K_p)}$$
$$\leqslant c \cdot \left(p^{-j(\alpha-\frac{1}{2})}\right) \|u\|_{W^\alpha(K_p)}, \quad j \in \mathbb{P},$$

with $c > 0$, this implies $u \in C^{\alpha-\frac{1}{2}}(K_p)$ and $\|u\|_{C^{\alpha-\frac{1}{2}}(K_p)} \leqslant c \|u\|_{W^\alpha(K_p)}$. Thus (iii) is proved.

Theorem 4.1.8 Suppose $\sigma \in S^m_{\rho,\delta}(K_p), m + 3(1-\rho) < 0, m < 0$, then $f \in L^r(K_p)$ implies $T_\sigma f \in L^r(K_p), 1 \leqslant r < +\infty$, and holds

$$\|T_\sigma f\|_{L^r(K_p)} \leqslant c \|f\|_{L^r(K_p)}, \quad 1 \leqslant r < +\infty.$$

Proof. For $f \in L^r(K_p)$, we deduce that

$$T_\sigma f(x) = \int_{\Gamma_p} \sigma(x,\xi) \int_{K_p} f(t) \chi_\xi(x-t)\, dt d\xi$$
$$= \int_{\Gamma_p} \sum_{j=0}^{+\infty} \sum_{k=0}^{+\infty} \omega_{kj}(x) \psi_j(\xi) \chi_{v(k)}(p^j \xi) \int_{K_p} f(t) \chi_\xi(x-t)\, dt d\xi$$
$$= \sum_{j=0}^{+\infty} \sum_{k=0}^{+\infty} \omega_{kj}(x) \int_{\Gamma_p} \psi_j(\xi) \chi_{v(k)}(p^j \xi) \int_{K_p} f(t) \chi_\xi(x-t)\, dt d\xi$$
$$= \sum_{j=0}^{+\infty} \sum_{k=0}^{+\infty} \omega_{kj}(x) \int_{K_p} f(x-t) \int_{\Gamma_p} \psi_j(\xi) \chi_{v(k)}(p^j \xi) \chi_\xi(t)\, d\xi dt$$

$$= \sum_{j=0}^{+\infty}\sum_{k=0}^{+\infty} \omega_{kj}(x) \int_{K_p} f(x-t) \int_{\Gamma_p} \psi_j(\xi)\chi_{p^j v(k)+t}(\xi) d\xi dt$$

$$= \sum_{j=0}^{+\infty}\sum_{k=0}^{+\infty} \omega_{kj}(x) g_{kj}(x),$$

with $g_{kj}(x) = f * h_{kj}(x)$, and $h_{kj}(x) = \int_{\Gamma_p} \psi_j(\xi)\chi_{p^j v(k)+x}(\xi) d\xi$. Thus

$$\|T_\sigma f\|_{L^r(K_p)} = \left\{\int_{K_p} |T_\sigma f(x)|^r dx\right\}^{\frac{1}{r}} \leqslant \sum_{j=0}^{+\infty}\sum_{k=0}^{+\infty} \left\{\int_{K_p} |\omega_{kj}(x) g_{kj}(x)|^r dx\right\}^{\frac{1}{r}}$$

$$= \|\omega_{00} g_{00}\|_{L^r(K_p)} + \sum_{k=1}^{+\infty} \|\omega_{k0} g_{k0}\|_{L^r(K_p)}$$

$$+ \sum_{j=1}^{+\infty}\sum_{k=0}^{+\infty} \|\omega_{kj} g_{kj}\|_{L^r(K_p)} + \sum_{j=1}^{+\infty}\sum_{k=2}^{+\infty} \|\omega_{kj} g_{kj}\|_{L^r(K_p)}$$

$$\leqslant c\left\{\|f * h_{00}\|_{L^r(K_p)} + \sum_{k=1}^{+\infty} |v(k)|^{-2} \|f * h_{k0}\|_{L^r(K_p)}\right.$$

$$+ \sum_{j=1}^{+\infty}\sum_{k=0}^{+\infty} p^{jm} \|f * h_{kj}\|_{L^r(K_p)}$$

$$\left.+ \sum_{j=1}^{+\infty}\sum_{k=2}^{+\infty} |v(k)|^{-2} p^{j(m+3(1-\rho))} \|f * h_{kj}\|_{L^r(K_p)}\right\}$$

$$\leqslant c \|f\|_{L^r(K_p)} \left\{1 + \sum_{k=1}^{+\infty} |v(k)|^{-2}\right.$$

$$\left.+ 2(1-p^{-1})\left[\sum_{j=1}^{+\infty}\sum_{k=0}^{+\infty} p^{jm} + \sum_{j=1}^{+\infty}\sum_{k=2}^{+\infty} |v(k)|^{-2} p^{j(m+3(1-\rho))}\right]\right\}$$

$$\leqslant C \|f\|_{L^r(K_p)}.$$

The proof is complete.

Theorem 4.1.9 Suppose $\sigma \in S^m_{\rho,\delta}(K_p), m + ([s]+1)\delta + 3(1-\rho) < 0$ and $m < 0$, then $f \in B^s_{rt}(K_p)$ implies $T_\sigma f \in B^s_{rt}(K_p), 1 \leqslant r, t < +\infty$, and holds

$$\|T_\sigma f\|_{B^s_{rt}(K_p)} \leqslant c \|f\|_{B^s_{rt}(K_p)}, \quad 1 \leqslant r, t < +\infty.$$

Specially, if $\rho = 1, \delta = 0, m < 0$ and $s > 0$, then $f \in B^s_{rt}(K_p)$ implies $T_\sigma f \in B^s_{rt}(K_p)$, $1 \leqslant r, t < +\infty$, and holds
$$\|T_\sigma f\|_{B^s_{rt}(K_p)} \leqslant c \|f\|_{B^s_{rt}(K_p)}, \quad 1 \leqslant r, t < +\infty.$$

Proof. We only give the line of proof: Estimate the $B^s_{rt}(K_p)$-norm of
$$T_\sigma f(x) = \sum_{j=0}^{+\infty} \sum_{k=0}^{+\infty} \omega_{kj}(x) g_{kj}(x),$$
with $g_{kj}(x) = f * h_{kj}(x)$, $h_{kj}(x) = \int_{\Gamma_p} \psi_j(\xi) \chi_{p^j v(k)+x}(\xi) d\xi$. It is

$\|T_\sigma f(x)\|_{B^s_{rt}(K_p)}$

$= \left\{ \int_{K_p} \|T_\sigma f(\cdot - y) - T_\sigma f(\cdot)\|^t_{L^r(K_p)} |y|^{-(st+1)} dy \right\}^{\frac{1}{t}}$

$= \left\{ \int_{K_p} \left\| \sum_{j=0}^{+\infty} \sum_{k=0}^{+\infty} [\omega_{kj}(\cdot - y) g_{kj}(\cdot - y) - \omega_{kj}(\cdot) g_{kj}(\cdot)] \right\|^t_{L^r(K_p)} |y|^{-(st+1)} dy \right\}^{\frac{1}{t}}$

$\leqslant \sum_{j=0}^{+\infty} \sum_{k=0}^{+\infty} \left\{ \int_{K_p} \|\omega_{kj}(\cdot - y) g_{kj}(\cdot - y) - \omega_{kj}(\cdot) g_{kj}(\cdot)\|^t_{L^r(K_p)} |y|^{-(st+1)} dy \right\}^{\frac{1}{t}}.$

Let
$$I_{kj} \equiv \left\{ \int_{K_p} \|\omega_{kj}(\cdot - y) g_{kj}(\cdot - y) - \omega_{kj}(\cdot) g_{kj}(\cdot)\|^t_{L^r(K_p)} |y|^{-(st+1)} dy \right\}^{\frac{1}{t}},$$
for $j, k \in \mathbb{P}$, then estimate I_{kj}, one may get the result of the theorem[84]. We omit the details.

Exercises

1. Establish the theory about the homogeneous B-type spaces and F-type spaces on a local field K_p.
2. How to establish the function space theory on the multiplication group K_p^* of K_p.
3. Prove Theorems 4.1.1~4.1.3.
4. Study some new special cases of B-type spaces and F-type spaces on K_p, and study their properties. Compare with those of $\mathbb{R}^n$ cases, propose some new open problems.
5. Can we generalize the Hölder spaces to the multiplication group K_p^* of K_p? What preparations are needed?

4.2 Lipschitz class on local fields

The Hölder spaces and Lipschitz class appear in lots of scientific areas, they play important roles, as well known. Scientists are also familiar with the properties of functions in the function spaces and they apply those functions to their scientific areas to solve the practical problems. When we study the function spaces underlying local fields, we hope that the functions defined on a local field and the distributions underlying local fields have more important applications.

Whether the hope of us can be achieved? The key step is to compare those properties of functions, function spaces, operators underlying Euclidean spaces and local fields, and so on, so that scientists can recognize the similar properties and the different properties of two underlying spaces. We emphasize that to study the topics on local fields and do comparison between those results on two underlying spaces are necessary and important.

4.2.1 Lipschitz classes on local fields

Let $f: K_p \to \mathbb{C}$ be a complex Haar measurable function on K_p.

Definition 4.2.1 (Lipschitz class) *Let $C(K_p)$ be the bounded continuous function space on K_p. For $\alpha > 0$,*

$$\mathrm{Lip}(C(K_p), \alpha) = \left\{ f \in C(K_p) : \|f(\cdot + h) - f(\cdot)\|_{C(K_p)} = O(|h|^\alpha),\ h \in K_p \right\}$$

*is said to be a **Lipschitz class** in $C(K_p)$ on local field K_p, simply, Lip class (compared with Definition 3.4.4).*

Theorem 4.2.1 *For a local field K_p, it holds*

$$\mathrm{Lip}(C(K_p), \alpha) = C^\alpha(K_p),\quad \alpha \in (0, +\infty).$$

Proof. Let $f \in \mathrm{Lip}(C(K_p), \alpha)$.

Since $\mathrm{Lip}(C(K_p), \alpha) \subset C(K_p) \subset \mathbb{S}^*(K_p)$, by unit decomposition (4.1.7) of the character group Γ_p of K_p,

$$1 = \Phi_{\Gamma^0}(\xi) + \sum_{j=1}^{+\infty} \Phi_{\Gamma^j \setminus \Gamma^{j-1}}(\xi),\quad \xi \in \Gamma_p,$$

it follows that

$$f^\wedge(\xi) = f^\wedge(\xi)\Phi_{\Gamma^0}(\xi) + \sum_{j=1}^{+\infty} f^\wedge(\xi)\Phi_{\Gamma^j \setminus \Gamma^{j-1}}(\xi) = f_0^\wedge(\xi) + \sum_{j=1}^{+\infty} f_j^\wedge(\xi),$$

with
$$f_0^\wedge(\xi) = f^\wedge(\xi)\Phi_{\Gamma^0}(\xi) = \left[(f^\wedge)^\vee * \Phi_{\Gamma^0}^\vee\right]^\wedge(\xi),$$
$$f_j^\wedge(\xi) = f^\wedge(\xi)\Phi_{\Gamma^j\backslash\Gamma^{j-1}}(\xi) = \left[(f^\wedge)^\vee * \Phi_{\Gamma^j\backslash\Gamma^{j-1}}^\vee\right]^\wedge(\xi), \quad j \in \mathbb{N}.$$

Thus
$$f_0(x) = f * \Phi_{\Gamma^0}^\vee(x) = \int_{K_p} f(x-t)\Phi_{\Gamma^0}^\vee(t)dt,$$
$$f_j(x) = f * \Phi_{\Gamma^j\backslash\Gamma^{j-1}}^\vee(x)$$
$$= \int_{K_p} f(x-t)\Phi_{\Gamma^j}^\vee(t)dt - \int_{K_p} f(x-t)\Phi_{\Gamma^{j-1}}^\vee(t)dt, \quad j \in \mathbb{N}.$$

Since
$$\Phi_{\Gamma^0}^\vee(x) = p^0\Phi_{B^0}(x),$$
$$\Phi_{\Gamma^j}^\vee(x) - \Phi_{\Gamma^{j-1}}^\vee(x) = p^j\Phi_{B^j}(x) - p^{j-1}\Phi_{B^{j-1}}(x), \quad j \in \mathbb{N},$$

thus
$$f_0(x) = \int_{K_p} p^0 f(x-t)\Phi_{B^0}(t)dt = \int_{B^0} f(x-\beta^0 y)\Phi_{B^0}(\beta^0 y)\, dy;$$
$$f_j(x) = \int_{K_p} p^j f(x-t)\Phi_{B^j}(t)dt - \int_{K_p} p^{j-1}f(x-t)\Phi_{B^{j-1}}(t)dt$$
$$= \int_{K_p} p^j f(x-\beta^j y)\Phi_{B^0}(\beta^j y)d(\beta^j y)$$
$$- \int_{K_p} p^{j-1}f(x-\beta^{j-1}y)\Phi_{B^0}(\beta^{j-1}y)d(\beta^{j-1}y), \quad j \in \mathbb{N},$$

with $|\beta| = p^{-1}$. Hence, by $f \in \mathrm{Lip}(C(K_p),\alpha) \Rightarrow \|\Delta_h f\|_{C(K_p)} = O(|h|^\alpha)$, and set
$$h = (x - \beta^j y) - (x - \beta^{j-1}y) = \beta^{j-1}y - \beta^j y,$$
the estimates of norms can be obtained
$$\|f_0\|_{L^\infty(K_p)} \leqslant \left\|\int_{B^0} |f(\cdot - \beta^0 y)\Phi_{B^0}(\beta^0 y)|\, dy\right\|_{L^\infty(K_p)} \leqslant cp^{-0\cdot\alpha},$$
$$\|f_j\|_{L^\infty(K_p)} \leqslant \left\|\int_{K_p} f(x-\beta^j y)\Phi_{B^0}(\beta^j y)\, dy\right.$$
$$\left. - \int_{K_p} f(x-\beta^{j-1}y)\Phi_{B^0}(\beta^{j-1}y)\, dy\right\|_{L^\infty(K_p)} \leqslant cp^{-j\alpha}, \quad j \in \mathbb{N}.$$

Note that $\|f\|_{C(K_p)} = \sup_j \{p^{j\alpha}\|f_j\|_{L^\infty(K_p)}\} < +\infty$, then
$$\mathrm{Lip}(C(K_p),\alpha) \subset C^\alpha(K_p), \quad \alpha \in (0,+\infty).$$

Conversely, for $f \in C^\alpha(K_p)$, then by unit decomposition theorem on K_p,
$$f(x) = f_0(x) + \sum_{j=1}^{\infty} f_j(x) = f * \Phi_{\Gamma^0}^{\vee}(x) + \sum_{j=1}^{\infty} f * \Phi_{\Gamma^j \backslash \Gamma^{j-1}}^{\vee}(x).$$
Without loss of generality, take $h \in K_p$ with $|h| < 1$, then there exists an integer $j_0 \in \mathbb{N}$ with $p^{-j_0-1} \leqslant |h| \leqslant p^{-j_0}$. Thus
$$f(x+h) - f(x) = \sum_{j=0}^{\infty} [f_j(x+h) - f_j(x)] \equiv \sum_{j \leqslant j_0} + \sum_{j > j_0}.$$
Estimate $\sum_{j > j_0}$ first. Since
$$\sum_{j > j_0} |f_j(x+h) - f_j(x)|$$
$$\leqslant \sum_{j > j_0} \int_{K_p} |f(x+h-t) - f(x-t)| \Phi_{\Gamma^j \backslash \Gamma^{j-1}}^{\vee}(t) dt$$
$$\leqslant c' p^j \sum_{j > j_0} \int_{K_p} |f(x+h-t) - f(x-t)| \Phi_{B^{j-1} \backslash B^j}(t) dt$$
$$\leqslant \sum_{j > j_0} c' p^j p^{-j\alpha} p^{-j} \leqslant c' \sum_{j > j_0} p^{-j\alpha} \leqslant c p^{-(j_0+1)\alpha} \leqslant c|h|^\alpha,$$
we get
$$\sum_{j > j_0} |f_j(x+h) - f_j(x)| \leqslant c|h|^\alpha.$$
Then, for $\sum_{j \leqslant j_0}$, it follows that
$$f_j(x+h) - f_j(x) = f * \Phi_{\Gamma^j \backslash \Gamma^{j-1}}^{\vee}(x+h) - f * \Phi_{\Gamma^j \backslash \Gamma^{j-1}}^{\vee}(x)$$
$$= \int_{K_p} f(t) \int_{\Gamma_p} \Phi_{\Gamma^j \backslash \Gamma^{j-1}}(\xi) \{\chi_{x+h-t}(\xi) - \chi_{x-t}(\xi)\} d\xi dt$$
$$= \int_{K_p} f(t) \int_{\Gamma^j \backslash \Gamma^{j-1}} \{\chi_{x+h-t}(\xi) - \chi_{x-t}(\xi)\} d\xi dt.$$
To compute $\chi_{x+h-t}(\xi) - \chi_{x-t}(\xi)$, let $\xi \in \Gamma^j \backslash \Gamma^{j-1}$, $j > 0$, $h \in K_p$, then
$$\xi = \xi_{-j} \beta^{-j} + \xi_{-j+1} \beta^{-j+1} + \cdots, \quad j > 0,$$
$$h = h_{j_0} \beta^{j_0} + h_{j_0+1} \beta^{j_0+1} + \cdots.$$
We set $x - t = y_k \beta^k + y_{k+1} \beta^{k+1} + \cdots$, $k \in \mathbb{Z}$. Then if j_0 is large enough, i.e., $|h|$ with $p^{-j_0-1} \leqslant |h| \leqslant p^{-j_0}$, is small enough, the points $x - t + h$ and

$x - t$ in the term $\chi_{x+h-t}(\xi) - \chi_{x-t}(\xi)$ belong to same B^k, $k \in \mathbb{Z}$, that is, for $k \in \mathbb{Z}$,

$$x - t = y_k \beta^k + y_{k+1}\beta^{k+1} + \cdots,$$

$$x - t + h = y_k\beta^k + y_{k+1}\beta^{k+1} + \cdots + y_{j_0-1}\beta^{j_0-1}$$
$$+ (y_{j_0} + h_{j_0})\beta^{j_0} + (y_{j_0+1} + h_{j_0+1})\beta^{j_0+1} + \cdots,$$

then $x - t + h$, $x - t \in B^k$. Thus, $\chi_{x+h-t}(\xi) - \chi_{x-t}(\xi) = 0$ for j_0 which is large enough. So

$$\left| \sum_{j \leqslant j_0} [f_j(x+h) - f_j(x)] \right|$$

$$\leqslant \sum_{j \leqslant j_0} \left| \int_{K_p} f(t) \int_{\Gamma_p} \Phi_{\Gamma^j \setminus \Gamma^{j-1}}(\xi) \{\chi_{x+h-t}(\xi) - \chi_{x-t}(\xi)\} d\xi dt \right| \leqslant c|h|^\alpha.$$

Combining the estimations of $\sum_{j > j_0}$ and $\sum_{j \leqslant j_0}$, it follows that

$$|f(x+h) - f(x)| \leqslant \sum_{j=0}^\infty |f_j(x+h) - f_j(x)|$$
$$= \sum_{j \leqslant j_0} |f_j(x+h) - f_j(x)| + \sum_{j > j_0} |f_j(x+h) - f_j(x)| \leqslant c|h|^\alpha.$$

This implies $\mathrm{Lip}(C(K_p), \alpha) \supset C^\alpha(K_p)$, $\alpha \in (0, +\infty)$. We conclude that

$$\mathrm{Lip}(C(K_p), \alpha) = C^\alpha(K_p), \quad \alpha \in (0, +\infty).$$

The proof is complete.

Theorem 4.2.2 *For a local field K_p, if $\alpha > \beta$ with $\alpha, \beta \in (0, +\infty)$, then*

$$\mathrm{Lip}(C(K_p), \alpha) \subset \mathrm{Lip}(C(K_p), \beta).$$

Proof. By the property of $C^\alpha(K_p)$ (Theorem 4.1.4).

Theorem 4.2.3 *Let $\alpha > 0$, then $\mathrm{Lip}(C(K_p), \alpha) \subset L^1_{\mathrm{loc}}(K_p)$.*

Proof. By Theorem 4.1.5, $C^\alpha(K_\alpha) = B^\alpha_{\infty\infty}(K_p), \alpha \in \mathbb{R}$, then for $\alpha > 0$, we have

$$\mathrm{Lip}(C(K_p), \alpha) = C^\alpha(K_p) = B^\alpha_{\infty\infty}(K_p) \subset L^1_{\mathrm{loc}}(K_p),$$

the last conclusion relationship $B^\alpha_{\infty\infty}(K_p) \subset L^1_{\mathrm{loc}}(K_p)$ comes from the following proposition.

Proposition If $0 < r \leqslant +\infty, 0 < t \leqslant +\infty, s > \sigma_r = \left(\frac{1}{r} - 1\right)_+$, then
$$B_{rt}^s(K_p) \subset L_{\text{loc}}^1(K_p).$$

Remark. This proposition shows that: each element in the B-type space $B_{rt}^s(K_p)$ is locally integrable under the conditions $0 < r \leqslant +\infty, 0 < t \leqslant +\infty, s > \sigma_r = \left(\frac{1}{r} - 1\right)_+$.

Proof. Firstly, let $1 \leqslant r \leqslant \infty$, then $\sigma_r = \left(\frac{1}{r} - 1\right)_+ = 0$ implies $s > 0 = \sigma_r$.

For $f \in B_{rt}^s(K_p) \subset \mathbb{S}^*(K_p)$, its L–P decomposition $f = \sum_{j=0}^{\infty} f * \varphi_j$ converges in the $\mathbb{S}^*(K_p)$ sense, where
$$\varphi_0(x) = p^0 \Phi_{B^0}(x), \varphi_j(x) = (p^j \Phi_{B^j} - p^{j-1} \Phi_{B^{j-1}})(x), j = 1, 2, \cdots.$$

We consider two cases

(a) If $1 \leqslant t \leqslant \infty$, then by Minkowski inequality and Hölder inequality, it follows

$$\left\| \sum_{j=0}^{\infty} f * \varphi_j \right\|_{L^r(K_p)} \leqslant \sum_{j=0}^{\infty} p^{-js} p^{js} \| f * \varphi_j \|_{L^r(K_p)}$$

$$\leqslant \left(\sum_{j=0}^{\infty} p^{-jst'} \right)^{1/t'} \left(\sum_{j=0}^{\infty} \left(p^{js} \| f * \varphi_j \|_{L^r(K_p)} \right)^t \right)^{1/t}$$

$$\leqslant c \| f \|_{B_{rt}^s(K_p)} < +\infty.$$

(b) If $0 < t < 1$, by $l_t \subset l_1$, it follows

$$\left\| \sum_{j=0}^{\infty} f * \varphi_j \right\|_{L^r(K_p)} \leqslant \sum_{j=0}^{\infty} \| f * \varphi_j \|_{L^r(K_p)} \leqslant \sum_{j=0}^{\infty} p^{js} \| f * \varphi_j \|_{L^r(K_p)}$$

$$\leqslant \left(\sum_{j=0}^{\infty} \left(p^{js} \| f * \varphi_j \|_{L^r(K_p)} \right)^t \right)^{1/t} \leqslant \| f \|_{B_{rt}^s(K_p)} < +\infty.$$

Thus for $0 < t \leqslant \infty$, it holds

$$f \in B_{rt}^s(K_p) \Rightarrow f \in L^r(K_p) \subset L^1_{\text{loc}}(K_p).$$

Secondly, let $0 < r < 1$, then $\sigma_r = \left(\frac{1}{r} - 1\right)_+ = \frac{1}{r} - 1$ implies $s > \sigma_r = \frac{1}{r} - 1$, i.e., $s - \frac{1}{r} + 1 > 0$. Thus, by the relationship of B-type spaces[59]: "if $0 < r_0 \leqslant r_1 < \infty, 0 < t \leqslant \infty, -\infty < s_1 \leqslant s_0 < \infty$, and $s_0 - \frac{1}{r_0} = s_1 - \frac{1}{r_1}$, then $B_{r_0 t}^{s_0}(K_p) \subset B_{r_1 t}^{s_1}(K_p)$", we get immediately

$$B_{rt}^s(K_p) \subset B_{1t}^{s-\frac{1}{r}+1}(K_p).$$

By the first step, we have $B_{1t}^{s-\frac{1}{r}+1}(K_p) \subset L^1(K_p)$, and thus $B_{rt}^s(K_p) \subset L^1(K_p)$.

4.2.2 Chains of function spaces on Euclidean spaces

We recall the function space chain underlying on Euclidean space, and refer to [71], [73], [102], [105].

An interesting example in the classical theory of construction theory of function:

$$f(x) = \sum_{k=1}^{+\infty} 2^{-2k} \cos 2^k x,$$

its best approximation is

$$E_n(C_{2\pi}, f) = \inf_{p \in T_n} \|f - p\|_{C_{2\pi}} = O(n^{-2}),$$

however, its second order continuous modulus is

$$\omega_2(C_{2\pi}, f, \delta) \geqslant \frac{4}{\pi^2} \delta^2 \frac{\ln \frac{1}{2\delta}}{\ln 2}.$$

In 1950s, mathematician S. B. Stechkin of Russia asserted that, we not only need the second order continuous modulus, but also need the third order, fourth order, $\cdots$, to describe the smoothness of functions. The above example shows this fact.

We show the results of functions defined on $\mathbb{R} = \mathbb{R}^1$ in this section, however, all results hold for $\mathbb{R}^n$.

1. m-order continuous modulus and m-order Lip class

Let $C_{2\pi}^m \equiv C^m([0, 2\pi])$ be the function space of all m-order continuous differentiable and 2π-periodic functions, $m \in \mathbb{N}$, and let

$$C_{2\pi} = \{f : f \text{ is } 2\pi\text{-periodic continuous on } \mathbb{R}\}.$$

Definition 4.2.2 (m-order difference) For $f \in C_{2\pi}$, $m \in \mathbb{N}$, the first order, the second order, $\cdots$, the mth-order difference are defined as

$$\Delta_h^1 f(x) \equiv \Delta_h f(x) = f(x + h) - f(x),$$
$$\Delta_h^2 f(x) = \Delta_h (\Delta_h f)(x) = \Delta_h (f(x + h) - f(x))$$
$$= f(x + 2h) - 2f(x + h) + f(x),$$
$$\cdots$$
$$\Delta_h^m f(x) = \Delta_h \left(\Delta_h^{m-1}\right) f(x).$$

Definition 4.2.3 (m^{th}-order continuous modulus) For $f \in C_{2\pi}$, $m \in \mathbb{N}$, then

$$\omega_m(C_{2\pi}, f, \delta) = \sup_{|h| \leq \delta} \|\Delta_h^m f(\cdot)\|_{C_{2\pi}}$$

is said to be m^{th}-**order continuous modulus** of $f \in C_{2\pi}$.

The second-order continuous modulus of f,

$$\omega_2(C_{2\pi}, f, \delta) \equiv \omega^*(C_{2\pi}, f, \delta) = \sup_{|h| \leq \delta} \|\Delta_h^2 f(\cdot)\|_{C_{2\pi}}$$

is said to be the **smooth modulus** also.

The mth-order continuous modulus has the following properties.

Theorem 4.2.4 Let $\omega_m(C_{2\pi}, f, \delta)$ be the mth-order continuous modulus of $f \in C_{2\pi}$. Then

(i) $\omega_m(C_{2\pi}, f, \delta)$ is a monotonic increasing function of δ ($\delta \geq 0$):

$$\delta_1 \leq \delta_2 \Rightarrow \omega_m(C_{2\pi}, f, \delta_1) \leq \omega_m(C_{2\pi}, f, \delta_2);$$

(ii) $\omega_m(C_{2\pi}, f, \delta)$ is a "decreasing function" of m ($m \in \mathbb{N}$):

$$j < m \Rightarrow \omega_m(C_{2\pi}, f, \delta) \leq 2^{m-j} \omega_j(C_{2\pi}, f, \delta);$$

(iii) $\omega_m(C_{2\pi}, f, \delta)$ has "dilation property" about δ ($\delta \geq 0$):

$$\lambda > 0 \Rightarrow \omega_m(C_{2\pi}, f, \lambda\delta) \leq (1 + \lambda)^m \omega_m(C_{2\pi}, f, \delta);$$

(iv) $\omega_m(C_{2\pi}, f, \delta)$ has a "scale property" about δ ($\delta \geq 0$):

$$\delta_1 < \delta_2 \Rightarrow \frac{\omega_m(C_{2\pi}, f, \delta_2)}{\delta_2^m} \leq \frac{2^m \omega_m(C_{2\pi}, f, \delta_1)}{\delta_1^m}.$$

Moreover, as $\delta \to 0$, it holds

$$\lim_{\delta \to 0+} \frac{\omega_m(C_{2\pi}, f, \delta)}{\delta^m} > 0, \quad \forall f \neq \text{const},$$

and

$$\lim_{\delta \to 0+} \omega_m(C_{2\pi}, f, \delta) = o(\delta^m) \Rightarrow f = \text{const}.$$

Definition 4.2.4 (m^{th}-*order Lipschitz class*) *For $0 < \alpha \leqslant m$, $m \in \mathbb{N}$, then*

$$\text{Lip}_m(C_{2\pi}, \alpha) = \left\{ f \in C_{2\pi} : \omega_m(C_{2\pi}, f, \delta) = O(\delta^\alpha), \delta \to 0^+ \right\}$$

*is said to be an mth-order **Lipschitz α-class**, simply, m^{th}-order **Lip class**, denoted by $\text{Lip}_m \alpha$.*

Theorem 4.2.5 *Let $\text{Lip}_m \alpha$ be the mth-order Lipschitz class, $0 < \alpha \leqslant m$, $m \in \mathbb{N}$. Then*

(i) There are the following relationships for m^{th}-order Lip class $\text{Lip}_m \alpha$:

$$0 < \alpha < m \Rightarrow \text{Lip}_m \alpha = \text{Lip}_{m+1} \alpha,$$
$$\alpha = m \Rightarrow \text{Lip}_m m \subsetneqq \text{Lip}_{m+1} m.$$

(ii) There are the following relationships for Lipschitz class $\text{Lip}_m \alpha$, $0 < \alpha \leqslant m$:

$$\beta < \alpha \Rightarrow \text{Lip}_m \alpha \subsetneqq \text{Lip}_m \beta;$$

(iii) For $f \in \text{Lip}_m \alpha$, $0 < \alpha \leqslant m$, it holds that: $\forall m \in \mathbb{N}, \exists M_m$, which is a constant such that for each $\delta > 0$, we have

$$f \in \text{Lip}_m \alpha \Rightarrow \omega_m(C_{2\pi}, f, \delta) \leqslant M_m \delta^\alpha.$$

2. Chain between C^{m-1} and C^m, $m \in \mathbb{N}$

One may be familiar with the relationships

$$\begin{aligned} 0 < \alpha < m &: \text{Lip}_m \alpha = \text{Lip}_{m+1} \alpha, \\ \alpha = m &: \text{Lip}_m m \subsetneqq \text{Lip}_{m+1} m; \end{aligned} \tag{4.2.1}$$

moreover, a "chain" between $C^1 \equiv C_{2\pi}^1$ and $C \equiv C_{2\pi}$,

$$C^1 \subsetneqq \text{Lip} 1 \subsetneqq \text{Lip} \alpha \subsetneqq \text{Lip} \beta \subsetneqq C, \tag{4.2.2}$$
$$\uparrow$$
$$1 > \alpha > \beta > 0$$

as well as

$$C^m \subsetneq \mathrm{Lip}_m m \subsetneq \mathrm{Lip}_{m+1} m \subsetneq \mathrm{Lip}_{m+1}\alpha \subsetneq \mathrm{Lip}_{m+1}\beta \subsetneq C^{m-1}.$$
$$\uparrow \qquad\qquad\qquad \uparrow$$
$$\alpha = m \qquad\qquad m > \alpha > \beta > 0$$
$$m \in \mathbb{N} = \{1, 2, 3, \cdots\} \qquad (4.2.3)$$

The Jackson theorem and the Bernstein theorem in the approximation theory hold for $m \in \mathbb{N}$:
$$0 < \alpha < m\colon f \in \mathrm{Lip}_m \alpha \Leftrightarrow E_n(C_{2\pi}, f) = O(n^{-\alpha}),$$
$$\alpha = m\colon f \in \mathrm{Lip}_m m \Rightarrow E_n(C_{2\pi}, f) = O(n^{-m}),$$
$$f \in \mathrm{Lip}_{m+1} m \Leftrightarrow E_n(C_{2\pi}, f) = O\left(n^{-m}\right). \qquad (4.2.4)$$

Let $\mathrm{Lip}^*\alpha \equiv \mathrm{Lip}^*(\mathbb{R}, \alpha)$ be the second-order Lipschitz class, then we have the chain for C^1 and C:
$$C^1 \subsetneq \mathrm{Lip}\,1 \subsetneq \mathrm{Lip}^*1 \subsetneq \mathrm{Lip}^*\alpha = \mathrm{Lip}\,\alpha \subsetneq \mathrm{Lip}\,\beta = \mathrm{Lip}^*\beta \subsetneq C$$
$$\text{not equal} \qquad \text{equal} \qquad\qquad \text{equal} \qquad\qquad (4.2.5)$$
$$1 > \alpha > 0 \qquad\qquad 1 > \alpha > \beta > 0$$

Note that, there is a "gap" between $\mathrm{Lip}\,\alpha$ and $\mathrm{Lip}^*\alpha$ at $\alpha = 1$:
$$0 < \alpha < 1 \Rightarrow \mathrm{Lip}\,\alpha = \mathrm{Lip}^*\alpha,$$
$$\alpha = 1 \Rightarrow \mathrm{Lip}\,1 \subsetneq \mathrm{Lip}^*1. \qquad (4.2.6)$$

The example $f(x) = \sin \ln |\sin x|$ shows that $\mathrm{Lip}\,1 \ne \mathrm{Lip}^*1$. The "gap" appears in the equivalent theorems also at $\alpha = 1$
$$0 < \alpha < 1\colon f \in \mathrm{Lip}\,\alpha \Leftrightarrow E_n(C_{2\pi}, f) = O(n^{-\alpha}),$$
$$\alpha = 1\colon f \in \mathrm{Lip}\,1 \Rightarrow E_n(C_{2\pi}, f) = O(n^{-1}), \qquad (4.2.7)$$

and
$$0 < \alpha \leqslant 1\colon f \in \mathrm{Lip}^*\alpha \Leftrightarrow E_n(C_{2\pi}, f) = O(n^{-\alpha}), \qquad (4.2.8)$$

where $E_n(C_{2\pi}, f)$ is the nth-degree trigonometric best approximation of $f \in C_{2\pi}$[9].

Moreover, the "gap" appears in the Hölder spaces $C^\alpha \equiv C^\alpha(\mathbb{R})$, $\alpha \in (0, +\infty) \setminus \mathbb{N}$, and the Zygmund classes $C_*^\alpha \equiv C_*^\alpha(\mathbb{R})$, $\alpha \in (0, +\infty) \setminus \mathbb{N}$,
$$0 < \alpha < 1\colon \qquad C^\alpha \leftrightarrow \mathrm{Lip}\,\alpha = \mathrm{Lip}^*\alpha \leftrightarrow C_*^\alpha$$
$$\alpha = 1\colon \qquad \mathrm{Lip}\,1 \subsetneq \mathrm{Lip}^*1$$
$$\uparrow \qquad\qquad \uparrow$$
$$\text{no Hölder space} \qquad \text{no Zygmund class}$$

$$1 < \alpha < 2 : C^\alpha \qquad \text{Lip}^*\alpha \leftrightarrow C^\alpha_* \qquad (4.2.9)$$

$$\uparrow$$

no Lip class

where $C^\alpha_* = \{f \in C_{2\pi} : |f(x+h) - 2f(x) + f(x-h)| = O(|h|)\}$ is the Zygmund class defined by Zygmund, that is Lip*1.

We note, for $\alpha > 1$ in case $\mathbb{R}$, $f \in \text{Lip}(\alpha, \mathbb{R})$ implies $f = \text{const.}$

$$|f(x+h) - f(x)| = o(h) \Rightarrow f = \text{const.} \qquad (4.2.10)$$

4.2.3 The cases on a local field K_p

Let K_p be a local field, $D = B^0 \subset K_p$ be a compact subgroup in K_p. The Lipschitz class on D is denoted by $\text{Lip}(C(D), \alpha)$ and $E_{p^n}(C(D), f)$ is the p^n-degree best approximation in D.

Theorem 4.2.6 *For $\alpha > 0, s \in \mathbb{P}$, it holds the equivalent relationship*

$$f^{(s)} \in \text{Lip}(C(D), \alpha) \Leftrightarrow E_{p^n}(C(D), f) = O\left((p^n)^{-(s+\alpha)}\right), \quad n \to +\infty. \qquad (4.2.11)$$

And for K_p, we also have

Theorem 4.2.7 *For $\alpha > 0, s \in \mathbb{P}$, it holds the equivalent relationship*

$$f^{(s)} \in \text{Lip}(C(K_p), \alpha) \Leftrightarrow E_{p^n}(C(K_p), f) = O\left((p^n)^{-(s+\alpha)}\right), \quad n \to +\infty. \qquad (4.2.12)$$

We may assert that: to describe the continuity and smoothness of functions defined on a local field K_p, it is enough to use the first order continuous modulus and first order Lipschitz class, but no need any higher order continuous module and higher order Lipschitz classes. In other words, the "gap"appeared in the case of $\mathbb{R}$ disappears in the case of K_p.

In the Theorem 4.1.6, the Holder type space $C^\sigma(K_p)$ points out the essential property of p-type calculus, combining with Theorem 4.2.1, a wonderful relationship is shown:

$$C^\alpha(K_p) \subset C^\beta(K_p), \quad 0 < \beta < \alpha < +\infty$$

$$\updownarrow \qquad \searrow$$

$$\text{Lip}(C(K_p), \alpha) \subset \text{Lip}(C(K_p), \beta), \quad 0 < \beta < \alpha < +\infty. \qquad (4.2.13)$$

The following example shows that (4.2.10) is failed on a local field K_p.

Example 4.2.1 Let $p = 2$, $x \in D \subset K_2$, $x = 0.x_1x_2x_3\cdots, x_j \in \{0,1\}, j \in \mathbb{N}$, the function $f : D \to \mathbb{R}$ (See Fig. 4.2.1):

$$f(x) = \begin{cases} 0, & 0 = x, \\ \vdots & \vdots \\ 3^{-4}, & 0.0001 < x \leqslant 0.001, \\ 3^{-3}, & 0.001 < x \leqslant 0.01, \\ 3^{-2}, & 0.01 < x \leqslant 0.1, \\ 3^{-1}, & 0.1 < x \leqslant 1. \end{cases} \quad (4.2.14)$$

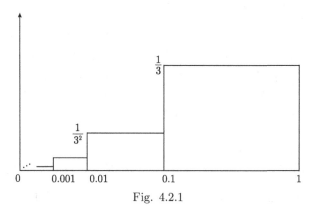

Fig. 4.2.1

Take $h = 0.\underbrace{0\cdots0}_{k-1}1 = \dfrac{1}{2^k}$, $k = 1, 2, \cdots$, then at $x = 0$,

$$|f(0 \oplus h) - f(0)| = |3^{-k} - 0| = 3^{-k} = (2^{-k})^{\log_2 3} = O\left(|h|^{\log_2 3}\right).$$

For $x \in (0,1]$, $x = 0.x_1x_2x_3\cdots, x_j \in \{0,1\}, j \in \mathbb{N}$, and $h = 0.\underbrace{0\cdots0}_{k-1}1 = \dfrac{1}{2^k}$, it holds

$$|f(x \oplus h) - f(x)| = |f(0.x_1\cdots x_{k-1}(x_k+1)x_{k+1}\cdots) - f(0.x_1\cdots x_{k-1}x_kx_{k+1}\cdots)| = O\left(|h|^{\log_2 3}\right).$$

This shows that $f \in \text{Lip}(\log_2 3, D)$, and $f(x)$ is not constant.

4.2.4 Comparison of Euclidean space analysis and local field analysis

As we have emphasized, the comparison of essential properties between those results on two underlying spaces, $\mathbb{R}$ and K_p, is necessary and important.

(4.2.1)~(4.2.10) describe the construction features of those functions defined on $\mathbb{R}^1 = \mathbb{R}$, whereas (4.2.11)~(4.1.13) describe the construction characters of those functions defined on K_p. Our construction features of $\mathbb{R}$ and K_p are quite different.

(i) **Operation structures of $\mathbb{R}$ and K_p.**

The addition $+$ and multiplication $\times$ on $\mathbb{R}$ are the usual operations on the Euclidean space, the operations $\oplus$ and $\otimes$ on K_p are term by term mod p, no carrying, or carrying from left to right, respectively. We denote by $(\mathbb{R}, +, \times)$ and $(K_p, \oplus, \otimes)$, then the two fields, $\mathbb{R}$ and K_p, have quite different operation structures.

(ii) **Topological structures of $\mathbb{R}$ and K_p.**

The topology of $\mathbb{R}$ is the usual topology τ on the Euclidean space; the topology of K_p is determined by the non-Archimedean valued norm $|x|$. Although the two topological spaces are locally compact topological fields, $\mathbb{R}$ is connected, whereas K_p is totally disconnected. We denote them by $(\mathbb{R}, +, \times, \tau)$ and $(K_p, \oplus, \otimes, |\cdot|)$.

We emphasize that, **an essential geometrical differences** is: two balls in $(\mathbb{R}, +, \times, \tau)$ have three positions, disjoint, intersect each other, one is contained in the other. However, two balls in $(K_p, \oplus, \otimes, |\cdot|)$ just have two positions, disjoint, one is contained in the other one.

(iii) **Character group structures of $\mathbb{R}$ and K_p.**

By the Pontryagin dual theorem, we have for $\mathbb{R}$ (or $[-1,1]$),

$$\Gamma_{\mathbb{R}} \xleftrightarrow{\text{iso.}} \mathbb{R} \quad (\Gamma_{[-1,1]} \xleftrightarrow{\text{iso.}} \mathbb{Z}).$$

For K_p (or D),

$$\Gamma_{K_p} \xleftrightarrow{\text{iso.}} K_p \quad (\Gamma_D \xleftrightarrow{\text{iso.}} \{0\} \cup \mathbb{N}).$$

Thus, $\Gamma_{\mathbb{R}}$ and Γ_K are connected and totally disconnected, respectively.

Table 4.2.1

underlying space	group	character group (character function)	character value
Euclidean space $\mathbb{R}$	compact $x \in [-1, 1]$	$\Gamma_{[-1,1]} = \{\exp 2\pi \mathrm{i} k x : k \in \mathbb{Z}\}$	$\lambda = 2\pi \mathrm{i} k$
	locally compact $x \in \mathbb{R}$	$\Gamma_\mathbb{R} = \{\exp 2\pi \mathrm{i} \xi x : \xi \in \mathbb{R}\}$	$\lambda = 2\pi \mathrm{i} \xi$
local field K_p	compact $x \in D$	$\Gamma_D = \{\chi_k(x) : k \in \{0\} \cup \mathbb{N}\}$	$\lambda = \langle k \rangle$
	locally compact $x \in K_p$	$\Gamma_{K_p} = \{\chi_\xi(x) : \xi \in K_p\}$	$\lambda = \langle \xi \rangle$

We emphasize that, **an essential analytical difference is**: the kernels of two Fourier transformations on $(\mathbb{R}, +, \times, \tau)$ and $(K_p, \oplus, \otimes, |\cdot|)$ are $\exp 2\pi \mathrm{i} \xi x$ and $\chi_\xi(x)$, (see ch2, §2.2), respectively.

Thus, the character equations, or eigen-equations of two spaces are quite different. They are $y' = \lambda y$ and $y^{\langle 1 \rangle} = \lambda y$, (see Example 3.3.1) respectively, where f' is the classical derivative of $f : \mathbb{R} \to \mathbb{R}$, on $\mathbb{R}$, and $f'(x)$ is the rate of change of f at point $x \in \mathbb{R}$ effected by those points in some neighborhoods of x, so it is a rate of change in local sense; whereas, $f^{\langle 1 \rangle}$ is the p-type derivative on K_p of $f : K_p \to \mathbb{R}$, and $f^{\langle 1 \rangle}(x)$ is the rate of change of f at $x \in K_p$ effected by all points in the domain K_p, so it is a rate of change in global sense.

Moreover, $\lambda = 2\pi \mathrm{i} \xi$, $\xi \in \mathbb{R}$, and $\lambda = \langle \xi \rangle$, $\xi \in K_p$, are character values, or eigen-values of $\mathbb{R}$ and K_p, respectively.

(iv) **Approximation equivalent theorems of $\mathbb{R}$ and K_p.**

From the point of view of approximation theory, the best approximation equivalent theorems describe the structures of functions. (4.2.7) and (4.2.8) describe the case of $\mathbb{R}$, and Theorem 4.2.7 shows the case of K_p. We refer to [25]~[29], [31], [32], [76]~[85], [90]~[96], [119]~[131].

It is clear that: (4.2.3) and (4.2.13) describe the smoothness of functions defined on $\mathbb{R}$ and K_p, respectively. Then we conclude that the classical calculus is suitable to that of $\mathbb{R}^n$, and the p-type calculus is suitable to the analysis of K_p. Correspondingly, the analysis of $\mathbb{R}$ needs to introduce Lip* class, and its approximation equivalent theorems need the second degree continuous modulus. However, the analysis of K_p needs only Lip class; moreover, its approximation equivalent theorems need only one degree continuous modulus, and relationship $C^\alpha(K_p) \leftrightarrow \text{Lip}(\alpha, K_p)$, $\alpha \in (0, +\infty)$, holds.

Recall Theorem 1.1 in [121]: for $f \in L^1([0, +\infty))$, and $\forall t \in \mathbb{R} = (-\infty, +\infty)$, it holds

$$\lim_{p \to +\infty} \int_0^{+\infty} f(x) \overline{w}_p(t, x)\, dx$$
$$= \int_0^{+\infty} f(x) \exp\left[-2\pi i \left(tx - (\text{singt})\{|t|\}\{x\}\right)\right] dx,$$

where $\{x\}$ is the decimal part of $x \in \mathbb{R}$, $w_p(t, x)$ is the p-adic Walsh function.

This Theorem shows the essential difference between the Fourier transformations on $\mathbb{R}$ and that of on K_p. It says that: the Fourier analysis on K_p is not a special case of that on $\mathbb{R}$ as $p \to +\infty$.

We may expect: the analysis on $\mathbb{R}^n$ is the powerful tool to describe the macro-universe (large scale), whereas, the analysis on K_p is the best tool to describe the micro-universe (small scale). Thus we may think that we have found a new idea, new method and new technique to study the non-linear problems, such as, chaos, fractals, and solitons.

Finally, we list some important function classes on $\mathbb{R}^n$ so that readers may find certain new topics on local field K_p to close this section.

Dini–Lip class: $DL = \left\{f \in C_{2\pi} : \lim_{\delta \to 0} \omega(\delta) \ln|\delta| = 0 \right\}$.

W-class: $W = \{f \in C_{2\pi} : \omega(\delta) \leqslant A\delta(1 + |\ln \delta|)\}, \delta > 0, A > 0$ independent of δ.

Zygmund class: $Z = \{f \in C_{2\pi} : |f(x+h) - 2f(x) + f(x-h)| \leqslant M|h|\} \equiv \text{Lip}^*1$. On the finite intervals, we have

$$\text{Lip } 1 \subset \text{Lip}^*1 \subset \text{Lip}^*\alpha = \text{Lip } \alpha \subset DL, \quad \text{for } 0 < \alpha < 1;$$
$$\text{Lip } 1 \subset W \subset \text{Lip } \alpha, \quad \text{for } 0 < \alpha < 1.$$

Since the inverse of "$f \in \text{Lip } 1 \Rightarrow E_n(f) = O(n^{-1})$" fails, then the "gap" happens such that there exists a hinder between approximation degrees and smoothness of functions, and the hinder can not be overcomed unless we introduce the second degree continuous modulus. However, for the Zygmund class, it holds

$$f \in \text{Lip}^*1 \Leftrightarrow E_n(f) = O(n^{-1}).$$

And for the W-class, if $f \in C_{[a,b]}$, it holds on any closed interval $[a', b'] \subset (a, b)$,

$$W \Leftarrow E_n(f) = O(n^{-1}).$$

Then we have Lip $1 \subset W \subset \text{Lip}^*1$ in any $[a', b'] \subset (a, b)$, but not holds on $[a, b]$.

Exercises

1. Establish the Lip theory on the multiplication group K_p^*.
2. Prove Theorem 4.2.4~4.2.5.
3. Compare the analysis on $\mathbb{R}$ with analysis on K_p, show the reasons of differences.
4. Show the essential properties of multiplication group K_p^*, and give a design to establish analysis on K_p^*.

4.3 Fractal spaces on local feilds

The study of mathematical theory in fractal geometry has been a quite longer history, such as those excellent work by Cantor and Weierstrass, except there was not the word "fractal". In 1960s, fractal sets and fractal functions had been applied to physics, signals, and other scientific fields as well as application fields; meanwhile fractal geometry has been developed quickly. In 1967, American mathematician B.B. Mandelbrot (was born in 1924, Warsaw, Poland) published *How long is the coast of Britain* on *Science*[36], such that fractal theory catches attentions of scientists, and makes an ascent in scientific and mathematical fields. Later, in 1977 and 1982, two books *Fractal: From, Chance and Dimension*[37] and *The Fractal Geometry of Nature*[38] have been published, as the fundamental literatures. The new idea and creative work of B.B. Mandelbrot have inaugurated new era of the modern fractal geometry. Due to his great and splendid research achievements, the Wolf Prize was awarded to him at his twilight age.

Then, from the last 30 years of 20th century, a lot of fractal theory and applications has appeared. Most excellent jobs[4],[5],[8],[13]~[15],[40] are on the underlying space $\mathbb{R}^n$. However, some mathematicians in the world have paid attentions to those of on the underlying space K_p, and are opening a new area of "fractal analysis on local fields", in which new algebraic, geometric, topological, analytic as well as physical idea and methods are mingled and matched each other, thus a quite new frame is established for studying fractals[87],[89],[97].

In this section, we establish the fractal spaces on local fields, and refer to [5], [8], [102].

4.3.1 Fractal spaces on K_p

Definition 4.3.1 (Fractal space) Let (X,d) be a complete metric space, denote by

$$\mathbb{K}(X) = \{A \subset X : A \text{ is compact in } X\},$$

and agree on $\emptyset \notin \mathbb{K}(X)$.

(i) $\forall x \in X, \forall B \in \mathbb{K}(X)$, the **distance** of x and B is defined as

$$d(x,B) = \min\{d(x,y) : y \in B\}.$$

(ii) $\forall A \in \mathbb{K}(X), \forall B \in \mathbb{K}(X)$, the **distance** of A to B is defined as

$$d(A,B) = \max\{d(x,B) : x \in A\},$$

and the **distance** of B to A is $d(B,A) = \max\{d(y,A) : y \in B\}$.

(iii) The **Hausdorff distance** $h(A,B)$ of A and B is defined as

$$h(A,B) = \max\{d(A,B), d(B,A)\}.$$

The set $(\mathbb{K}(X), h)$ is said to be a **fractal space** on X, or **space of fractals** on X; it is a metric space on the space (X,d); an element in $(\mathbb{K}(X), h)$ is said to be a **fractal**.

For a complete metric space (X,d), denote by

$$\mathbb{LK}(X) = \{A \subset X : A \text{ is locally compact in } X\},$$

and agree on $\emptyset \notin \mathbb{LK}(X)$.

Correspondingly, $h(A,B) = \max\{d(A,B), d(B,A)\}$ is said to be the **generalized Hausdorff distance** of A and B for $A, B \in \mathbb{LK}(X)$; and $(\mathbb{LK}(X), h)$ is said to be a **generalized fractal space** on X.

Remark. It is clear that $d(A,B) \neq d(B,A)$, and the Hausdorff distance $h(A,B)$ is an ultra-distance on $\mathbb{K}(X)$ such that $(\mathbb{K}(X), h)$ is an ultra-metric space, and also is a metric space on (X,d) (see [8] for proof).

Example 4.3.1 In the one dimension Euclidean space $(\mathbb{R}, d)$, the distance is $d(x,y) = |x-y|$ (absolute value), then $\mathbb{K}(\mathbb{R}) = \{A \subset \mathbb{R} : A \text{ is compact in } \mathbb{R}\}$. With the **Hausdorff distance** $h(A, B)$, $A, B \in \mathbb{K}(\mathbb{R})$, we have the fractal space $(\mathbb{K}(\mathbb{R}), h)$ on $\mathbb{R}$.

Take $(X, d) = (K_p, d)$ with the distance

$$d(x,y) = \begin{cases} ||x| - |y||, & |x| \neq |y|, \\ |x-y|, & |x| = |y|, \ x \neq y, \\ 0, & x = y, \end{cases} \quad (4.3.1)$$

where $|\cdot|$ is the non-Archimedean valued norm on local field K_p, thus for

$$\mathbb{K}(K_p) = \{A \subset K_p : A \text{ is compact in } K_p\},$$

then $(\mathbb{K}(K_p), h)$ is a fractal space on K_p.

Correspondingly, $(\mathbb{LK}(K_p), h)$ is the generalized fractal space on K_p.

Sometimes, mathematicians take distance $d(x,y) = |x-y|$, the non-Archimedean valued of x and y, then $(\mathbb{K}(K_p), h)$ is also an ultra-metric fractal space on K_p.

Example 4.3.2 The Cantor type set C_p on K_p.

Let K_p be the p-series field with prime $p \geqslant 2$. Let

$$C_p = D \Big\backslash \Big(\bigcup_{j=1}^{+\infty} V_j \Big) = B^0 \Big\backslash \Big(\bigcup_{j=1}^{+\infty} V_j \Big), \quad (4.3.2)$$

where

$$V_1 = B^1 \cup (2\beta^0 + B^1) \cup \cdots \cup ((p-2)\beta^0 + B^1),$$
$$V_2 = (\beta^0 + B^2) \cup (\beta^0 + 2\beta^1 + B^2) \cup \cdots \cup (\beta^0 + (p-2)\beta^1 + B^2)$$
$$\cdots .$$

Compare with Chapter 1, Section 1.2.8, the Cantor type set C_3 on K_3. The Figure 4.3.1 is the structure draft of $D = B^0$ in K_5 with $p = 5$.

4.3.2 Completeness of $(\mathbb{K}(K_p), h)$ on K_p

Definition 4.3.2 (Dilation of set S) (i) Let (X, d) be a metric space, for a subset $S \subset X$, the following set with $r \geqslant 0$,

$$S + r = \{y \in X : d(x, y) \leqslant r, x \in S\}$$

is said to be a **dilation of S with radius r for $S \subset X$**.

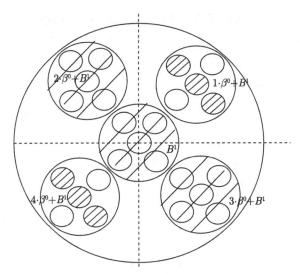

Figure 4.3.1

(ii) Let (K_p, d) be the metric space on local field K_p with d in (4.3.1), and let $S \subset K_p$ be a subset. Then a **dilation of S with radius** $r \geqslant 0$ for $S \subset K_p$ is defined by

$$S + r = \{y \in X : d(x, y) \leqslant r, x \in S \subset K_p\}. \tag{4.3.3}$$

Definition 4.3.3 (Cauchy sequence in $\mathbb{K}(X)$) Let (X, d) be a complete metric space, $(\mathbb{K}(X), h)$ be the fractal space on X.

(i) The Cauchy sequence $\{A_n\}_{n=1}^{+\infty} \subset \mathbb{K}(X)$ in $(\mathbb{K}(X), h)$ is defined as: $\forall \varepsilon > 0, \exists N \in \mathbb{N}$, such that for $n, m > N$, hold

$$A_n \subset A_m + \varepsilon \quad \text{and} \quad A_m \subset A_n + \varepsilon.$$

(ii) The Cauchy sequence $\{A_n\}_{n=1}^{+\infty} \subset \mathbb{K}(K_p)$ in $(\mathbb{K}(K_p), h)$ on local field K_p is defined as: $\forall \varepsilon > 0, \exists N \in \mathbb{N}$, such that for $n, m > N$, hold

$$A_n \subset A_m + \varepsilon \quad \text{and} \quad A_m \subset A_n + \varepsilon. \tag{4.3.4}$$

Theorem 4.3.1 (i) If (X, d) be a complete metric space, then space $(\mathbb{K}(X), h)$ is complete; and if $\{A_n\}_{n=1}^{+\infty} \subset \mathbb{K}(X)$ is a Cauchy sequence in $(\mathbb{K}(X), h)$, then

$$A = \lim_{n \to +\infty} A_n \in \mathbb{K}(X)$$

can be written as

$$A = \left\{ x \in X : \exists \text{ Cauchy sequence } \{x_n \in A_n\}, \text{ s.t. } \lim_{n \to +\infty} x_n = x \right\}.$$

(ii) *The fractal space* $(\mathbb{K}(K_p), h)$ *on* (K_p, d) *is a complete metric space; and if* $\{A_n\}_{n=1}^{+\infty} \subset \mathbb{K}(K_p)$ *is a Cauchy sequence in* $(\mathbb{K}(K_p), h)$, *then*
$$A = \lim_{n \to +\infty} A_n \in \mathbb{K}(K_p)$$
can be expressed as
$$A = \left\{ x \in K_p : \exists \text{ Cauchy sequence } \{x_n \in A_n\}, \text{ s.t. } \lim_{n \to +\infty} x_n = x \right\}.$$

Proof. We only prove (ii).

① Prove $h(A, B)$ is a metric on $\mathbb{K}(K_p)$.

The first and second conditions of metric holds, clearly. We check the third one. By
$$d(A, B) \leqslant d(A, C) + d(C, B) \quad \text{and} \quad d(B, A) \leqslant d(B, C) + d(C, A),$$
thus
$$\begin{aligned} h(A, B) &= \max\{d(A, B), d(B, A)\} \\ &\leqslant \max\{d(A, C), d(C, B)\} + \max\{d(B, C), d(C, A)\} \\ &\leqslant \max\{d(A, C), d(C, A)\} + \max\{d(B, C), d(C, B)\} \\ &\leqslant h(A, C) + h(C, B). \end{aligned}$$

② Prove the completeness of fractal space $(\mathbb{K}(K_p), h)$.

By two properties (for simply, we use $+$ instead of the addition operator $\oplus$ on K_p):

1° *Let* $A, B \in \mathbb{K}(K_p)$, *for any* $\varepsilon > 0$, *it holds*
$$h(A, B) \leqslant \varepsilon \Leftrightarrow A \subset B + \varepsilon \text{ and } B \subset A + \varepsilon.$$

Since
$$d(A, B) \leqslant \varepsilon \Rightarrow \max\{d(a, B) : a \in A\} \leqslant \varepsilon \Rightarrow d(a, B) \leqslant \varepsilon, \quad \forall a \in A$$
$$\Rightarrow a \in B + \varepsilon, \forall a \in A \Rightarrow A \subset B + \varepsilon.$$

Conversely, by $d(a, B) = \min\{d(a, b) : b \in B\}$, it holds
$$A \subset B + \varepsilon \Rightarrow \exists b \in B, \text{ s.t. } d(a, b) \leqslant \varepsilon, \quad \forall a \in A$$
$$\Rightarrow d(a, B) \leqslant \varepsilon, \quad \forall a \in A$$
$$\Rightarrow d(A, B) \leqslant \varepsilon.$$

Then, $d(A, B) \leqslant \varepsilon \Leftrightarrow A \subset B + \varepsilon$.

Similarly, $d(B, A) \leqslant \varepsilon \Leftrightarrow B \subset A + \varepsilon$. Thus,
$$h(A, B) = \max\{d(A, B), d(B, A)\} \leqslant \varepsilon \Leftrightarrow A \subset B + \varepsilon \quad \text{and} \quad B \subset A + \varepsilon.$$

Hence 1° is proved.

2° Let $\{A_n\}_{n=1}^{+\infty} \subset \mathbb{K}(K_p)$ be a Cauchy sequence in $\mathbb{K}(K_p)$, and $\{n_j\}_{j=1}^{+\infty}$ be a sub-sequence in $\mathbb{N}: 0 < n_1 < n_2 < \cdots < n_j < \cdots$. If there is a Cauchy sequence $\left\{x_{n_j}^0 \in A_{n_j}\right\}_{j=1}^{+\infty}$ in (K_p, d), then there exists a Cauchy sequence $\{\tilde{x}_n \in A_n\}_{n=1}^{+\infty}$ such that $\tilde{x}_{n_j} = x_{n_j}^0$, $j \in \mathbb{N}$.

This is an extension theorem, it plays an important role in our theory.

Let $\{A_n\}_{n=1}^{+\infty} \subset \mathbb{K}(K_p)$ be a Cauchy sequence in $\mathbb{K}(K_p)$, then by Definition 4.3.3 (ii), $\forall \varepsilon > 0, \exists N > 0$, s.t. $n, m > N$, hold $A_n + \varepsilon \supset A_m$ and $A_m + \varepsilon \supset A_n$. So that by 1°, it holds $h(A_n, A_m) \leqslant \varepsilon$.

Now, we construct a Cauchy sequence $\{\tilde{x}_n \in A_n\}_{n=1}^{+\infty}$ in (K_p, d) by virtue of the given Cauchy sequence $\{A_n\}_{n=1}^{+\infty} \subset \mathbb{K}(K_p)$.

Since $\{A_n\}_{n=1}^{+\infty}$ is a Cauchy sequence in $\mathbb{K}(K_p)$, then $\forall \varepsilon > 0, \exists N > 0$, s.t. $n, m > N$, holds $h(A_n, A_m) < \dfrac{\varepsilon}{3}$. Thus, correspondingly, take $x_n^0 \in A_n$, $n \in \mathbb{N}$, for $\{A_n\}_{n=1}^{+\infty} \subset \mathbb{K}(K_p)$.

Let $\{n_j\}_{j=1}^{+\infty}$ be an increasing subsequence in $\mathbb{N}$, $0 < n_1 < n_2 < \cdots < n_j < \cdots$, and for the above $x_n^0 \in A_n$, let $\left\{x_{n_j}^0 \in A_{n_j}\right\}_{j=1}^{+\infty} \subset (K_p, d)$, it is a Cauchy sequence in (K_p, d). Now we prove: there exists a Cauchy sequence $\{\tilde{x}_n \in A_n\}_{n=1}^{+\infty}$, such that $\tilde{x}_{n_j} = x_{n_j}^0$, $j = 1, 2, \cdots$.

In fact, we construct $\{\tilde{x}_n\}_{n=1}^{+\infty} \subset K_p$: for $n \in \{1, 2, \cdots, n_1\}$, selecting

$$\tilde{x}_n \in \left\{x_n \in A_n : d\left(x_n, x_{n_1}^0\right) = d\left(x_{n_1}^0, A_n\right)\right\},$$

that is, $\tilde{x}_n$ is the one of the nearest points from $x_{n_1}^0$ to A_n, this $\tilde{x}_n$ exists since the compactness of A_n. Continuously, for $j \in \{2, 3, \cdots\}$, $\forall n \in \{n_j + 1, \cdots, n_{j+1}\}$, selecting

$$\tilde{x}_n \in \left\{x_n \in A_n : d\left(x_n, x_{n_j}^0\right) = d\left(x_{n_j}^0, A_n\right)\right\}.$$

Then we prove the sequence $\{\tilde{x}_n\}_{n=1}^{+\infty}$ satisfies 2°, i.e., it is the extension of sequence $\left\{x_{n_j}^0 \in A_{n_j}\right\}_{j=1}^{+\infty}$ in $\{A_n\}_{j=1}^{+\infty}$. By the constructive method above, $\tilde{x}_{n_j} = x_{n_j}^0$, and $\tilde{x}_n \in A_n$, so we only need to prove that $\{\tilde{x}_n\}_{n=1}^{+\infty}$ is a Cauchy sequence.

Since $\left\{x_{n_j}^0 \in A_{n_j}\right\}_{j=1}^{+\infty} \subset (K_p, d)$, and $\{A_n\}_{n=1}^{+\infty} \subset (\mathbb{K}(K_p), h)$ is a Cauchy sequence, then,

$$\forall \varepsilon > 0, \quad \exists N_1 > 0, \text{ s.t. } n_k, n_j \geqslant N_1, \text{ holds } d\left(x_{n_k}, x_{n_j}\right) \leqslant \dfrac{\varepsilon}{3},$$

$\exists N_2 > 0$, s.t. $m, n \geqslant N_2$, holds $d(A_m, A_n) \leqslant \frac{\varepsilon}{3}$.

Let $N = \max\{N_1, N_2\}$, so that for $m, n \geqslant N$, holds
$$d(\tilde{x}_m, \tilde{x}_n) \leqslant d(\tilde{x}_m, x_{n_j}) + d(x_{n_j}, x_{n_k}) + d(x_{n_k}, \tilde{x}_n), \quad (4.3.5)$$
where $m \in \{n_{j-1}+1, n_{j-1}+2, \cdots, n_j\}$, $n \in \{n_{k-1}+1, n_{k-1}+2, \cdots, n_k\}$.

Since $h(A_m, A_n) < \frac{\varepsilon}{3}$, thus $\tilde{x}_m \in A_m$, $x_{n_j}^0 \in \{x_{n_j}^0\} + \frac{\varepsilon}{3}$, such that $d(\tilde{x}_m, x_{n_j}^0) \leqslant \frac{\varepsilon}{3}$; similarly, $d(x_{n_k}^0, \tilde{x}_n) \leqslant \frac{\varepsilon}{3}$. Hence, if $m, n \geqslant N$, (4.3.5) implies
$$d(\tilde{x}_m, \tilde{x}_n) \leqslant \frac{\varepsilon}{3} + \frac{\varepsilon}{3} + \frac{\varepsilon}{3} = \varepsilon.$$
Then, 2° is proved.

We turn to prove the completeness of $(\mathbb{K}(K_p), h)$.

Let $A = \left\{ x \in K_p : \exists \text{ Cauchy sequence } \{x_n \in A_n\}, \text{ s.t. } \lim_{n \to +\infty} x_n = x \right\}$, where $\{A_n\} \subset \mathbb{K}(K_p)$ is the given Cauchy sequence. We prove:

(a) $A \neq \varnothing$;
(b) A is a closed subset, thus it is complete (by completeness of (K_p, d));
(c) $\forall \varepsilon > 0$, $\exists N > 0$, such that for $n > N$, holds $A \subset A_n + \varepsilon$;
(d) A is totally bounded, thus it is compact;
(e) $\lim_{n \to +\infty} A_n = A$.

For (a), we prove: for the given Cauchy sequence $\{A_n\}_{n=1}^{+\infty} \subset \mathbb{K}(K_p)$, there exists a Cauchy sequence $\{a_j \in A_j\}$ in (K_p, d), it converges to $a \in A$, i.e., $\lim_{j \to +\infty} a_j = a$.

Taking $N_1 < N_2 < \cdots < N_j < \cdots$, such that
$$h(A_m, A_n) < \frac{1}{2^j}, \quad m, n > N_j, \quad j = 1, 2, \cdots.$$

Selecting $x_{N_1} \in A_{N_1}$, by $h(A_{N_1}, A_{N_2}) \leqslant \frac{1}{2}$, then there exists $x_{N_2} \in A_{N_2}$, s.t. $d(x_{N_1}, x_{N_2}) \leqslant \frac{1}{2}$.

Inductively, if select $x_{N_j} \in A_{N_j}$, $j = 1, 2, \cdots, k$, s.t. $d(x_{N_{j-1}}, x_{N_j}) \leqslant \frac{1}{2^{j-1}}$, then hold $h(A_{N_k}, A_{N_{k+1}}) \leqslant \frac{1}{2^k}$ and $x_{N_k} \in A_{N_k}$. So that, $\exists x_{N_{k+1}} \in A_{N_{k+1}}$, s.t. $d(x_{N_k}, x_{N_{k+1}}) \leqslant \frac{1}{2^k}$. Take one of the nearest points $x_{N_{k+1}}$ (it exists by compactness of $A_{N_{k+1}}$) from x_{N_k} to $A_{N_{k+1}}$. So that by the above induction, $\exists \{x_{N_j} \in A_{N_j}\}$, s.t. $d(x_{N_j}, x_{N_{j+1}}) \leqslant \dfrac{1}{2^j}$, $j = 1, 2, \cdots$.

Moreover, $\{x_{N_j} \in A_{N_j}\}$ is a Cauchy sequence in K_p because: $\forall \varepsilon > 0$, take $N_\varepsilon > 0$, s.t. $\sum\limits_{j=N_\varepsilon}^{+\infty} \dfrac{1}{2^j} < \varepsilon$. Thus, for $m > n > N_\varepsilon$, it holds

$$d(x_{N_m}, x_{N_n}) \leqslant d(x_{N_m}, x_{N_{m+1}}) + d(x_{N_{m+1}}, x_{N_{m+2}}) + \cdots + d(x_{N_{n-1}}, x_{N_n})$$
$$\leqslant \sum_{j=N_\varepsilon}^{+\infty} \dfrac{1}{2^j} < \varepsilon.$$

Till now, for the given Cauchy sequence $\{A_n\}_{n=1}^{+\infty} \subset \mathbb{K}(K_p)$ and a sequence of positive integers $N_1 < N_2 < \cdots < N_j < \cdots$, the Cauchy sequence $\{x_{N_j} \in A_{N_j}\}$ in (K_p, d) has been constructed by induction, and $\lim\limits_{j \to +\infty} x_{N_j} = x \in K_p$. By 2°, there exists a Cauchy sequence $\{a_j \in A_j\}$, s.t. $a_{N_j} = x_{N_j}$ and $\lim\limits_{j \to +\infty} a_j = x \in A \subset K_p$. Thus, $x \in A$ implies $A \neq \emptyset$. (a) is proved.

For (b), take a sequence $\{a_j \in A\}$ in A and converges to a, $\lim\limits_{j \to +\infty} a_j = a$. We prove $a \in A$. In fact, $\forall j \in \mathbb{N}$, for $a_j \in A$, $\exists \{x_{j,n} \in A_n\}$, s.t. $\lim\limits_{n \to +\infty} x_{j,n} = a_j$. Then there is an increasing integer sequence $\{N_j\}_{j=1}^{+\infty}$, s.t. $d(a_{N_j}, a) < \dfrac{1}{j}$. Moreover, there is an increasing integer subsequence $\{m_j\}$, s.t. $d(x_{N_j, m_j}, a_{N_j}) \leqslant \dfrac{1}{2^j}$. Thus, these imply that $d(x_{N_j, m_j}, a) \leqslant \dfrac{1}{2^j}$. Let $y_{m_j} = x_{N_j, m_j}$, then $y_{m_j} \in A_{m_j}$, and $\lim\limits_{j \to +\infty} y_{m_j} = a$. By 2°, $y_{m_j} \in A_{m_j}$ can be extended to a Cauchy sequence $\{\tilde{y}_j \in A_j\}$ with $\lim\limits_{j \to +\infty} \tilde{y}_j = a$, and $a \in A$ by definition of A. This implies that A is a closed set. (b) is proved.

For (c), take any Cauchy sequence $\{A_n\} \subset \mathbb{K}(K_p)$; then $\forall \varepsilon > 0$, $\exists N > 0$, s.t. for $m, n \geqslant N$, holds $h(A_m, A_n) \leqslant \varepsilon$. Thus, for $n \geqslant N$, and $m \geqslant n$, holds $A_m \subset A_n + \varepsilon$. Next, to prove $A \subset A_n + \varepsilon$, we deduce
$a \in A \Rightarrow \exists \{a_j \in A_j\}$, s.t. $\lim\limits_{j \to +\infty} a_j = a$,

$\Rightarrow$ take N large enough, s.t. $m \geqslant N$ implies $d(a_m, a) < \varepsilon$,

$\Rightarrow$ $m \geqslant N$ implies $a_m \in A_n + \varepsilon \Rightarrow m \geqslant N$ implies $A_m \subset A_n + \varepsilon$,

$\Rightarrow$ $a \in A_n + \varepsilon$ (compactness of A_n implies $A_n + \varepsilon$ is a closed set),

$\Rightarrow$ $A \subset A_n + \varepsilon$ for n large enough $\Rightarrow$ (c) is proved.

For (d), if A is not totally bounded, then for some Cauchy sequence $\{A_n\} \subset \mathbb{K}(K_p)$, $\exists \varepsilon_0 > 0$, s.t. there is no ε_0-net, in other words, there exists a sequence $\{x_j\}_{j=1}^{+\infty} \subset A$, s.t. $d(x_j, x_k) \geqslant \varepsilon_0$, $j \neq k$,

$\Rightarrow \exists n \in \mathbb{N}$ large enough, s.t. $A \subset A_n + \dfrac{\varepsilon_0}{3}$ by (c),

$\Rightarrow \forall x_j, \exists y_j \in A_n$, s.t. $d(x_j, y_j) \leqslant \dfrac{\varepsilon_0}{3}$,

$\Rightarrow \exists$convergent subsequence $\{y_{n_j}\}$ of $\{y_j\} \subset A_n$ by compactness of A_n,

$\Rightarrow \exists y_{n_j}, y_{n_k} \in \{y_{n_j}\}$, s.t. $d(y_{n_j}, y_{n_k}) < \dfrac{\varepsilon_0}{3}$,

$\Rightarrow d(x_{n_j}, x_{n_k}) \leqslant d(x_{n_j}, y_{n_j}) + d(y_{n_j}, y_{n_k}) + d(y_{n_k}, x_{n_k}) < \dfrac{\varepsilon_0}{3} + \dfrac{\varepsilon_0}{3} + \dfrac{\varepsilon_0}{3} = \varepsilon_0$,

$\Rightarrow$ contrary to the assumption of induction $\Rightarrow$ (d) is proved.

For (e), take any Cauchy sequence $\{A_j\} \subset \mathbb{K}(K_p)$, we prove $\lim\limits_{j \to +\infty} A_j = A$. Since $A \in \mathbb{K}(K_p)$ by (d) then by (c) and 1°, we only need to prove: $\forall \varepsilon > 0$, $\exists N > 0$, s.t. for $n \geqslant N$, holds $A_n \subset A + \varepsilon$. We deduce

$\forall \varepsilon > 0$, $\exists N > 0$, s.t. for $m, n \geqslant N$, holds $h(A_m, A_n) \leqslant \varepsilon$,

$\Rightarrow$ for $m, n \geqslant N$, holds $A_m \subset A_n + \dfrac{\varepsilon}{2} \Rightarrow$ for $n \geqslant N$, take $y \in A_n$,

$\Rightarrow \exists \{N_j\}_{j=1}^{+\infty}$, s.t. $n < N_1 < \cdots < N_j < \cdots$; also for $m, k \geqslant N_j$, hold $A_m \subset A_k + \dfrac{\varepsilon}{2^{j+1}}$ and $A_n \subset A_{N_1} + \dfrac{\varepsilon}{2}$,

$\Rightarrow \exists x_{N_1} \in A_{N_1}$, s.t. $d(y, x_{N_1}) \leqslant \dfrac{\varepsilon}{2}$, by $y \in A_n$,

$\Rightarrow \exists x_{N_2} \in A_{N_2}$, s.t. $d(x_{N_1}, x_{N_2}) \leqslant \dfrac{\varepsilon}{2^2}$, by $x_{N_1} \in A_{N_1}$,

......

$\Rightarrow \exists x_{N_1}, x_{N_2}, \cdots$, s.t. $x_{N_j} \in A_{N_j}$, and $d(x_{N_j}, x_{N_{j+1}}) \leqslant \dfrac{\varepsilon}{2^{j+1}}$,

$\Rightarrow d(y, x_{N_j}) \leqslant \varepsilon, \forall j \in \mathbb{N}$; and $\{x_{N_j}\}$ is a Cauchy sequence,

$\Rightarrow \{x_{N_j}\}$ converges to $x \in A$; and $d(y, x_{N_j}) \leqslant \varepsilon$ implies $d(y, x) \leqslant \varepsilon$,

$\Rightarrow \forall n \geqslant N$ implies $A_n \subset A + \varepsilon$,

$\Rightarrow$ (e) is proved.

Combining the proofs of (a)$\sim$(e), it follows that $(\mathbb{K}(K_p), h)$ is a complete space.

The following relations for (X, d) and $(\mathbb{K}(X), h)$ are useful.

Theorem 4.3.2 Let (X,d) be a complete metric space. For $A, B \in \mathbb{K}(X)$,
(i) $\exists x \in A$, $\exists y \in B$, s.t. $h(A,B) = d(x,y)$.
(ii) If $B \subset A \subset X$, $x \in X$, then $d(x,B) \geqslant d(x,A)$.
(iii) For $A, B \in \mathbb{K}(X)$, it holds $d(A,B) \neq d(B,A)$, usually.
(iv) For $A, B, C \in \mathbb{K}(X)$, it holds $d(A \cup B, C) = \max\{d(A,C), d(B,C)\}$.
(v) For $A, B, C, D \in \mathbb{K}(X)$, it holds

$$h(A \cup B, C \cup D) \leqslant \max\{h(A,C), h(B,D)\}.$$

4.3.3 Some useful transformations on K_p

Definition 4.3.4 (transformation) Let (X,d) and $(Y,\tilde{d})$ be two metric spaces. A mapping $f : X \to Y$ from X to Y is said to be a **transformation**, if for any $x \in X$, there is unique $y = f(x) \in Y$ corresponding to x; denoted by $y = f(x)$.

A transformation $f : X \to Y$ is said to be **one to one**, if

$$f(x_1) = f(x_2) \Leftrightarrow x_1 = x_2, \quad \forall x_1, x_2 \in X.$$

The set $f(S) = \{f(x) \in X : x \in S\} \subset Y$ is said to be an **image of set** $S \subset X$ under the transformation $f : X \to Y$;

If $f : X \to Y$ satisfies $f(X) = Y$, then f is said to be an **onto mapping**;

If $f : X \to Y$ is an one to one, onto, then it is said to be an **invertible mapping**; the inverse transformation denotes by f^{-1} with $f^{-1} : X \to X$.

In the above, metric spaces (X,d) and $(Y,\tilde{d})$ can be taken as $\mathbb{R}, \mathbb{R}^n, \mathbb{C}, K_p, \cdots$, and so on.

1. Useful transformations on $\mathbb{R}^n$ and $\mathbb{C}$

(1) Cases on $\mathbb{R}$
  (a) **Polynomial transformation.** $f : \mathbb{R} \to \mathbb{R}$

$$f(x) = a_0 + a_1 x + a_2 x^2 + \cdots + a_N x^N$$

with coefficients $a_j \in \mathbb{R}, j = 0, 1, 2, \cdots, N$; $N \in \mathbb{P}$ is a fixed non-negative integer, $a_N \neq 0$; then f is said to be an N-*degree polynomial* on $\mathbb{R}$.

  (b) **Affine transformation.** $f : \mathbb{R} \to \mathbb{R}$

$$f(x) = ax + b$$

with $a, b \in \mathbb{R}$, is said to be an **affine transformation**. If $a = 1$, $f(x) = x + b$ is said to be a **translation**. If $b = 0$, $f(x) = ax$ is said to be a **linear transformation**.

(c) **Linear fractional transformation** $f : \mathbb{R} \to \mathbb{R}$

$$f(x) = \frac{ax+b}{cx+d}, \quad ad \neq bc$$

with $a, b, c, d \in \mathbb{R}$ is said to be a **linear fractional transformation**, or a **Möbius transformation**.

For convention, if $c \neq 0$, then set $f(x)|_{x=-\frac{d}{c}} = \infty$; if $c = 0$, then $f(x)|_{x \to \infty} = \infty$.

2) Cases on $\mathbb{R}^2$

2-dimension affine transformation $w : \mathbb{R}^2 \to \mathbb{R}^2$

$$w(x_1, x_2) = \begin{bmatrix} a & b \\ c & d \end{bmatrix} \begin{bmatrix} x_1 \\ x_2 \end{bmatrix} + \begin{bmatrix} e \\ f \end{bmatrix} = Ax + t,$$

is said to be a **2-dimension affine transformation**, or **2-dimension similar transformation**, where $A = \begin{bmatrix} a & b \\ c & d \end{bmatrix} \in \mathfrak{M}_{2 \times 2}$, $x = \begin{bmatrix} x_1 \\ x_2 \end{bmatrix}$, $t = \begin{bmatrix} e \\ f \end{bmatrix} \in \mathfrak{M}_{2 \times 1}$, where $\mathfrak{M}_{2 \times 2}$ and $\mathfrak{M}_{2 \times 1}$ are 2×2 and 2×1 matrix spaces, respectively.

The geometric sense of 2-dimension affine transformation is: let (a, c), $(b, d) \in \mathbb{R}^2$ have the pole coordinate expresses

$$\begin{cases} a = r_1 \cos \theta_1, \\ b = r_2 \cos \left(\theta_2 + \frac{\pi}{2} \right), \end{cases} \quad \begin{cases} c = r_1 \sin \theta_1, \\ d = r_2 \sin \left(\theta_2 + \frac{\pi}{2} \right). \end{cases}$$

Then the matrix A becomes

$$A = \begin{bmatrix} a & b \\ c & d \end{bmatrix} = \begin{bmatrix} r_1 \cos \theta_1 & -r_2 \sin \theta_2 \\ r_1 \sin \theta_1 & r_2 \cos \theta_2 \end{bmatrix}.$$

Thus, the 2-dimension transformation w rotates x_1 axis and x_2 axis with angles θ_1 and θ_2, respectively, then translates them to the new position t.

We Agree with: for $r_1 = r_2 = 1$, $\theta_1 = \theta_2 = \theta$, i.e.,

$$A = \begin{bmatrix} \cos \theta & -\sin \theta \\ \sin \theta & \cos \theta \end{bmatrix};$$

or $r_1 = r_2 = 1$, $\theta_1 = \theta_2 + \pi$, i.e.,

$$A = \begin{bmatrix} \cos\theta & \sin\theta \\ \sin\theta & -\cos\theta \end{bmatrix},$$

is said to be an **orthogonal transformation** with determinant $|A| = \pm 1$.

An orthogonal transformation is said to be a **rotation transformation**, they keep the length of an vector in $\mathbb{R}^2$. For $\theta = 0$, then $A = \begin{bmatrix} 1 & 0 \\ 0 & -1 \end{bmatrix}$, is said to be a **reflection transformation**.

3) Cases on $\mathbb{C}$

(a) **Mobius transformation on Riemann surfaces** For $f : \mathbb{C} \to \mathbb{C}$,
$$f(z) = \frac{az+b}{cz+d}, \quad ad \neq bc, \quad a,b,c,d \in \mathbb{C}.$$

(b) **Analytic transformation on Riemann surfaces** For $f : \mathbb{C} \to \mathbb{C}$, if the transformation on $\mathbb{C}$
$$f(z) = \frac{az+b}{cz+d}, \quad ad \neq bc, \quad a,b,c,d \in \mathbb{C}$$
satisfies: $\forall z_0 \in \mathbb{C}$, there exists $w(z) = a_z + b$, $a, b \in \mathbb{C}$, with $a = a(z_0)$, $b = b(z_0)$, such that
$$\lim_{z \to z_0} \frac{d(f(z), w(z))}{d(z, z_0)} = 0,$$
then f is said to be an **analytic transformation**.

2. Transformation sets in a metric space (X, d)

Let (X, d) be a metric space, the set
$$\mathbb{F} \equiv \mathbb{F}(X) = \{f : f \text{ is a transformation from } X \text{ to } X\}$$
is said to be a **transformation set** $\mathbb{F} \equiv \mathbb{F}(X)$ **on metric space** (X, d).

Introduce an operation as composition $\circ$ of two transformations on $\mathbb{F} \equiv \mathbb{F}(X)$,
$$f, g \in \mathbb{F} \Rightarrow f \circ g(x) = f(g(x)), \quad x \in X,$$
and the **identity transformation** $I : x \to I(x) = x$ as an unit element, such that $\mathbb{F}$ becomes a **semi-group** $(\mathbb{F}, \circ)$. This $(\mathbb{F}, \circ)$ has a subset $\widetilde{\mathbb{F}} = \{f \in \mathbb{F}(X) : \exists f^{-1} \text{ of } f\}$, with the operation $\circ$, $(\widetilde{\mathbb{F}}, \circ)$ becomes a group. Then $(\widetilde{\mathbb{F}}, \circ)$ is said to be a **transformation group**.

There are three important cases of the transformations: **contraction mapping, iterate, iterated function system**.

(1) Contraction mapping

Definition 4.3.5 (Contraction mapping) *Let (X,d) be a metric space. A mapping $f : X \to X$ is said to be a* **contraction mapping from X to X**, *if there exists a constant $s, 0 < s < 1$, such that*
$$d(f(x), f(y)) \leqslant s d(x,y), \quad x, y \in X,$$
where s is called a **contraction factor**.

Theorem 4.3.3 *Let $w : X \to X$ be a mapping on metric space (X,d).*

(i) *If w is continuous, then w maps (X,d) to (X,d).*

(ii) *If w is contraction mapping, then w is continuous.*

(iii) *If w is contractive with $0 < s < 1$, then the mapping $W : \mathbb{K}(X) \to \mathbb{K}(X)$ defined from w by*
$$W(B) = \{w(x) : x \in B\}, \quad \forall B \in \mathbb{K}(X),$$
where W is a contraction mapping on $(\mathbb{K}(X), h)$ with contraction factor s.

(iv) *If $\left(\{w_n\}_{n=1}^N, s_n : n = 1, \cdots, N\right)$ is a set of contraction mappings on $(\mathbb{K}(X), h)$, $W : \mathbb{K}(X) \to \mathbb{K}(X)$ is defined by*
$$W(B) = \bigcup_{n=1}^N w_n(B), \quad \forall B \in \mathbb{K}(X),$$
where W is a contraction mapping with contraction factor $s = \max\{s_n : 1 \leqslant n \leqslant N\}$ on the fractal space $(\mathbb{K}(X), h)$.

To study the contraction mappings, we list the properties of continuous mappings.

Theorem 4.3.4 *Let (X_1, d_1) and (X_2, d_2) be two metric spaces, $f : X_1 \to X_2$ a continuous mapping.*

(i) *If the sequence $\{x_k\}_{k=1}^{+\infty} \subset X_1$ converges to $x \in X_1$, then*
$$\lim_{k \to +\infty} f(x_k) = f(x).$$

(ii) *$\forall E \in \mathbb{K}(X_1)$ compact, then $f : E \to X_2$ is uniformly continuous, i.e., $\forall \varepsilon > 0, \exists \delta(\varepsilon) > 0$, such that $\forall x, y \in E$ with $d_1(x, y) < \delta$, implies $d_2(f(x), f(y)) < \varepsilon$.*

(iii) *If (X_1, d_1) and (X_2, d_2) are complete, and $f : X_1 \to X_2$ is continuous, one to one, onto mapping, then f is a homeomorphic mapping from X_1 to X_2.*

(iv) *If $(X, d) = (X_1 \times X_2, \max(d_1, d_2))$ is a product space, and $E_1 \in \mathbb{K}(X_1), E_2 \in \mathbb{K}(X_2)$, then $E_1 \times E_2 \in \mathbb{K}(X) \equiv \mathbb{K}(X_1 \times X_2)$.*

(v) If (X_3, d_3) is complete metric space, $f : X_1 \times X_2 \to X_3$ satisfies: $\forall \varepsilon > 0$, $\exists \delta > 0$, s.t. $d_1(x_1, y_1) < \delta$ implies $d_3(f(x_1, x_2), f(y_1, y_2)) < \varepsilon$, and $\forall \varepsilon > 0$, $\exists \delta > 0$, s.t. $d_2(x_2, y_2) < \delta$ implies $d_3(f(x_1, x_2), f(y_1, y_2)) < \varepsilon$. Then the mapping $f : X_1 \times X_2 \to X_3$ is continuous on the metric space $(X_3, d_3) = (X_1 \times X_2, d_3)$ with $d_3(x, y) = d((x_1, x_2), (y_1, y_2)) = \max\{d_1(x_1, y_1), d_2(x_2, y_2)\}$.

Definition 4.3.6 (Fixed point of a mapping) Let (X, d) be a complete metric space. If a mapping $f : X \to X$ satisfies: there exists point $x_f \in X$, such that $f(x_f) = x_f$, then x_f is said to be a **fixed point** of f.

Theorem 4.3.5 Let $f : X \to X$ be a contraction mapping with contraction factor s on a complete metric space (X, d).

(i) If $F(B) = \{f(x) : x \in B\}$, $B \in \mathbb{K}(X)$, then $F : \mathbb{K}(X) \to \mathbb{K}(X)$ is a contraction mapping with contraction factor s from $\mathbb{K}(X)$ to $\mathbb{K}(X)$.

(ii) There exists an unique fixed point $x_f \in X$ of f, such that
$$\lim_{k \to +\infty} f^{0\ k}(x) = x_f, \quad x \in X,$$
with $f^{0\ k}(x) = f \circ f^{0\ k-1}(x) = f(f^{0\ k-1}(x))$, $f^{0\ 0}(x) = f(x)$, $k \in \mathbb{N}$.

Proof. The proof of existence is based on the inequality
$$d(f^{0\ k}(x), f^{0\ m}(x)) \leq s^{m \wedge k} d(0, f^{0\ |m-k|}(x)), \quad m, k \in \mathbb{P},$$
with $m \wedge k = \min(m, k)$, and
$$d(x, f^{0\ k}(x)) \leq d(x, f^{0\ 1}(x)) + d(f^{0\ 1}(x), f^{0\ 2}(x))$$
$$+ \cdots + d(f^{0\ (k-1)}(x), f^{0\ k}(x))$$
$$\leq (1 + s + s^2 + \cdots + s^{k-1}) d(x, f(x)) \leq \frac{1}{1-s} d(x, f(x)),$$
thus
$$d(f^{0\ k}(x), f^{0\ m}(x)) \leq s^{m \wedge k} \frac{1}{1-s} d(x, f(x)).$$

It concludes that $\{f^{0\ k}(x)\}_{k=1}^{+\infty} \subset X$ is a Cauchy sequence in X, and by the completeness of X, a point $x_f \in X$ exists, and by the continuousness of contraction mapping, it follows
$$f(x_f) = f\left(\lim_{k \to +\infty} f^{0\ k}(x)\right) = \lim_{k \to +\infty} f^{0\ (k+1)}(x) = x_f.$$
Uniqueness can be obtained by
$$d(x_f, y_f) = d(f(x_f), f(y_f)) \leq s d(x_f, y_f)$$
and $0 < s < 1$.

(2) Forward iterate, backward iterate of a function

Definition 4.3.7 (Forward k-iterate, backward k-iterate)   Let $f \in \mathbb{F}$, denoted by

$$f^{0\ k} : X \to X, \quad k \in \mathbb{P}$$

the transformation: $\forall x \in X$,

$$f^{0\ 0}(x) = x,$$
$$f^{0\ 1}(x) = f(x),$$
$$\cdots\cdots$$
$$f^{0\ k}(x) = f \circ f^{0\ k-1}(x) = f\left(f^{0\ k-1}(x)\right),$$

then the transformation $f^{0\ k} : X \to X$ is said to be a **forward k-iterate of f on X**; or simply, **forward k-iterate of f**.

If $f \in \tilde{\mathbb{F}}$, and let $f^{0\ -k}(x) = \left(f^{0\ k}\right)^{-1}(x)$, $k \in \mathbb{P}$, then the transformation

$$f^{0\ -k} : X \to X, \quad k \in \mathbb{P}$$

is said to be a **backward k-iterate of f**.

Theorem 4.3.6   The forward k-iterate $f^{0\ k} : X \to X$ on the metric space (X, d) of $f \in \mathbb{F}$ satisfies semi-group property

$$f^{0\ m} \circ f^{0\ k} = f^{0(m+k)}, \quad m, k \in \mathbb{P}.$$

Example 4.3.3   Let $f : \mathbb{R} \to \mathbb{R}$, $f(x) = 2x$, $x \in \mathbb{R}$, find the forward n-iterate.

Solution.   By definition, it follows

$$f^{00}(x) = x, f^{01}(x) = 2x, f^{02}(x) = 2^2 x, \cdots, f^{0k}(x) = 2^k x, \cdots, \quad k \in \mathbb{N}.$$

Example 4.3.4   Let $f : \mathbb{R}^2 \to \mathbb{R}^2$, $f(x_1, x_2) = \left(2x_1, x_2^2 + x_1\right)$, find the forward 2-iterate.

Solution.   By definition, it follows

$$f^{00}(x_1, x_2) = (x_1, x_2), \quad f^{01}(x_1, x_2) = \left(2x_1, x_2^2 + x_1\right),$$
$$f^{02}(x_1, x_2) = \left(4x_1, \left(x_2^2 + x_1\right)^2 + 2x_1\right).$$

Example 4.3.5   Let $\sum_{n=0}^{+\infty} ba^n = b + ba + ba^2 + ba^3 + \cdots$ be a geometric series with $0 < a < 1$ and $b > 0$. For the interval $I_0 = [0, b]$ and $f(x) = ax + b$, find the forward k-iterate of I_0 under f.

Solution. The forward k-iterates are

$$f^{00}(I_0) = I_0, f^{01}(I_0) = f(I_0) = I_1 = [b, ab+b], \cdots, f^{0k}(I_0) = I_k.$$

Note that, the left and right end points of I_k are the partial sums s_{k-1} and s_k of the geometric series, respectively, i.e.,

$$I_k = [s_{k-1}, s_k], \quad k \in \mathbb{N},$$

where $s_0 = b$, $s_k = \sum_{j=0}^{k} ba^j$, $k \in \mathbb{N}$. Let $I = \bigcup_{k=0}^{+\infty} I_k$, then

$$f(I) = \left[b, \frac{b}{1-a}\right) = I - I_0.$$

(3) Iterated function system

Definition 4.3.8 (Iterated function system) *Let $w_n : X \to X$ be contraction functions with contraction factors s_n, $n = 1, 2, \cdots, N$, on complete metric space (X, d). Then $\{X; w_n, s_n\}_{n=1}^{N}$ is said to be an **iterated function system on** (X, d), or simply, **IFS**. $s = \max\{s_n : 1 \leqslant n \leqslant N\}$ is said to be the **contraction factor** of the IFS.*

For $B \in \mathbb{K}(X)$, denoted by $w(B)$ the union set $w(B) \equiv \bigcup_{n=1}^{N} w_n(B)$.

Since the compactness of sets in $\mathbb{K}(X)$, and the continuity of w_n, we have

$$w(B) \in \mathbb{K}(X);$$

and the contraction mapping $w : X \to X$ with the contraction factor s. Thus, w has unique "fixed point" $A \in \mathbb{K}(X)$, satisfies

$$A = w(A) = \bigcup_{n=1}^{N} w_n(A), \qquad (4.3.6)$$

which is determined by

$$A = \lim_{k \to +\infty} w^{0k}(B), \quad B \in \mathbb{K}(X).$$

Definition 4.3.9 (Attractor of an IFS) *The fixed point A of IFS $\{X; w_n, s_n\}_{n=1}^{N}$ in (4.3.6) is said to be an **attractor of IFS**.*

Example 4.3.6 Take $X = \mathbb{R}$, elements of $\{w_1, w_2\}$ are

$$w_1 = \frac{1}{3}x, \quad w_2 = \frac{1}{3}x + \frac{2}{3},$$

then $w(B) = w_1(B) \cup w_2(B)$, $B \in \mathbb{K}(\mathbb{R})$, and w is a contraction mapping with contraction factor $s = \dfrac{1}{3}$.

If we take $B = [0,1]$, then $B^{0k} = w^{0k}(B)$, $k \in \mathbb{N}$, is the k-th lever of the Cantor set C_3.

It is easy to prove that $\lim\limits_{k \to +\infty} w^{0k}(B) = A$ is the Cantor set C_3. And

$$A = \left(\frac{1}{3}A\right) \cup \left(\frac{1}{3}A + \frac{2}{3}\right),$$

with $\alpha A = \{\alpha x : \alpha \in \mathbb{C}, x \in A\}$, $A + \beta = \{x + \beta : x \in A, \beta \in \mathbb{C}\}$.

Theorem 4.3.7 *Let $\{X; w_n, s_n\}_{n=1}^{N}$ be an IFS on a complete metric space. Then the mapping determined by*

$$w(B) \equiv \bigcup_{n=1}^{N} w_n(B), \quad B \in \mathbb{K}(X)$$

satisfies $h(w(B), w(C)) \leqslant s\, h(B,C)$, $0 \leqslant s < 1$, $B, C \in \mathbb{K}(X)$, with an unique fixed point $A \in \mathbb{K}(X)$. And the attractor $A = w(A) \equiv \bigcup\limits_{n=1}^{N} w_n(A)$ is determined by

$$A = \lim_{k \to +\infty} w^{0k}(B), \quad \forall B \in \mathbb{K}(X).$$

Definition 4.3.10 (Condensation mapping, condensation set) *Let (X, d) be a complete metric space, and $C \in \mathbb{K}(X)$, define a mapping*

$$w_0(B) = C, \quad B \in \mathbb{K}(X).$$

*Then $w_0 : \mathbb{K}(X) \to \mathbb{K}(X)$ is said to be a **condensation mapping**, and the set $C \in \mathbb{K}(X)$ is said to be the **associated condensation set**.*

The condensation is a contraction mapping with contraction factor $s = 0$.

If $\{X; w_n, 0 \leqslant s_n < 1\}_{n=0}^{N}$ is an IFS on (X, d) with condensation mapping w_0, then the IFS $\{X; w_n, s_n\}_{n=0}^{N}$ is said to be a **hyperbolic** IFS **with condensation w_0 and contraction factor** $s = \max\{s_n : 0 \leqslant s_n < 1, n = 0, 1, \cdots, N\}$ **on** (X, d).

Theorem 4.3.8 *Let $\{X; w_n, s_n\}_{n=0}^{N}$ be a hyperbolic IFS with condensation w_0 and contraction factor $s = \max\{s_n : 0 \leqslant s_n < 1, n = 0, 1, \cdots, N\}$ on (X, d). Then the mapping $W : \mathbb{K}(X) \to \mathbb{K}(X)$ determined by the IFS $\{X; w_n, s_n\}_{n=0}^{N}$*

$$W(B) \equiv \bigcup_{n=0}^{N} w_n(B), \quad \forall B \in \mathbb{K}(X)$$

is a contraction mapping with contraction factor s on the complete metric space $(\mathbb{K}(X), h)$ satisfying

$$h(W(B), W(C)) \leqslant s h(B, C), \quad 0 \leqslant s < 1, \quad B, C \in \mathbb{K}(X),$$

and W has an unique fixed point $A = W(A) = \bigcup_{n=0}^{N} w_n(A) \in \mathbb{K}(X)$ given by

$$A = \lim_{k \to +\infty} W^{0k}(B), \quad \forall B \in \mathbb{K}(X).$$

Definition 4.3.11 (Collage of IFS) Let $L \in \mathbb{K}(X)$ be a given compact set, and $\{X; w_n, s_n\}_{n=0}^{N}$ an hyperbolic IFS on complete metric space (X, d). If for $\varepsilon > 0$,

$$h\left(L, \bigcup_{n=0}^{N} w_n(L)\right) \leqslant \varepsilon,$$

then L is said to be a **collage of** IFS $\{X; w_n, s_n\}_{n=0}^{N}$.

Theorem 4.3.9 Let (X, d) be a complete metric space, $L \in \mathbb{K}(X)$ a given compact set, $\{X; w_n, s_n\}_{n=0}^{N}$ a hyperbolic IFS on (X, d). If $L \in \mathbb{K}(X)$ is a collage of the IFS $\{X; w_n, s_n\}_{n=0}^{N}$, then

$$h(L, A) \leqslant \frac{\varepsilon}{1-s},$$

where $A \in \mathbb{K}(X)$ is the attractor of $\{X; w_n, s_n\}_{n=0}^{N}$; or equivalently,

$$h(L, A) \leqslant \frac{1}{1-s} h\left(L, \bigcup_{n=0}^{N} w_n(L)\right).$$

Proof. By four steps.

① If $f : X \to X$ is a contraction mapping with contraction factor s, and $x_f \in X$ is its fixed point. Then

$$d(x, x_f) \leqslant \frac{1}{1-s} d(x, f(x)), \quad \forall x \in X.$$

In fact, by the continuity of metric d, we deduce

$$d(x, x_f) \leqslant d\left(x, \lim_{n \to +\infty} f^{0n}(x)\right) = \lim_{n \to +\infty} d(x, f^{0n}(x))$$

$$\leqslant \lim_{n \to +\infty} \sum_{m=1}^{n} d\left(f^{0\ m-1}, f^{0m}(x)\right)$$

$$\leqslant \lim_{n \to +\infty} d(x, f(x)) \{1 + s + \cdots + s^{n-1}\}$$

$$\leqslant \frac{1}{1-s} d(x, f(x)).$$

② If (P, d_P) is a metric space, and $w: P \times X \to X$ is a contraction mapping with contraction factor s on X for $\forall p \in P$, that is, $\forall p \in P$, $w(p, x)$ is a contraction mapping on X. Suppose that for each fixed $x \in X$, the mapping w is continuous with respect to $p \in P$. Then the fixed point x_w of $w(p, x)$ is depending on p continuously, that is the mapping $x_w: P \to X$ is continuous.

In fact, for the fixed point $x_w(p)$ of $w(p, x)$, $p \in P$, we deduce: $\forall \varepsilon > 0$, $\forall q \in P$, it follows

$$d(x_w(p), x_w(q))$$
$$= d(w(p, x_w(p)), w(q, x_w(q)))$$
$$\leqslant d(w(p, x_w(p)), w(q, x_w(p))) + d(w(q, x_w(p)), w(q, x_w(q)))$$
$$\leqslant d(w(p, x_w(p)), w(q, x_w(p))) + s d(x_w(p), x_w(q)),$$

thus

$$d(x_w(p), x_w(q)) \leqslant \frac{1}{1-s} d(w(p, x_w(p)), w(q, x_w(p))).$$

Let $d_P(p, q) \to 0$, then

$$d(w(p, x), w(q, x)) \leqslant c d_P(p, q), \quad \forall p, q \in P, x \in X$$

and $d(x_w(p), x_w(q)) \leqslant \dfrac{c}{1-s} d_P(p, q)$. This implies the continuity of $x_w: P \to X$.

③ Let $w_n : \mathbb{N} \times X \to X$ be a continuous mapping of $x \in X$ and depending on $n \in \mathbb{N}$ continuously. i.e., take $P = \{1, \cdots, N\}, N \in \mathbb{N}$ in ② as a finite set. Since $\{P = \{1, \cdots, N\}, d_P\}$ is a compact metric space, we define

$$W(p, B) = \bigcup_{n=1}^{N} w_n(p, B), \quad \forall B \in \mathbb{K}(X),$$

then $W: \mathbb{K}(X) \to \mathbb{K}(X)$ is continuous on $p \in P$, i.e., $W(p, B)$ is continuous on $p \in P$ for all $B \in \mathbb{K}(X)$ in the metric space $(\mathbb{K}(X), h)$.

In fact, we only need to prove ③ for $N = 1$. Since $\forall B \in \mathbb{K}(X)$, take $p, q \in P$, and $\forall \varepsilon > 0$, then

$$d(w_1(p, B), w_1(q, B))$$
$$= \max_{x \in B} \min_{y \in B} d(w_1(p, x), w_1(q, y))$$
$$\leqslant \max_{x \in B} \min_{y \in B} \{d(w_1(p, x), w_1(p, y)) + d(w_1(p, y), w_1(q, y))\}.$$

By the compactness of $P \times B$, and by the continuity of $w_1 : P \times B \to X$, we conclude that w_1 is continuous uniformly, thus, $\forall \varepsilon > 0$, $\exists \delta > 0$, s.t. for $d_P(p,q) < \delta$, holds
$$d(w_1(p,x), w_1(q,y)) < \varepsilon, \quad \forall y \in B.$$
Then, if $d_P(p,q) < \delta$, it holds
$$d(w_1(p,B), w_1(q,B)) < \max_{x \in B} \min_{y \in B} \{d(w_1(p,x), w_1(p,y)) + \varepsilon\}$$
$$\leqslant d(w_1(p,B), w_1(p,B)) + \varepsilon = \varepsilon.$$

Similarly, if $d_P(p,q) < \delta$, then it holds $d(w_1(q,B), w_1(p,B)) < \varepsilon$. Combining the above results, for $d_P(p,q) < \delta$, it holds
$$h(w_1(p,B), w_1(q,B)) < \varepsilon.$$

④ Let $\{X; w_n, s_n\}_{n=0}^N$ be an IFS on (X,d), and mappings w_n depending on $n \in P$ continuously, where P is a compact metric space. Then the attractor $A(p) \in \mathbb{K}(X)$ is depending on parameter $p \in P$ with respect to Hausdorff metric h.

The collage theorem is proved (for more details, see [8]).

The dependent property of the attractor of IFS on parameter continuously is shown in the collage theorem, this is important for applications, since one can control the attractor of IFS by controlling the parameter.

The concepts of self-similar functions, contraction mapping, iterate, IFS on the fractal space $(\mathbb{K}(K_p), h)$ on local field (K_p, d) was developed recently. We refer to [123], [132]~[138]. The interpolation theory on local fields is also studied, but it is quite young.

Exercises

1. Prove: $h(A,B) = \max\{d(A,B), d(B,A)\}$ is a distance.
2. Prove: $d(x,y)$ defined in (4.3.1) is a distance on local field K_p, and then (K_p, d) becomes a complete metric space, and $(\mathbb{K}(K_p), h)$ is a complete fractal space on K_p.
3. Prove: the Cantor type set C_p belongs to fractal space $(\mathbb{K}(\mathbb{R}), d)$.
4. Let $(\mathbb{K}(K_p), h)$ be a fractal space on local field K_p, prove: Theorem 4.3.4 and 4.3.5.
5. Let $(\mathbb{K}(K_p), h)$ be a fractal space on K_p, prove: Theorem 4.3.7 and 4.3.8.

Chapter 5

Fractal Analysis on Local Fields

The study of fractal analysis based on local fields was started at the end of last century. However, the scientific results show that this area has a bright future.

We begin with the basic knowledge about fractal analysis on local fields, and then introduce new results and some open problems.

Suppose that K_p is a local field K_p with prime $p \geqslant 2$, and K_p is taken as a p-series field S_p or p-adic field A_p in this chapter, and the main references are [34], [40], [54]~[59], [102], [132]-[138].

5.1 Fractal dimensions on local fields

5.1.1 Hausdorff measure and dimension

1. Hausdorff measure

For $E \subset K_p$, the value $|E|_d = \sup\{d(x,y) : x, y \in E\}$ is said to be a **diameter of the set** E, where $d(x,y) = |x-y|$ in (1.2.8) is an ultra-distance on a metric space (K_p, d).

For $E \subset K_p$ and $n \in \mathbb{Z}$, the family of sets $\{U_j\}_{j=1}^{+\infty}$ is said to be an **n-covering of set** E, if $E \subset \bigcup_{j=1}^{+\infty} U_j$, where $\forall U_j \subset K_p$ is open with diameter $|U_j|_d \leqslant p^{-n}$.

Definition 5.1.1 (s-dimensional Hausdorff measure) *For $s \geqslant 0$ and $n \in \mathbb{Z}$,*

$$H_n^s(E) = \inf\left\{\sum_{j=1}^{+\infty} |U_j|_d^s : \bigcup_j U_j \supset E \text{ is } n\text{-covering of } E\right\}$$

is said to be an **approximate s-dimensional Hausdorff measure of** E.

It is clear that the limit $\lim\limits_{n\to+\infty} H_n^s(E) = H^s(E)$ exists, and $H^s(E)$ is said to be the **s-dimensional Hausdorff measure of** E. Simply, Hausdorff measure of E.

Similar to that on Euclidean spaces, we have

Theorem 5.1.1 $H_n^s(E)$ and $H^s(E)$ are outer-measures; moreover, $H^s(E)$ is an ultra-measure.

Theorem 5.1.2 $H^s(E)$ is translation invariant and dilation invariant: for $\lambda \in K_p$,
 (i) $H^s(E + \lambda) = H^s(E)$, where $E + \lambda = \{x + \lambda : x \in E\}$;
 (ii) $H^s(\lambda E) = |\lambda|^s H^s(E)$, where $\lambda E = \{\lambda x : x \in E\}$.

Proof. For (i). Let $\{U_j\}$ be an n-covering of E, then $\{U_j + \lambda\}$ is an n-covering of $E + \lambda$. Thus, $H_n^s(E + \lambda) \leqslant H_n^s(E)$, so that

$$H^s(E + \lambda) \leqslant H^s(E). \tag{5.1.1}$$

On the other hand, take $\lambda = -\mu \in K_p$ in (5.1.1), then $H^s(E - \mu) \leqslant H^s(E)$. Thus,

$$H^s((E - \mu) + \mu) \leqslant H^s(E + \mu) \text{ implies } H^s(E) \leqslant H^s(E + \mu).$$

By arbitrariness of λ, μ, (i) is proved.

The property (ii) can be obtained by $|\lambda U|_d = |\lambda| \sup\{|x - y| : x, y \in U\}$.

2. Hausdorff dimension

The s-dimensional Hausdorff measure $H^s(E)$ has the following important property.

Lemma 5.1.1 Let $E \subset K_p$ be a non-empty Borel set. If there exists $s_0 \in (0, +\infty)$, such that $H^{s_0}(E) < +\infty$, then $\forall s > s_0$, $H^s(E) = 0$ holds; moreover, if there exists $s_1 \in (0, +\infty)$, such that $H^{s_1}(E) > 0$, then $\forall s < s_1$, $H^s(E) = +\infty$ holds.

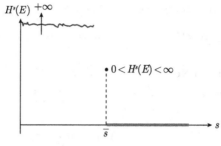

Figure. 5.1.1

Thus, $\forall E \subset K_p$, $\exists\, \bar{s} \in \mathbb{R}$ with $0 \leqslant \bar{s} \equiv D_H(E) < +\infty$, such that

for $0 \leqslant s < \bar{s} \equiv D_H(E) < +\infty$, $H^s(E) = +\infty$ holds;

for $0 \leqslant D_H(E) \equiv \bar{s} < s < +\infty$, $H^s(E) = 0$ holds.

And for $0 \leqslant s < t < +\infty$,

$$H^s(E) < +\infty \quad \text{implies} \quad H^t(E) = 0;$$

$$H^t(E) > 0 \quad \text{implies} \quad H^s(E) = +\infty.$$

Definition 5.1.2 (Hausdorff dimension) *The Hausdorff dimension of $E \subset K_p$ is defined as*

$$\dim_H E = \sup\{s : H^s(E) > 0\} = \sup\{s : H^s(E) = +\infty\}$$
$$= \inf\{t : H^s(E) < +\infty\} = \inf\{t : H^t(E) = 0\}. \quad (5.1.2)$$

Theorem 5.1.3 *Hausdorff dimension $\dim_H E$ has the following properties:*

(i) $E \subset K_p \Rightarrow 0 \leqslant \dim_H E \leqslant 1$;
(ii) $E_1 \subset E_2 \Rightarrow \dim_H E_1 \leqslant \dim_H E_2$;
(iii) $\dim_H \left(\bigcup_{k \geqslant 1} E_k \right) = \sup_{k \geqslant 1} \{\dim_H E_k\}$;
(iv) $\dim_H E = \sup\{\dim_H F : F \subset E, F \in \mathbb{K}(K_p)\}$;
(v) $|E| > 0 \Rightarrow \dim_H E = 1$, *where $|E|$ is the Haar measure of $E \subset K_p$;*
(vi) $E \subset K_p$ *is a countable set* $\Rightarrow \dim_H E = 0$.
Proofs are left to exercises.

3. Hausdorff net measure

Hausdorff net measure $H_{\mathfrak{F}}^s(E)$ plays important role in the fractal analysis on local fields.

Definition 5.1.3 (Net in K_p) *Let $\mathfrak{F} = \{A \subset K_p\}$ be a family of subsets in K_p. If $\forall \varepsilon > 0$ and $\forall x \in K_p$, $\exists A \in \mathfrak{F}$, s.t. $x \in A$, and $|A|_d \leqslant \varepsilon$, then $\mathfrak{F}$ is said to be a net in K_p. Denoted by*

$$\mathfrak{M} = \{\mathfrak{F} : \mathfrak{F} = \{A \subset K_p\} \text{ is a net in } K_p\}, \quad (5.1.3)$$

and the set of all subsets of K_p by

$$2^{K_p} = \{A \subset K_p : A \text{ is a subset in } K_p\}. \quad (5.1.4)$$

Definition 5.1.4 (s-dimensional Hausdorff net measure) *For $s \geqslant 0$, $\forall E \in 2^{K_p}$, an **s-dimensional Hausdorff net measure of E in net $\mathfrak{F}$** is defined as*

$$H_{\mathfrak{F}}^s(E) = \lim_{n \to +\infty} \inf \left\{ \sum_{j=1}^{+\infty} |U_j|_d^s : E \subset \bigcup_j U_j, U_j \in \mathfrak{F}, |U_j|_d \leqslant p^{-n} \right\}. \quad (5.1.5)$$

For $\mathfrak{F}_1, \mathfrak{F}_2 \in \mathfrak{M}$, if there exist constants $c_1, c_2 > 0$, s.t. $\forall E \in 2^{K_p}$, $\forall s \geqslant 0$, it holds

$$c_1 H_{\mathfrak{F}_1}^s(E) \leqslant H_{\mathfrak{F}_2}^s(E) \leqslant c_2 H_{\mathfrak{F}_1}^s(E), \quad (5.1.6)$$

*then it is said that the **nets $\mathfrak{F}_1$ and $\mathfrak{F}_2$ are equivalent**, denoted by $\mathfrak{F}_1 \approx \mathfrak{F}_2$; also, the **$s$-dimensional Hausdorff net measures** $H_{\mathfrak{F}_1}^s(E)$ and $H_{\mathfrak{F}_2}^s(E)$ are said to be equivalent, and denoted by $H_{\mathfrak{F}_1}^s(E) \approx H_{\mathfrak{F}_2}^s(E)$.*

*Moreover, if $\forall E \in 2^{K_p}$, $\forall s \geqslant 0$, $H_{\mathfrak{F}_1}^s(E) = H_{\mathfrak{F}_2}^s(E)$ holds, then it is said that the **nets $\mathfrak{F}_1$ and $\mathfrak{F}_2$ are strongly equivalent**, denoted by $\mathfrak{F}_1 \equiv \mathfrak{F}_2$.*

Theorem 5.1.4 *Let $\mathfrak{M} = \{\mathfrak{F}\}$ be a net set of K_p, and 2^{K_p} the set of all subsets of K_p. If $\mathfrak{F}_1, \mathfrak{F}_2, \mathfrak{F}_3, \mathfrak{F}_4, \mathfrak{F}_5 \in \mathfrak{M}$ are the **closed set net, open set net, compact set net, closed ball net, and open ball net**, respectively, then $2^{K_p} \equiv \mathfrak{F}_1 \equiv \mathfrak{F}_2 \equiv \mathfrak{F}_3 \equiv \mathfrak{F}_4 \equiv \mathfrak{F}_5$.*

Proof. By the structure of local field, it is clear that $\mathfrak{F}_3 \equiv \mathfrak{F}_4 \equiv \mathfrak{F}_5$.

To prove $2^{K_p} \approx \mathfrak{F}_4$, take any n-covering $\{E_j\}$ of $E \in 2^{K_p}$, it is at most countable. Suppose that $|E_j|_d = p^{-n_j}$, $n_j \geqslant n$. Take $x_j \in E_j$ with $U_j = x_j + B_{n_j}$, then by the structure of a local field, $U_j \supset E_j$, and $|U_j|_d = |E_j|_d = p^{-n_j}$. Thus, $\sum_j |E_j|_d^s = \sum_j |U_j|_d^s$. Since $U_j \in \mathfrak{F}_4$, and by the arbitrariness of $\{E_j\}$, then $\inf\left\{\sum_j |E_j|_d^s\right\} = \inf\left\{\sum_j |U_j|_d^s\right\}$, this implies $H^s(E) \geqslant H^s_{\mathfrak{F}_4}(E)$. On the other hand, by $2^{K_p} \supset \mathfrak{F}_4$, it follows that $H^s(E) \leqslant H^s_{\mathfrak{F}_4}(E)$. Thus we have $H^s(E) = H^s_{\mathfrak{F}_4}(E)$. Then $2^{K_p} \equiv \mathfrak{F}_4$. So

$$2^{K_p} \equiv \mathfrak{F}_3 \equiv \mathfrak{F}_4 \equiv \mathfrak{F}_5. \tag{5.1.7}$$

Moreover, since $2^{K_p} \supset \mathfrak{F}_1 \supset \mathfrak{F}_4$, we have $H^s(E) \leqslant H^s_{\mathfrak{F}_1}(E) \leqslant H^s_{\mathfrak{F}_4}(E)$; thus by $H^s(E) = H^s_{\mathfrak{F}_4}(E)$, it holds $H^s(E) = H^s_{\mathfrak{F}_1}(E) = H^s_{\mathfrak{F}_4}(E)$. This implies $2^{K_p} \equiv \mathfrak{F}_1 \equiv \mathfrak{F}_4$. Thus,

$$2^{K_p} \equiv \mathfrak{F}_1 \equiv \mathfrak{F}_3 \equiv \mathfrak{F}_4 \equiv \mathfrak{F}_5. \tag{5.1.8}$$

Finally, $2^{K_p} \supset \mathfrak{F}_2 \supset \mathfrak{F}_5$ implies $H^s(E) \leqslant H^s_{\mathfrak{F}_2}(E) \leqslant H^s_{\mathfrak{F}_5}(E)$, then by $H^s(E) = H^s_{\mathfrak{F}_5}(E)$, we get $H^s(E) = H^s_{\mathfrak{F}_2}(E) = H^s_{\mathfrak{F}_5}(E)$, and this implies $2^{K_p} \equiv \mathfrak{F}_2 \equiv \mathfrak{F}_5$. Thus,

$$2^{K_p} \equiv \mathfrak{F}_1 \equiv \mathfrak{F}_2 \equiv \mathfrak{F}_3 \equiv \mathfrak{F}_4 \equiv \mathfrak{F}_5. \tag{5.1.9}$$

This theorem has important meaning: to evaluate the Hausdorff measure, we may choose any strong equivalent net in $2^{K_p}, \mathfrak{F}_1, \mathfrak{F}_2, \mathfrak{F}_3, \mathfrak{F}_4, \mathfrak{F}_5$. For example, take $\mathfrak{F}_4$ or $\mathfrak{F}_5$, then we have a convenient and simple form of the Hausdorff measure on K_p

$$H^s(E) = \lim_{n\to+\infty} \inf\left\{\sum_{j=1}^{+\infty} |U_j|_d^s : \bigcup_j U_j \supset E \text{ is an open ball covering of } E, |U_j|_d < p^{-n}\right\}.$$

Example 5.1.1 Cantor type set C_p^q on local field $K_p, p \geqslant 2, 0 \leqslant q \leqslant p-1$.

We defined the Cantor type set C_3 on K_3 in the Subsection 1.2, and C_p on K_p in Example 4.3.2, now we generalize this set on a p-series field, K_p with prime element $\beta \in K_p, |\beta| = p^{-1}$. For $0 \leqslant q \leqslant p-1$, let

$$V_0 = D = \{x \in K_p : |x| \leqslant 1\},$$
$$V_1 = (0 \cdot \beta^0 + B^1) \cup (2\beta^0 + B^1) \cup \cdots \cup ((p-q)\beta^0 + B^1),$$
$$V_2 = (1 \cdot \beta^0 + 0 \cdot \beta^1 + B^2) \cup (1 \cdot \beta^0 + 2 \cdot \beta^1 + B^2)$$
$$\cup \cdots \cup (1 \cdot \beta^0 + (p-q) \cdot \beta^1 + B^2),$$
$$\cdots$$

Denoted by

$$C_p^q = D \setminus \left(\bigcup_{j=1}^{+\infty} V_j \right),$$

and C_p^q is said to be a **Cantor type set on local field K_p**. We rewrite C_p^q as

$$C_p^q = \bigcap_{n=1}^{+\infty} \bigcup_{i=1}^{q^n} I_{n,i},$$

where

$$I_{1,1} = 1 \cdot \beta^0 + B^1, \quad I_{1,2} = (p-q+1)\beta^0 + B^1, \cdots, I_{1,q} = (p-1)\beta^0 + B^1;$$
$$I_{2,1} = 1 \cdot \beta^0 + 1 \cdot \beta^1 + B^2, \cdots, I_{2,q^2} = (p-1)\beta^0 + (p-1)\beta^1 + B^2;$$
$$\cdots \tag{5.1.10}$$

The construction process of Cantor type set C_p^q is: take away balls $0 \cdot \beta^0 + B^1$, $2\beta^0 + B^1$, $\cdots$, $(p-q)\beta^0 + B^1$ from $D = B^0$, keep $I_{1,1}$, $I_{1,2}$, $\cdots$, $I_{1,q}$. Repeating the process, after n-th step, there are $I_{n,1}$, $I_{n,2}$, $\cdots$, I_{n,q^n} left, and $|I_{n,i}|_d = p^{-n}$. The sets $I_{n,1}$, $I_{n,2}$, $\cdots$, I_{n,q^n} are said to be in the n-th lever. See Fig. 5.1.2.

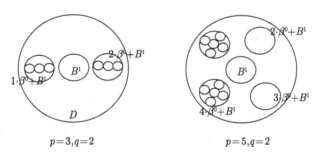

$p=3, q=2$ $\qquad\qquad$ $p=5, q=2$

Figure 5.1.2

We now determine the Hausdorff measure and Hausdorff dimension of the Cantor type set C_p^q.

Lemma 5.1.2 *For the Cantor type set $C_p^q = \bigcap_{n=1}^{+\infty} \bigcup_{i=1}^{q^n} I_{n,i}$, and any open set U with $U \subset D \subset K_p$, suppose that:* ① $U \cap C_p^q \neq \varnothing$; ② $|U|_d = p^{-t}$; ③ *for any fixed $l \in \mathbb{N}$, there exists $s > 0$, independent of U and t, such that*

$$\sum_{I_{l,i} \subset U} |I_{l,i}|_d^s = |U|_d^s, \qquad (5.1.11)$$

where $I_{l,i} \subset U$ satisfying: $I_{l,i} \cap I_{l,k} = \varnothing$, $i \neq k$, $|I_{l,i}|_d = p^{-l}$. Then $s = \dfrac{\ln q}{\ln p}$.

Proof. For any fixed $l \in \mathbb{N}$, by the construction of C_p^q, without lose of generality, we suppose that there are q^{l-t} balls $I_{t,i}$, disjoint each other with diameter p^{-l}, and contained in U, satisfy (5.1.11). Thus, since

$$\left(p^{-t}\right)^s = |U|_d^s = \sum_{I_{l,i} \subset U} |I_{l,i}|_d^s = \sum_{i=1}^{q^{l-t}} |I_{l,i}|_d^s = q^{l-t} \left(p^{-l}\right)^s,$$

it follows that $q^{l-t} p^{-ls} = p^{-ts}$, this is $q^{l-t} = p^{(l-t)s}$, it implies $s = \dfrac{\ln q}{\ln p}$.

Theorem 5.1.5 *Let K_p be a p-series field. Then the Cantor type set* $C_p^q = \bigcap\limits_{n=1}^{+\infty} \bigcup\limits_{i=1}^{q^n} I_{n,i}$ *has the following properties:*

(i) *For Haar measure of C_p^q, denoted by $|C_p^q|$, then $|C_p^q| = 0$.*

(ii) *For $s = \dfrac{\ln q}{\ln p}$-dimension Hausdorff measure of C_p^q, then $H^s\left(C_p^q\right) = 1$.*

(iii) *For Hausdorff dimension $\dim_H C_p^q$ of C_p^q, then $\dim_H C_p^q = \dfrac{\ln q}{\ln p}$.*

Proof. We prove (ii), $H^s\left(C_p^q\right) = 1$, $s = \dfrac{\ln q}{\ln p}$, firstly.

For this purpose, we only need to prove: for any ball covering $\{U_1, U_2, \cdots\}$ of C_p^q with $\bigcup\limits_j U_j \supset C_p^q$, $U_j \cap C_p^q \neq \varnothing$, then for $s = \dfrac{\ln q}{\ln p}$,

$$\sum_j |U_j|_d^s = 1.$$

Since each ball $I_{n,i}$ is open, closed and compact in K_p for $i = 1, \cdots, q^n$, $n = 1, 2, \cdots$, thus, the set $\bigcup\limits_{i=1}^{q^n} I_{n,i}$ is compact for any $n \in \mathbb{N}$, and then $C_p^q = \bigcap\limits_{n=1}^{+\infty} \bigcup\limits_{i=1}^{q^n} I_{n,i}$ is compact. So there exists finite ball covering $\{U_1, U_2, \cdots, U_m\}$, $U_j \cap U_k = \varnothing$, $j \neq k$, such that

$$C_p^q \subset \bigcup_{j=1}^m U_j.$$

Let N be the largest integer such that $|U_j|_d \leqslant p^{-N}$, $1 \leqslant j \leqslant m$. And choose $k > N$, such that each $I_{k,i}$ is contained in certain U_j. Then, by the Lemma 5.1.2,

$$H^s(C_p^q) = \sum_j |U_j|_d^s = \sum_j \sum_{I_{k,i} \subset U_j} |I_{k,i}|_d^s = \sum_{i=1}^{q^k} |I_{k,i}|_d^s = 1,$$

with $s = \dfrac{\ln q}{\ln p}$, $|I_{k,i}|_d = p^{-k}$. Thus, (ii) has been shown. And (iii) is the corollary of (ii).

The result (i) can be obtained by evaluation for Cantor type set C_p^q directly, $|C_p^q| = 0$.

Remark. The methods of proofs in Theorem 5.1.5 for local fields are definitely different from those for Euclidean spaces, and they are very delicate, technical, and utilizable, and suitable for fractal analysis on local fields.

5.1.2 Box dimension

1. Preparation lemmas

We use the following notations. For $E \subset K_p$, $n \in \mathbb{N}$,

$N_n(E)$: the smallest number of balls of covering of E in K_p with diameters p^{-n}.

$N_n^*(E)$: the smallest number of balls of covering of E in K_p with diameters $\leqslant p^{-n}$;

$M_n(E)$: the largest number of disjoint balls with centers in E and diameters p^{-n}.

As we know, the balls in the definitions of $N_n(E)$ and $N_n^*(E)$ are all disjoint each other by the well known structure property of local fields.

Lemma 5.1.3 *Let $E \subset K_p$, $n \in \mathbb{N}$. Then*

$$N_n(E) = N_n^*(E) = M_n(E). \tag{5.1.12}$$

Proof. We prove $N_n(E) = N_n^*(E)$, firstly.

It is clear that $N_n^*(E) \leqslant N_n(E)$.

To prove $N_n^*(E) \geqslant N_n(E)$, take $\{U_1, U_2, \cdots, U_{N_n^*(E)}\}$, a covering of E with diameters $\leqslant p^{-n}$, $U_j \cap E \neq \varnothing, 1 \leqslant j \leqslant N_n^*(E)$. Without lose of generality, let

$$|U_1|_d = |U_2|_d = \cdots = |U_k|_d = p^{-n},$$

and $|U_j|_d < p^{-n}, k < j \leqslant N_n^*(E)$, with $1 \leqslant k \leqslant N_n^*(E)$.

Then we assert that: for all i and j satisfying $i \neq j$ and $k < i, j \leqslant N_n^*(E)$,

$$d(U_i, U_j) > p^{-n}. \tag{5.1.13}$$

In fact, if there would be a pair i and j satisfying $i \neq j, k < i, j \leqslant N_n^*(E)$, such that $d(U_i, U_j) \leqslant p^{-n}$, then by $|U_i|_d < p^{-n}, |U_j|_d < p^{-n}$, and by the structure of local fields, it holds $|U_i \cup U_j|_d \leqslant p^{-n}$. Thus, there exists a new ball, denoted by U_0 with $|U_0|_d = p^{-n}$, such that $U_i \cup U_j \subset U_0$. This contradicts that $N_n^*(E)$ is the smallest number of balls with diameters $\leqslant p^{-n}$ in all coverings of E. Thus, (5.1.13) holds.

Now, for any i satisfying $k < i \leqslant N_n^*(E)$, we fix certain $x_i \in U_i \cap E$, then for some $j \neq i$ and $k < i, j \leqslant N_n^*(E)$, it holds $d(x_i + B^n, U_j) > 0$. Thus, $d(x_i + B^n, x_j) > 0$. So we conclude that $x_j \notin x_i + B^n$, and then $(x_i + B^n) \cap (x_j + B^n) = \varnothing$. Hence,

$$d(x_i + B^n, x_j + B^n) > 0$$

holds for all i and j with $k < i, j \leqslant N_n^*(E)$. This implies the set of balls

$$U_1, U_2, \cdots, U_k, x_{k+1} + B^n, \cdots, x_{N_n^*(E)} + B^n \tag{5.1.14}$$

becomes a new covering of E in which balls are disjoint each other and with diameters p^{-n} in (5.1.14). Then we conclude $N_n(E) \leqslant N_n^*(E)$. Combining with $N_n^*(E) \leqslant N_n(E)$, it follows that $N_n^*(E) = N_n(E)$.

Then we prove $N_n(E) = M_n(E)$.

To prove $N_n(E) \leqslant M_n(E)$, let $\{A_j\}_{j=1}^{+\infty}$ be a covering of E with diameters p^{-n}. Without lose of generality, suppose that $A_j \cap E \neq \varnothing, \forall j$. Thus the centers of A_j are in E. Since any two balls in a local field only have two positions: disjoint, or one is contained in the other one, so that by the definition of $M_n(E)$, it follows that $N_n(E) \leqslant M_n(E)$.

Conversely, to prove $N_n(E) \geqslant M_n(E)$, let $\{E_1, E_2, \cdots, E_{N_n(E)}\}$ be a covering of E with diameters p^{-n}, and $E_j \cap E \neq \varnothing, j = 1, \cdots, N_n(E)$. Suppose that $\{F_1, F_2, \cdots, F_{M_n(E)}\}$ is a set of disjoint balls with diameters p^{-n} and centers in E. Then $\forall i, 1 \leqslant i \leqslant M_n(E)$, there exists $x_i \in F_i \cap E$. Thus, there exists a corresponding set E_j to $x_i \in F_i \cap E$, such that $x_i \in E_j \in \{E_1, E_2, \cdots, E_{N_n(E)}\}$, this implies $E_j = F_i$.

Then, $\forall i$ with $1 \leqslant i \leqslant M_n(E)$ determines a $j, 1 \leqslant j \leqslant N_n(E)$, such that $E_j = F_i$. On the other hand, the balls in $\{E_1, E_2, \cdots, E_{N_n(E)}\}$ are disjoint each other, and so are balls in $\{F_1, F_2, \cdots, F_{M_n(E)}\}$, thus, we have $M_n(E) \leqslant N_n(E)$.

The proof is complete.

Remark. The above Lemma fails in the case of Euclidean spaces. By using this Lemma, we may evaluate the measures and dimensions for fractals, underlying on local fields conveniently.

2. Box dimension

Definition 5.1.5 (Box dimension) *Let $E \in 2^{K_p}$ be a bounded set in K_p. Then*

$$\overline{\dim}_B E = \limsup_{n \to +\infty} \frac{\ln N_n(E)}{n \ln p} \quad \text{and} \quad \underline{\dim}_B E = \liminf_{n \to +\infty} \frac{\ln N_n(E)}{n \ln p} \quad (5.1.15)$$

*are said to be **upper box dimension** and **lower box dimension** of E, respectively. If*

$$\overline{\dim}_B E = \underline{\dim}_B E = \dim_B E,$$

*then $\dim_B E$ is said to be the **box dimension** of E (sometimes called box-counting dimension).*

Theorem 5.1.6 *Let $E \in 2^{K_p}$ be a non-empty set in K_p. Then the upper and lower box dimensions have formulae*

$$\overline{\dim}_B E = \limsup_{n \to +\infty} \left\{ 1 + \frac{\ln |E(p^{-n})|}{n \ln p} \right\}$$

and

$$\underline{\dim}_B E = \liminf_{n \to +\infty} \left\{ 1 + \frac{\ln |E(p^{-n})|}{n \ln p} \right\},$$

where $|E(\varepsilon)|$ is the Haar measure of $E(\varepsilon) = \{x \in K_p : d(x, E) < \varepsilon\}$, an ε-neighborhood of E.

Proof. Only prove for the case of upper box dimension.

Suppose that a set E is covered by $N_n(E)$ disjoint balls with diameters p^{-n}, then its $\varepsilon = p^{-n}$-neighborhood $E(p^{-n})$ is covered by the $N_n(E)$ balls too. Thus,

$$|E(p^{-n})| \leqslant N_n(E) p^{-n}.$$

On the other hand, all $N_n(E)$ disjoint balls are covered by p^{-n}-neighborhood of $E(p^{-n})$, so that

$$|E(p^{-n})| \geqslant N_n(E) p^{-n}.$$

Then, $|E(p^{-n})| = N_n(E)p^{-n}$. Substitute in $\overline{\dim}_B E$, it follows that

$$\overline{\dim}_B E = \limsup_{n\to+\infty} \frac{\ln N_n(E)}{n\ln p} = \limsup_{n\to+\infty} \frac{\ln p^n |E(p^{-n})|}{n\ln p}$$
$$= \limsup_{n\to+\infty} \left\{ 1 + \frac{\ln|E(p^{-n})|}{n\ln p} \right\}.$$

The proof is complete.

Theorem 5.1.7 Let $E \in 2^{K_p}$ be a non-empty bounded set in K_p. Then

(i) $\dim_H E \leqslant \underline{\dim}_B E \leqslant \overline{\dim}_B E$;

(ii) $\underline{\dim}_B E$ and $\overline{\dim}_B E$ are monotonic increasing functions of E;

(iii) $0 \leqslant \underline{\dim}_B E \leqslant \overline{\dim}_B E \leqslant 1$; and $|E| > 0$ implies $\dim_B E = 1$;

(iv) $\overline{\dim}_B (E_1 \cup E_2) = \max\{\overline{\dim}_B E_1, \overline{\dim}_B E_2\}$;

(v) $\overline{\dim}_B E = \overline{\dim}_B \overline{E}$, $\underline{\dim}_B E = \underline{\dim}_B \overline{E}$, where $\overline{E}$ is the closure of E.

Proof. For (i):

$s < \dim_H E \Rightarrow H^s(E) = +\infty \Rightarrow \exists n_0 \in \mathbb{N}$, s.t. $H_n^s(E) > 1, \forall n > n_0$;

on the other hand,

$$H_n^s(E) \leqslant N_n(E)p^{-ns} \Rightarrow 1 \leqslant H_n^s(E) \leqslant N_n(E)p^{-ns}, \quad \forall n > n_0;$$

combining the above, it follows that $\dim_H E \leqslant \underline{\dim}_B E$.

By definition, it follows that $\underline{\dim}_B E \leqslant \overline{\dim}_B E$. So (i) is proved.

To prove (ii):

$E_1 \subset E_2 \Rightarrow N_n(E_1) \leqslant N_n(E_2), \forall n \in \mathbb{N} \Rightarrow$ the monotonicity is proved.

To prove (iii):

$$\overline{\dim}_B E = \limsup_{n\to+\infty} \left\{ 1 + \frac{\ln|E(p^{-n})|}{n\ln p} \right\}$$
$$= \limsup_{n\to+\infty} \left\{ 1 - \frac{\ln|E(p^{-n})|}{\ln p^{-n}} \right\} \leqslant 1 \Rightarrow \overline{\dim}_B E \leqslant 1;$$

moreover,

$|E| > 0 \Rightarrow \dim_H E = 1 \Rightarrow 1 = \dim_H E \leqslant \overline{\dim}_B E \leqslant 1 \Rightarrow \overline{\dim}_B E = 1$.

To prove (iv): by (ii),

$$\overline{\dim}_B (E_1 \cup E_2) \geqslant \overline{\dim}_B E_1, \quad \overline{\dim}_B (E_1 \cup E_2) \geqslant \overline{\dim}_B E_2;$$

$$\Rightarrow \overline{\dim}_B (E_1 \cup E_2) \geqslant \max \{\overline{\dim}_B E_1, \overline{\dim}_B E_2\};$$

on the other hand,

$$\forall \alpha > \max \{\overline{\dim}_B E_1, \overline{\dim}_B E_2\},$$
$$\Rightarrow N_n(E_j) \leqslant p^{n\alpha}, j = 1, 2, \text{ for } n \in \mathbb{N} \text{ large enough,}$$
$$\Rightarrow N_n(E_1 \cup E_2) \leqslant N_n(E_1) + N_n(E_2) \leqslant 2p^{n\alpha},$$
$$\Rightarrow \overline{\dim}_B (E_1 \cup E_2) \leqslant \alpha,$$
$$\Rightarrow \overline{\dim}_B (E_1 \cup E_2) \geqslant \max \{\overline{\dim}_B E_1, \overline{\dim}_B E_2\} \Rightarrow \text{(iv)}.$$

To prove (v):

$E \subset \overline{E} \Rightarrow N_n(E) \leqslant N_n(\overline{E})$ (by monotonicity);

conversely, if $\{U_1, U_2, \cdots, U_k\}$ is a set of balls with diameters p^{-n},

and $\bigcup_{j=1}^{k} U_j \supset E$,

$$\Rightarrow \bigcup_{j=1}^{k} U_j \supset \overline{E} \text{ (balls are open, closed, and compact),}$$
$$\Rightarrow N_n(\overline{E}) \leqslant N_n(E),$$
$$\Rightarrow N_n(\overline{E}) = N_n(E) \text{ (by } N_n(E) \leqslant N_n(\overline{E}) \text{ and } N_n(\overline{E})) \leqslant N_n(E),$$
$$\Rightarrow \text{(v)}.$$

3. Important example

We construct an example which upper box dimension and lower dimension are not equal, denoted by C, it is compact set in K_p.

Take q_1, q_2 with $2 \leqslant q_1 < q_2 \leqslant p-1$; then take s, t such that

$$\frac{\ln q_1}{\ln p} < s < t < \frac{\ln q_2}{\ln p}. \tag{5.1.16}$$

① For q_1 and p, construct Cantor type set $C_p^{q_1} = \bigcap_{n=1}^{+\infty} \bigcup_{j=1}^{q_1^n} I_{n,j}$ (see Example 5.1.1). Thus, q_1 balls $I_{1,1}, I_{1,2}, \cdots, I_{1,q_1}$ in D have built with diameters p^{-1} satisfying

$$q_1 (p^{-1})^s \leqslant 1. \tag{5.1.17}$$

② In each ball $I_{1,m}, m = 1, \cdots, q_1$, with diameter p^{-1}, construct k_1 times the Cantor type sets $C_p^{q_2}$, as the 1th lever of C, where k_1 is large enough satisfying

$$q_1 q_2^{k_1} \left(p^{-1-k_1}\right)^t \geqslant 1; \qquad (5.1.18)$$

by the assumption of q_1, q_2, s, t in (5.1.17), then the natural number k_1 exists.

③ In each ball $I_{2,m}, m = 1, 2, \cdots, q_1^{k_1}$, with diameters p^{-1-k_1} in the 1st lever of C, construct k_2 times the Cantor type sets $C_p^{q_1}$, as the 2nd lever of C with $k_2 (k_2 > k_1)$ large enough satisfying

$$q_1 q_2^{k_1} q_1^{k_2} \left(p^{-1-k_1-k_2}\right)^s \leqslant 1; \qquad (5.1.19)$$

by the assumption of q_1, q_2, s, t in (5.1.17), then the natural numbers k_1, k_2 exist.

④ Continue the above process, then the natural number sequence $\{k_j\}$, $j \in \mathbb{N}$, is determined so that the jth lever of the Cantor type set C is constructed correspondingly. Without ending, the Cantor type set $C \subset K_p$ is constructed.

Theorem 5.1.8 Let $2 \leqslant q_1 < q_2 \leqslant p - 1$, $\dfrac{\ln q_1}{\ln p} < s < t < \dfrac{\ln q_2}{\ln p}$. Then

$$\underline{\dim}_B C \leqslant s < t \leqslant \overline{\dim}_B C$$

for the Cantor type set $C \subset K_p$ constructed as above.

Proof. By the construction above, we have

$$N_1(C) = q_1, \ N_{1+k_1}(C) = q_1 q_2^{k_1}, \ N_{1+k_1+k_2}(C) = q_1 q_2^{k_1} q_1^{k_2}, \ \cdots,$$

$$N_{1+k_1+k_2+\cdots+k_{2j-1}}(C) = q_1 q_2^{k_1} q_1^{k_2} \cdots q_2^{k_{2j-1}},$$

$$N_{1+k_1+k_2+\cdots+k_{2j-1}+k_{2j}}(C) = q_1 q_2^{k_1} q_1^{k_2} \cdots q_2^{k_{2j-1}} q_1^{k_{2j}},$$

$$\cdots.$$

Thus,

$$\underline{\dim}_B C \leqslant \lim_{j \to +\infty} \frac{\ln N_{1+k_1+k_2+\cdots+k_{2j}}(C)}{(1+k_1+k_2+\cdots+k_{2j})\ln p}$$

$$= \lim_{j \to +\infty} \frac{\ln q_1 q_2^{k_1} q_1^{k_2} \cdots q_1^{k_{2j}}}{(1+k_1+k_2+\cdots+k_{2j})\ln p} \leqslant s;$$

$$\overline{\dim}_B C \geqslant \lim_{j \to +\infty} \frac{\ln N_{1+k_1+k_2+\cdots+k_{2j-1}}(C)}{(1+k_1+k_2+\cdots+k_{2j-1})\ln p}$$

$$= \lim_{m \to +\infty} \frac{\ln q_1 q_2^{k_1} q_1^{k_2} \cdots q_2^{k_{2j-1}}}{(1 + k_1 + k_2 + \cdots + k_{2j-1}) \ln p} \geqslant t.$$

The proof is complete.

5.1.3 Packing measure and dimension

1. Pre-Packing measure, Packing measure

Definition 5.1.6 (s-dimension pre-Packing outer measure) *Let $E \in 2^{K_p}$ be a non-empty set in K_p. If $U_j \cap E \neq \varnothing$ for $\forall j$, then the ball family $\{U_j\}$ with diameters $\leqslant p^{-n}$, disjoint each other, is said to be n-Packing family.*

Let $s \geqslant 0$ and $n \in \mathbb{N}$, and denote

$$P_n^s(E) = \sup \left\{ \sum_{j=1}^{+\infty} |U_j|_d^s : \{U_j\} \text{ is } n\text{-Packing family of } E \right\}.$$

Then $\text{pre}P^s(E) \equiv \lim_{n \to +\infty} P_n^s(E)$ *is said to be an s-**dimension pre-Packing outer measure** of E.*

It is easy to check: for $0 \leqslant s < t < +\infty$, s-dimension pre-Packing outer measure holds

$$\text{pre}P^s(E) < +\infty \Rightarrow \text{pre}P^t(E) = 0;$$

$$\text{pre}P^s(E) = +\infty \Rightarrow \text{pre}P^t(E) > 0.$$

Definition 5.1.7 (s-dimension Packing measure, pre-Packing dimension) *Let $E \in 2^{K_p}$ be a non-empty set in K_p. Then*

$$P^s(E) = \inf \left\{ \sum_{i=1}^{+\infty} \text{pre}P^s(E_i) : E = \bigcup_i E_i \right\}$$

*is said to be an s-**dimension Packing measure** of E. Simply, **Packing measure** of E. And*

$$\Delta(E) \equiv \text{pre}\dim_P(E) = \sup \{s : \text{pre}P^s(E) = +\infty\} = \inf \{s : \text{pre}P^s(E) = 0\}$$

*is said to be a **pre-Packing dimension** of E.*

2. Packing dimension

Definition 5.1.8 (Packing dimension) Let $E \in 2^{K_p}$ be a non-empty set in K_p. Then

$$\dim_P(E) = \inf\left\{\sup\left\{\operatorname{predim}_P(E_i) : E = \bigcup_i E_i\right\}\right\}$$

$$= \inf\left\{\sup\left\{\Delta(E_i) : E = \bigcup_i E_i\right\}\right\}$$

is said to be a **Packing dimension** of E.

Example 5.1.2 s-dimension pre-Packing outer measures of some sets in K_p.

Let $\tilde{S} = \{\beta^0, \beta^1, \cdots, \beta^k, \cdots\} \subset D$, and

$$S = \left\{A \subset \tilde{S} : A \text{ is a set of finite sum of elements in } \tilde{S}\right\}.$$

Take $s = 1$, then for the set $\{\beta^k\}$ with single element β^k, we have

$$\operatorname{pre}P^1(\{\beta^k\}) = 0, \quad k = 1, 2, \cdots.$$

Moreover, by the density of S in D, it follows that

$$\operatorname{pre}P^1(S) = 1.$$

Note that, the s-dimension pre-Packing outer measure is not outer measure, since by $\operatorname{pre}P^1(S) = 1$ and $\operatorname{pre}P^1(\{\beta^k\}) = 0$, we have

$$1 = \operatorname{pre}P^1(S) \geqslant \sum_{k=0}^{+\infty} \operatorname{pre}P^1(\{\beta^k\}) = 0,$$

this contradicts the property of the outer measure.

Example 5.1.3 The pre-Packing dimension $\operatorname{predim}_P(C_p^q)$ of the Cantor type set C_p^q.

We see that the n-covering of E in Theorem 5.1.5 is also an n-Packing of E, thus

$$\operatorname{predim}_P(C_p^q) = s = \frac{\ln q}{\ln p}.$$

Next, we show the properties of s-dimension pre-Packing outer measure $\operatorname{pre}P^s(E)$, s-dimension Packing measure $P^s(E)$, pre-Packing dimension $\operatorname{predim}_P(E)$ and the Packing dimension $\dim_P(E)$.

Theorem 5.1.9 Let $s \geqslant 0$. For a non-empty set $E \subset K_p$, it holds that

(i) $H^s(E) \leqslant P^s(E) \leqslant \mathrm{pre}P^s(E)$; $\dim_H(E) \leqslant \dim_P(E) \leqslant \mathrm{pre}\dim_P(E)$.

(ii) $\overline{\dim}_B E = \mathrm{predim}_P E$; If $\dim_B E$ exists, then $\dim_B E = \mathrm{predim}_P E$.

Proof. For (i): since the n-covering of E is its n-Packing family, for $\forall n \in \mathbb{N}$,

$$H_n^s(E) \leqslant P_n^s(E),$$

this implies $H^s(E) \leqslant \mathrm{pre}P^s(E)$ in (i). The second inequality $H^s(E) \leqslant P^s(E)$ in (i) can be proved as follows. By a property of Hausdorff measure

$$H^s(E) = \inf \left\{ \sum_i H^s(E_i) : E = \bigcup_i E_i \right\},$$

and by $H^s(E) \leqslant \mathrm{pre}P^s(E)$, we have

$$H^s(E) = \inf \left\{ \sum_i H^s(E_i) : E = \bigcup_i E_i \right\}$$

$$\leqslant \inf \left\{ \sum_i \mathrm{pre}P^s(E_i) : E = \bigcup_i E_i \right\} = P^s(E).$$

For (ii): let $s \geqslant 0$, by $M_n(E) p^{-ns} \leqslant P_n^s(E)$, we have

$$\limsup_{n \to +\infty} M_n(E) p^{-ns} \leqslant \mathrm{pre}P^s(E).$$

Thus,

$$s > \mathrm{predim}_P(E) \Rightarrow \mathrm{pre}P^s(E) = 0 \Rightarrow \limsup_{n \to +\infty} M_n(E) p^{-ns} = 0,$$

$$\Rightarrow s \geqslant \overline{\dim}_B E \Rightarrow \mathrm{predim}_P(E) \geqslant \overline{\dim}_B E.$$

Conversely, for $\mathrm{predim}_P(E) = 0$, it is clear that $0 = \mathrm{predim}_P(E) \leqslant \overline{\dim}_B E$; we only need to prove the inequality for $\mathrm{predim}_P(E) > 0$.

Let $\mathrm{predim}_P(E) > 0$, and take $0 < \alpha < s < \mathrm{predim}_P(E)$, we define a nature number subsequence $\{n_j\}$:

$s < \mathrm{predim}_P(E)$,

$\Rightarrow \mathrm{pre}P^s(E) = +\infty$,

$\Rightarrow$ take n_0, s.t. $P_{n_0}^s(E) = \sup \left\{ \sum_{j=1}^{+\infty} |U_j|_d^s : \{U_j\} \text{ is } n_0\text{-Packing of } E \right\} > 1$,

$\to$ if n_j has been defined, choose n_j-family $\{U_j\}$, s.t. $\sum |U_j|_d^s \geqslant 1$,

$\Rightarrow$ let $j_k = \#\{U_j : |U_j|_d = p^{-k}\}$ be the numbers of U_j with diameter p^{-k}, then

$$\sum_{k=0}^{+\infty} p^{-ks} j_k = \sum |U_j|_d^s \geqslant 1,$$

$\Rightarrow \exists k \in \mathbb{N} \cup \{0\}$, s.t. $p^{-ks} j_k \geqslant p^{-k\alpha} \left(1 - p^{-\alpha}\right)$ $\left(\text{otherwise}, \sum_{k \geqslant 0} p^{-ks} j_k < 1\right)$,

$\Rightarrow n_{j+1} = k + 1$ is determined, so does the natural number subsequence $\{n_j\}$,

$\Rightarrow \forall n_j$, define $\{V_i : 1 \leqslant i \leqslant j_k\}$, s.t. $|V_i|_d = p^{-k-1}, j_k \geqslant p^{k(s-\alpha)} \left(1 - p^{-\alpha}\right)$,

$\Rightarrow M_{k+1}(E) \geqslant j_k \geqslant p^{k(s-\alpha)} \left(1 - p^{-\alpha}\right) \Rightarrow \overline{\dim}_B E \geqslant s - \alpha$,

$\Rightarrow \text{predim}_P(E) \leqslant \overline{\dim}_B E \Rightarrow$ the proof is complete.

The Packing measure is a kind of outer measure, it has some similar properties as the Hausdorff measure, such as monotonicity, countable stability (see Theorem 5.1.3 (i), (ii)).

Example 5.1.4 Packing dimension $\dim_P \left(C_p^q\right)$ of Cantor type set C_p^q. We have by Theorem 5.1.9 and Example 5.1.3,
$$\dim_P \left(C_p^q\right) = \frac{\ln q}{\ln p}.$$

3. Relationship between dimensions

Summarize as follows:
$$\dim_H(E) \leqslant \underline{\dim}_B(E) \leqslant \overline{\dim}_B(E) = \text{pre}\dim_P(E);$$
$$\dim_H(E) \leqslant \dim_P(E) \leqslant \overline{\dim}_B(E) = \text{pre}\dim_P(E).$$

We concern that whether there are examples to show the inequality $<$ holds? The example in Section 5.1.2 shows $\underline{\dim}_B(C) < \overline{\dim}_B(C)$ for the Cantor type set C. The following example shows that $\dim_H(E) < \dim_B(E)$.

Example 5.1.5 For $\tilde{S}$ and S in Example 5.1.2, then
$$\dim_H S = 0 < 1 = \dim_B S.$$
In fact, $\dim_H S = 0$ since S is countable. By $\overline{S} = D$, then
$$\dim_B S = \dim_B \overline{S} = \dim_B D = 1.$$

There are many open problems about measures and dimensions on local fields, such as, one may consider or define: similarity dimension, Fourier dimension, spectrum dimension, capacity dimension, information dimension, Liapunov dimension, and corresponding measures, and so on. Moreover, one can study the mathematical properties, physics senses and applications of various kinds of measures and dimensions on local fields.

Exercises

1. Prove Theorem 5.1.3.
2. Study the invariant properties of various dimensions under the transformations: translation, Lip transform, self-similar transform, affine transformation.
3. Study the properties of the Hausdorff dimensions of open sets, closed sets, compact sets in local fields.
4. If $E \subset K_p$ with positive Haar measure $|E| > 0$, how about its Hausdorff measure, and its Hausdorff dimension?
5. For $\dim_H(E) \leqslant \underline{\dim}_B(E) \leqslant \overline{\dim}_B(E) = \operatorname{pre dim}_P(E)$, find or construct some sets such that $<$ holds, and some sets that $=$ holds.
6. For $\dim_H(E) \leqslant \dim_P(E) \leqslant \overline{\dim}_B(E) = \operatorname{pre dim}_P(E)$, find or construct some sets such that $<$ holds, and some sets that $=$ holds.

5.2 Analytic expressions of dimensions of sets in local fields

The main aim of this section is to establish analytic expressions of the Hausdorrf dimensions and Fourier dimensions of sets in local fields[59],[102].

5.2.1 Borel measure and Borel measurable sets

Based upon Section 5.1, we study Borel sets and Borel set class in $\mathfrak{B}(K_p)$ on a local field K_p, as well as corresponding measures on $\mathfrak{B}(K_p)$ and measurable sets.

Let $\mu : \mathfrak{B}(K_p) \to [0, +\infty)$ be a Borel measure on K_p, and μ is in the Borel measure set

$$\mathfrak{N}(K_p) = \{\mu : \mu \text{ a Borel measure on } K_p\}.$$

Denoted by $(K_p, \mathfrak{B}(K_p), \mu)$ the corresponding measure space, $\mu \in \mathfrak{N}(K_p)$.

The Haar measure of subset $E \subset K_p$ is denoted by $|E|$, $0 \leqslant |E| \leqslant \infty$. Then we regard $(K_p, \mathfrak{B}(K_p), \mu(E) = |E|)$ as a special measure space $(K_p, \mathfrak{B}(K_p), \mu)$, and for the fractal space $(\mathbb{K}(K_p), h)$, it holds $\mathbb{K}(K_p) \subset \mathfrak{B}(K_p)$.

The Borel measures of compact sets and open sets holds
(i) For compact set $E \in \mathbb{K}(K_p) \Rightarrow 0 \leqslant \mu(E) < +\infty$.
(ii) For an open set $E \in \mathfrak{B}(K_p), E \neq \varnothing \Rightarrow 0 < \mu(E) < +\infty$.

Definition 5.2.1 (support of a Borel measure) Let $\mu \in \mathfrak{N}(K_p)$ be a Borel measure on K_p. Then

$$\operatorname{supp} \mu = K_p \setminus \left\{ \bigcup_j U_j : \mu|_{U_j}(E) = 0, U_j \subset K_p \text{ is open}, \forall E \subset K_p \right\}$$

is said to be the **support of Borel measure** μ.

By definition, the support of Borel measure μ is a closed set.

5.2.2 Distribution dimension

The s-dimension Hausdorff measure and Hausdorff dimension for a non-empty set E in a local field K_p have been defined in the Section 5.1 by virtue of the point of view of geometry in covering, however, in this section, we will define so called the distribution dimension which is an equivalent one of Hausdorff dimension as an analytic definition, and this new dimension will play important role in the fractal analysis.

1. Distribution dimension

Let $\mathbb{K}(K_p)$ be the fractal space on K_p. The following lemma is a key one.

Lemma 5.2.1 *Let $E \subset K_p$ be a non-empty Borel set, and $H^d(E) = +\infty$ for some $0 < d < +\infty$. Then there exist a compact set $F \in \mathbb{K}(K_p)$ with $F \subset E$ and $0 < H^d(F) < +\infty$, and a constant $b > 0$, such that $\forall l \in \mathbb{Z}^+$, $i \in \mathbb{Z}$,*

$$H^d\left(F \cap B^{l,i}\right) \leqslant b p^{-id},$$

where $B^{l,i} = z_{l,i} + B^i$, $z_{l,i} \in K_p$, and $B^{l,i} \cap B^{k,i} = \varnothing, k \neq l$.

The proof of Lemma is similar to that of the classical case, we refer to [13] and [40].

Definition 5.2.2 (Space $B^{s,\Theta}_{rt}(K_p)$) *Let $\Theta \subset K_p$ be a non-empty closed set with Haar measure $|\Theta| = 0$. For $s \in \mathbb{R}$, $0 < r,t \leqslant +\infty$, the subspace of B-type space $B^s_{rt}(K_p)$ is defined by*

$$B^{s,\Theta}_{rt}(K_p) = \{ f \in B^s_{rt}(K_p) : \langle f, \varphi \rangle = 0, \forall \varphi \in \mathbb{S}(K_p), \varphi|_\Theta = 0 \},$$

where $\varphi|_\Theta$ is the restriction of φ on Θ.

Specially, when $s = \sigma, r = t = \infty$, we have the Hölder type space $C^\sigma(K_p) = B^\sigma_{\infty\infty}(K_p)$ and subspace $C^{\sigma,\Theta}(K_p) = B^{\sigma,\Theta}_{\infty\infty}(K_p)$.

Lemma 5.2.2 *If $s \leqslant \sigma_r = \left(\dfrac{1}{r} - 1\right)_+$, $0 < r,t \leqslant +\infty$, then the space*

$B_{rt}^{s,\Theta}(K_p)$ consists of distributions in $\mathbb{S}^*(K_p)$.

Proof. For $0 < r, t \leqslant +\infty$, $s > \sigma_r = \left(\dfrac{1}{r} - 1\right)_+$, we have $B_{rt}^s(K_p) \subset L_{\text{loc}}^1(K_p)$ by Theorem 4.1.3 (v), so that $B_{rt}^{s,\Theta}(K_p)$ consists of regular distributions. Thus, for closed set $\Theta \subset K_p$ with $|\Theta| = 0$, if $0 < r, t \leqslant +\infty$, $s > \sigma_r$, then it holds

$$\operatorname{supp} f \subset \Theta, \quad \forall f \in B_{rt}^{s,\Theta}(K_p).$$

This implies $B_{rt}^{s,\Theta}(K_p) = \{0\}$. In other words, if $0 < r, t \leqslant +\infty$, $s \leqslant \sigma_r$, then $B_{rt}^{s,\Theta}(K_p)$ consists of singular distributions.

Definition 5.2.3 (Distribution dimension) Let $E \subset K_p$ be a nonempty Borel set with Haar measure $|E| = 0$. The distribution dimension of E is defined as

$$\dim_D E = \sup\left\{d : C^{-1+d,\Theta}(K_p) \neq \{0\} \text{ for some } \Theta \in \mathbb{K}(K_p), \Theta \subset E\right\},$$

where $C^{-1+d,\Theta}(K_p) \neq \{0\}$ means that the space $C^{-1+d,\Theta}(K_p)$ is nontrivial.

Theorem 5.2.1 The distribution dimension $\dim_D E$ has the following properties:

(i) $E_1 \subset E_2 \Rightarrow \dim_D E_1 \leqslant \dim_D E_2$.

(ii) If $E \in \mathbb{K}(K_p)$ is a compact set, then

$$\dim_D E = \sup\{d : C^{-1+d,E}(K_p) \neq \{0\}\}.$$

(iii) If $E \subset K_p$, $|E| = 0$, then

$$\dim_D E = \sup\{\dim_D \Theta : \Theta \in \mathbb{K}(K_p), \Theta \subset E\}.$$

(iv) $0 \leqslant \dim_D E \leqslant 1$.

Proof. We only prove for (iv).

If $-1 + d > 0$, for a compact set $\Theta \in \mathbb{K}(K_p)$, then $C^{-1+d,\Theta}(K_p)$ becomes trivial, i.e., $C^{-1+d,\Theta}(K_p) = B_{\infty\infty}^{-1+d,\Theta}(K_p) = \{0\}$. However, $C^{-1+d,\Theta}(K_p) \neq \{0\}$ implies $-1 + d \leqslant 0$, this is $d \leqslant 1$.

To prove $\dim_D E \geqslant 0$, we show that Dirac distribution $\delta \in C^{-1}(K_p)$. Because $s = -1$, $r = t = \infty$ with $s < \sigma_r = \left(\dfrac{1}{r} - 1\right)_+ = 0$, thus, if $\{0\} = \Theta \subset E$, then singular distribution δ must be in $C^{-1}(K_p) =$

$B_{\infty\infty}^{-1}(K_p)$. This implies $C^{-1,\Theta}(K_p) \neq \{0\}$. But the single point set $\{0\}$ can be replaced by any point in E, thus

$\dim_E = \sup\{d : C^{-1+d,\Theta}(K_p) \neq \{0\}$ for some set $\Theta \in \mathbb{K}(K_p), \Theta \subset E\} \geq 0$.

Then, (iv) is proved.

The sense of the distribution dimension is: it is an analytic expression of Hausdorff dimension. To prove this assertion, we need two lemmas.

Lemma 5.2.3 Let $\Theta \subset K_p$ be a non-empty Borel set with Haar measure $|\Theta| = 0$. Then

(i) $\forall s \leq 0$, $C^{s,\Theta}(K_p) \neq \{0\} \Leftrightarrow \{\varphi \in \mathbb{S}(K_p) : \varphi|_\Theta = 0\}$ is not dense in $B_{11}^{-s}(K_p)$;

(ii) $\forall s \leq 0, 1 < r \leq +\infty$, $B_{rr}^{s,\Theta}(K_p) \neq \{0\} \Leftrightarrow \{\varphi \in \mathbb{S}(K_p) : \varphi|_\Theta = 0\}$ is not dense in $B_{r'r'}^{-s}(K_p)$.

Proof. For (i): by $\left(B_{11}^{-s}(K_p)\right)^* = B_{\infty\infty}^s(K_p) = C^s(K_p)$ and by the Hahn–Banach Theorem, there exists non-zero element $f \in C^{s,\Theta}(K_p)$, so that $C^{s,\Theta}(K_p) \neq \{0\}$, hence

$$\overline{\{\varphi \in \mathbb{S}(K_p) : \varphi|_\Theta = 0\}} \subsetneqq B_{11}^{-s}(K_p).$$

Conversely, if there exists a non-zero element $f \in C^{s,\Theta}(K_p)$, then it must follow

$$\overline{\{\varphi \in \mathbb{S}(K_p) : \varphi|_\Theta = 0\}} \subsetneqq B_{11}^{-s}(K_p),$$

otherwise, it will lead to contradiction. Then, (i) is proved.

The proof of (ii) can be obtained by $\left(B_{r'r'}^{-s}(K_p)\right)^* = B_{rr}^s(K_p)$, similarly.

Lemma 5.2.4 Let $\Theta \subset K_p$ be a non-empty compact set with the Hausdorff dimension $\dim_H \Theta < 1$. Then, for $\forall \rho$, with $\dim_H \Theta < \rho < 1$, it holds

(i) $\{\varphi \in \mathbb{S}(K_p) : \varphi|_\Theta = 0\}$ is dense in $B_{11}^{-\rho+1}(K_p)$;

(ii) $\{\varphi \in \mathbb{S}(K_p) : \varphi|_\Theta = 0\}$ is dense in $B_{r'r'}^{\frac{-\rho+1}{r'}}(K_p), 1 < r \leq +\infty$.

Proof. For (i), by three steps.

① Without lose of generality, we suppose that $\Theta \subset D$. By the compactness of Θ, for each $\eta > 0$ and $i \in \mathbb{N}$, there exists a finite ball covering $\{U_j\}$ of Θ with diameters $|U_j|_d \leq p^{-i}$, $j = 1, 2, \cdots, N$, such that $\sum_{j=1}^{N} |U_j|_d^\rho < \eta$. Thus, there exist N positive integers $k_1, k_2, \cdots, k_N$, $k_j \geq i$, such that $U_j \cap D = x_j + B^{k_j}$ with diameter $\left|B^{k_j}\right|_d = p^{-k_j} < p^{-i}$.

② For the characteristic function $\Phi_0(x) = \begin{cases} 1, & x \in D \\ 0, & x \notin D \end{cases}$ of $D = \{x \in K_p : |x| \leq 1\}$, and $\forall \varepsilon > 0$, there exists a function $\varphi \in \mathbb{S}(K_p)$ with $\varphi|_\Theta = 0$, such that

$$\|\varphi - \Phi_0\|_{B_{11}^{-\rho+1}(K_p)} < \varepsilon.$$

In fact, let $\varphi(x) = \left(1 - \sum_{j=1}^{N} \Phi_{U_j}(x)\right)\Phi_0(x)$, Φ_{U_j} be the characteristic function of U_j. It is clear that $\varphi|_\Theta = 0$ and $\varphi \in \mathbb{S}(K_p)$. So that $\varphi - \Phi_0 = \sum_{j=1}^{N} \Phi_{U_j}\Phi_0$.

To estimate the norm, by virtue of equality

$$\Phi_{x_j+B^{k_j}} * \varphi_k(x) = \begin{cases} 0, & k > k_j, \\ p^{k-k_j}\Phi_{x_j+B^k}(x) - p^{k-1-k_j}\Phi_{x_j+B^{k-1}}(x), & 1 \leq k \leq k_j, \\ p^{-k_j}\Phi_{x_j+D}(x), & k = 0, \end{cases}$$

with $\varphi_k = p^k\Phi_{B^k} - p^{k-1}\Phi_{B^{k-1}}$, $k = 1, 2, \cdots$, and $\varphi_0 = \Phi_0$, it follows that

$$\|\varphi - \Phi_0\|_{B_{11}^{-\rho+1}(K_p)} = \left\|\sum_{j=1}^{N}\Phi_{U_j}\Phi_0\right\|_{B_{11}^{-\rho+1}(K_p)} = \left\|\sum_{j=1}^{N}\Phi_{U_j \cap D}\right\|_{B_{11}^{-\rho+1}(K_p)}$$

$$= \left\|\sum_{j=1}^{N}\Phi_{x_j+B^{k_j}}\right\|_{B_{11}^{-\rho+1}(K_p)}$$

$$\leq \sum_{j=1}^{N}\left(\sum_{k=0}^{+\infty} p^{(-\rho+1)k}\int_{K_p}\left|\Phi_{x_j+B^{k_j}} * \varphi_k(x)\right|dx\right)$$

$$\leq \sum_{j=1}^{N}\left(p^{-k_j} + 2\sum_{k=1}^{k_j} p^{(-\rho+1)k}p^{-k_j}\right)$$

$$\leq c\sum_{j=1}^{N}\left(p^{-k_j}p^{(-\rho+1)k_j}\right)$$

$$= c\sum_{j=1}^{N} p^{-\rho k_j} \leq c\eta,$$

where the constant c is independent of η.

③ Prove $\overline{\{\varphi \in \mathbb{S}(K_p) : \varphi|_\Theta = 0\}} = B_{11}^{-\rho+1}(K_p)$.

Since $\mathbb{S}(K_p)$ is dense in $B_{11}^{-\rho+1}(K_p)$, we only need to prove: $\forall g \in \mathbb{S}(K_p)$, $\forall \varepsilon > 0$, there exists a function $\varphi \in \mathbb{S}(K_p)$ with $\varphi|_\Theta = 0$, such that $\|g - \varphi\|_{B_{11}^{-\rho+1}(K_p)} < \varepsilon$.

In fact, $\forall \varphi \in \mathbb{S}(K_p)$ is a combination of translations and dilations of the characteristic function of D, thus, by ②, it follows that
$$\|\varphi(\cdot - h) - \Phi_0(\cdot - h)\|_{B_{11}^{-\rho+1}(K_p)} < \varepsilon, \quad \forall h \in K_p$$
and
$$\|\varphi(h\cdot) - \Phi_0(h\cdot)\|_{B_{11}^{-\rho+1}(K_p)} < \varepsilon, \quad \forall h \in K_p.$$
Then, this implies ③, and thus (i) holds.

For (ii), also by three steps.

① is similar to that of ① in (i).

② Prove: for $\Phi_0(x)$, and $1 < r \leqslant +\infty$, $\forall \varepsilon > 0$, there exists $\varphi \in \mathbb{S}(K_p)$, such that $\|\varphi - \Phi_0\|_{B_{r',r'}^{\frac{-\rho+1}{r'}}(K_p)} < \varepsilon$.

In fact, $1 < r \leqslant +\infty$ implies $r' > 1$, thus there exists an unique $l \in \mathbb{N}$, such that $l < r' \leqslant l+1$. Moreover, for a ball covering $\{U_j\}_{j=1}^{N}$ of compact set $\Theta \subset K_p$, there exists balls V_j contained in U_j, such that $|V_j|_d = |U_j|_d^{l+1}$. Thus, we have $V_j \cap D = x_j + B^{k_j(l+1)}$.

Let $\varphi(x) = \left(1 - \sum_{j=1}^{N} \Phi_{V_j}(x)\right) \Phi_0(x)$, then

$$\|\varphi - \Phi_0\|_{B_{r',r'}^{\frac{-\rho+1}{r'}}(K_p)}$$

$$= \left\|\sum_{j=1}^{N} \xi_{V_j} \Phi_0\right\|_{B_{r',r'}^{\frac{-\rho+1}{r'}}(K_p)} = \left\|\sum_{j=1}^{N} \Phi_{x_j + B^{k_j(l+1)}}\right\|_{B_{r',r'}^{\frac{-\rho+1}{r'}}(K_p)}$$

$$\leqslant \sum_{j=1}^{N} \left(\sum_{k=0}^{+\infty} p^{(-\rho+1)k} \int_{K_p} \left|\Phi_{x_j + B^{k_j(l+1)}} * \varphi_k(x)\right|^{r'} dx\right)^{\frac{1}{r'}}$$

$$\leqslant \sum_{j=1}^{N} \left(p^{-k_j(l+1)r'} + c_1 \sum_{k=1}^{k_j(l+1)} p^{(-\rho+1)k} p^{(k-k_j(l+1))r'} p^{-k}\right)^{\frac{1}{r'}}$$

$$\leqslant c_2 \sum_{j=1}^{N} \left(p^{-k_j(l+1)r'} p^{(r'-\rho)k_j(l+1)}\right)^{\frac{1}{r'}} \leqslant c_2 \sum_{j=1}^{N} p^{\frac{-k_j(l+1)\rho}{l+1}} \leqslant c_3 \eta,$$

where constants c_1, c_2, c_3 are independent of η.

③ The proof is similar to that of ③ in (i), and thus
$$\overline{\{\varphi \in \mathbb{S}(K_p) : \varphi|_{\Theta} = 0\}} = B_{r',r'}^{\frac{-\rho+1}{r'}}(K_p).$$
The proof is complete.

2. Relationship between distribution dimension and Hausdorff dimension

Theorem 5.2.2 Let $E \subset K_p$ be a non-empty Borel set, and the Haar measure $|E| = 0$. Then
$$\dim_D E = \dim_H E. \tag{5.2.1}$$

Proof. By three steps.

First step. Prove: if (5.2.1) holds for each compact set $\Theta \in \mathbb{K}(K_p), \Theta \subset E$, then it holds for every Borel set E with $|E| = 0$.

In fact, $\forall \Theta \in \mathbb{K}(K_p)$ compact, $\Theta \subset E$, (5.2.1) holds, $\dim_D \Theta = \dim_H \Theta$. Then by Theorem 5.2.1(ii), from $\dim_D \Theta = \sup\{d : C^{-1+d,\Theta}(K_p) \neq \{0\}\}$, we get
$$\dim_H \Theta = \sup\{d : C^{-1+d,\Theta}(K_p) \neq \{0\}\}.$$

On the other hand, by Theorem 5.1.3 (iv), for the Hausdorff dimension, it follows
$$\dim_H E = \sup\{\dim_H \Theta : \Theta \in \mathbb{K}(K_p), \Theta \subset E\}. \tag{5.2.2}$$

Moreover, by Theorem 5.2.1 (iii), if $E \subset K_p$ with $|E| = 0$, then
$$\dim_D E = \sup\{\dim_D \Theta : \Theta \in \mathbb{K}(K_p), \Theta \subset E\}. \tag{5.2.3}$$

By the assumption, (5.2.1) holds for compact set, that is
$$\dim_D \Theta = \dim_H \Theta, \quad \forall \Theta \in \mathbb{K}(K_p), \Theta \subset E,$$
so that $\dim_H \Theta$ in (5.2.2) and $\dim_H \Theta$ in (5.2.3) are equal for $\Theta \in \mathbb{K}(K_p)$, $\Theta \subset E$. Thus, the proof of first step is complete.

Second step. Prove: for non-empty compact set $\Theta \in \mathbb{K}(K_p), \Theta \subset E$ with $|\Theta| = 0$, then it holds $\dim_H \Theta \leqslant \dim_D \Theta$.

In fact, if $\dim_H \Theta = 0$, then by Theorem 5.2.1 (iv), $0 \leqslant \dim_D \Theta \leqslant 1$, thus
$$\dim_H \Theta = 0 \leqslant \dim_D \Theta.$$

If $\dim_H \Theta > 0$, we prove $\dim_H \Theta \leqslant \dim_D \Theta$ as follows.

$\dim_H \Theta > 0 \Rightarrow \forall \rho > 0$, then $0 < \rho < \dim_H \Theta$ implies $H^\rho(\Theta) = +\infty$

$\Rightarrow \exists \Lambda \in \mathbb{K}(K_p), \Lambda \subset \Theta$, and $\exists b > 0$, s.t. $0 < H^\rho(\Lambda) < +\infty$,

so for $\forall l \in \mathbb{Z}^+$ and $i \in \mathbb{Z}$, holds $H^\rho\left(\Lambda \cap B^{l,i}\right) \leqslant bq^{-i\rho}$

$\Rightarrow$ define a distribution $f \in \mathbb{S}^*(K_p)$ by the functional expression

$$\langle f, \varphi \rangle = \int_\Lambda \varphi(x) dH^\rho(x), \quad \forall \varphi \in \mathbb{S}(K_p). \tag{5.2.4}$$

(Note, for a compact set Λ with $0 < H^\rho(\Lambda) < +\infty$, the distribution $f \in \mathbb{S}^*(K_p)$ in (5.2.4) satisfies

① $\varphi|_\Theta = 0$ implies $\langle f, \varphi \rangle = 0$;

② $f \in C^{-1+\delta}(K_p)$; and $f \in \mathbb{S}^*(K_p)$ has the L-P decomposition

$$f = \sum_{j=0}^{+\infty} f * \varphi_j, \quad \varphi_j = \Delta_j - \Delta_{j-1}, \Delta_j(x) = p^j \Phi_j(x), \Delta_{-1}(x) = 0.$$

③ $\forall \varphi \in \mathbb{S}(K_p) \Rightarrow f * \varphi(x) = \langle f, \varphi(x - \cdot) \rangle$ and

$$(f * \varphi_j)(x) = \langle f, \varphi_j(x - \cdot) \rangle = \int_\Lambda \varphi_j(x - \lambda) dH^\rho(\lambda)$$

$$\leqslant c p^{jn} p^{-j\rho} = c p^{j(n-\rho)}, \quad \forall j \in \mathbb{Z}^+,$$

c is independent of j (by Theorem 3.1.25),

$$\Rightarrow \|f\|_{C^{-1+\rho}(K_p)} = \sup_j \left(p^{j(-1+\rho)} \sup_x |(f * \varphi_j)(x)| \right)$$

$$\leqslant c \sup_j \left(p^{j(-1+\rho)} p^{j(1-\rho)} \right) = c,$$

$\Rightarrow f \in C^{-1+\rho,\Theta}(K_p) \quad (f \neq 0$ by the construction of $f)$,

$\Rightarrow \dim_D \Theta = \sup\{d : C^{-1+d,\Theta}(K_p) \neq 0\} \geqslant \rho$,

$\Rightarrow \dim_H \Theta \leqslant \dim_D \Theta$.

The proof of second step is complete.

Third step. Prove: for a non-empty compact set $\Theta \in \mathbb{K}(K_p), \Theta \subset E$, $|\Theta| = 0$, it holds $\dim_H \Theta \geqslant \dim_D \Theta$.

In fact, if $\dim_H \Theta = 1$, by Theorem 5.2.1 (iv), $0 \leqslant \dim_D \Theta \leqslant 1$, thus

$$0 \leqslant \dim_D \Theta \leqslant 1 = \dim_H \Theta.$$

If $\dim_H \Theta < 1$, we prove $\dim_D \Theta \leqslant \dim_H \Theta$ as follows.

$\dim_H \Theta < 1 \Rightarrow \forall \rho > 0$, then $\dim_H \Theta < \rho < 1$ implies $H^\rho(\Theta) = 0$,

$$\Rightarrow \overline{\{\varphi \in \mathbb{S}(K_p) : \varphi|_\Theta = 0\}} = B_{11}^{1-\rho}(K_p), \text{ Lemma 5.2.4 (i)}$$
$$\Rightarrow C^{-1+\rho}(K_p) = \{0\}, \text{ Lemma 5.2.3 (i)}$$
$$\Rightarrow \dim_D \Theta \leqslant \rho,$$
$$\Rightarrow \dim_D \Theta \leqslant \dim_H \Theta.$$

The proof of the third step is complete, and so is for the Theorem. There is the other analytic expression of Hausdorff dimension.

Theorem 5.2.3 *Let $E \subset K_p$ be a non-empty Borel set, with Haar measure $|E| = 0$, if $1 < r \leqslant +\infty, 0 < t \leqslant +\infty$. Then*

$$\dim_H E = \sup\left\{d : B_{rt}^{\frac{-1+d}{r'},\Theta}(K_p) \neq \{0\}, \text{ for some } \Theta \in \mathbb{K}(K_p), \Theta \subset E\right\}. \tag{5.2.5}$$

Proof. By three steps.

First step. Prove: if (5.2.5) holds for each compact set $\Theta \in \mathbb{K}(K_p)$, $\Theta \subset E$, then it holds for every Borel set E with $|E| = 0$.

In fact, if (5.2.5) holds for each compact set $\Theta \in \mathbb{K}(K_p), \Theta \subset E$, then

$$\sup\left\{d : B_{rt}^{\frac{-1+d}{r'},\Theta}(K_p) \neq \{0\}, \text{ for some } \Lambda \in \mathbb{K}(K_p), \Lambda \subset \Theta\right\}$$
$$= \sup\left\{d : B_{rt}^{\frac{-1+d}{r'},\Theta}(K_p) \neq \{0\}\right\}.$$

Thus, by $\dim_H E = \sup\{\dim_H \Theta : \Theta \in \mathbb{K}(K_p), \Theta \subset E\}$, thus, (5.2.5) holds for each Borel set E with $|E| = 0$.

Second step. Prove: for each non-empty compact $\Theta \in \mathbb{K}(K_p), |\Theta| = 0$, it holds

$$\dim_H \Theta \leqslant \sup\left\{d : B_{rt}^{\frac{-1+d}{r'},\Theta}(K_p) \neq \{0\}\right\}.$$

In fact, we only need to prove the inequality

$$\dim_H \Theta \leqslant \sup\left\{d : B_{rt}^{\frac{-1+d}{r'},\Theta}(K_p) \neq \{0\}\right\} \tag{5.2.6}$$

holds for non-empty compact set Θ with $\dim_H \Theta > 0$.

For ρ with $0 < \rho < \dim_H \Theta$, $\exists \Lambda \in \mathbb{K}(K_p), \Lambda \subset \Theta$, and $\exists b > 0$, s.t.

$$0 < H^\rho(\Lambda) < +\infty;$$

and $\forall l \in \mathbb{Z}^+, i \in \mathbb{Z}$, it holds

$$H^\rho\left(\Lambda \cap B^{l,i}\right) \leqslant bq^{-i\rho}.$$

Define a distribution $f \in \mathbb{S}^*(K_p)$ by the functional expression as in (5.2.4)

$$\langle f, \varphi \rangle = \int_\Lambda \varphi(x) dH^\rho, \quad \forall \varphi \in \mathbb{S}(K_p).$$

It is clear that, $f \in \mathbb{S}^*(K_p)$ satisfies
① $0 < H^\rho(\Lambda) < +\infty$; ② $\varphi|_\Theta = 0$ implies $\langle f, \varphi \rangle = 0$.
Then, we deduce

$$\|f\|_{B_{r\infty}^{-\frac{1-\rho}{r'}}}$$

$$= \sup_j \left(p^{j(-\frac{1-\rho}{r'})} \|f * \varphi_j\|_{L^r(K_p)} \right)$$

$$= \sup_j \left(p^{j(-\frac{1-\rho}{r'})} \left\| \int_\Lambda \varphi_j(x-\lambda) dH^\rho(\lambda) \right\|_{L^r(K_p)} \right)$$

$$= \sup_j \left(p^{j(-\frac{1-\rho}{r'})} \left\| \left(\int_\Lambda \varphi_j(x-\lambda) dH^\rho(\lambda) \right)^{\frac{1}{r}} \left(\int_\Lambda \varphi_j(x-\lambda) dH^\rho(\lambda) \right)^{\frac{1}{r'}} \right\|_{L^r(K_p)} \right)$$

$$\leqslant c_1 \sup_j \left(p^{j(-\frac{1-\rho}{r'})} p^{j(\frac{1-\rho}{r'})} \left\| \left(\int_\Lambda \varphi_j(x-\lambda) dH^\rho(\lambda) \right)^{\frac{1}{r}} \right\|_{L^r(K_p)} \right)$$

$$= c_1 \sup_j \left(\int_{K_p} \int_\Lambda \varphi_j(x-\lambda) dH^\rho(\lambda) dx \right)^{\frac{1}{r}}$$

$$= c_1 \sup_j \left(\int_\Lambda \int_{K_p} \varphi_j(x-\lambda) dx dH^\rho(\lambda) \right)^{\frac{1}{r}} = c,$$

where c_1, c are independent of j. Thus, $f \in B_{r\infty}^{-\frac{1-\rho}{r'},\Theta}(K_p)$. Moreover, f is non-zero, so that (5.2.6) holds for $t = \infty$.

By the Proposition 2.2.1 in [139], the above estimation of $\|f\|_{B_{r\infty}^{-\frac{1-\rho}{r'}}}$ holds for $0 < t < +\infty$, thus (5.2.6) holds for $0 < t < +\infty$. The proof of second step is complete.

Third step. Prove: for non-empty compact set $\Theta \in \mathbb{K}(K_p), \Theta \subset E$ with $|\Theta| = 0$,

$$\dim_H \Theta \geqslant \sup \left\{ d : B_{rt}^{-\frac{-1+d}{r'},\Theta}(K_p) \neq \{0\} \right\}. \tag{5.2.7}$$

Similarly, we only need to prove (5.2.7) for $\dim_H \Theta < 1$.

If $\dim_H \Theta < \rho < 1$, we have $H^\rho(\Theta) = 0$. By Lemma 5.2.4 (ii), it follows
$$\overline{\{\varphi \in \mathbb{S}(K_p) : \varphi|_\Theta = 0\}} = B_{r'r'}^{\frac{1-\rho}{r'}}(K_p),$$
thus, for $1 < r \leqslant +\infty$, holds $B_{rr}^{\frac{-1+\rho}{r}}(K_p) = \{0\}$, so (5.2.7) holds for $r = t$. Then by the Proposition 2.2.1 in [139], (5.2.7) holds for $1 < r \leqslant +\infty, 0 < t \leqslant \infty$.

The proof is complete.

Theorem 5.2.4 *Let $E \subset K_p$ be a non-empty Borel set, with Haar measure $|E| = 0$, if $1 < r \leqslant +\infty, 0 < t \leqslant +\infty$. Then*

$$\dim_D E = \dim_H E$$
$$= \sup\left\{d : F_{rt}^{\frac{-1+d}{r'},\Theta}(K_p) \neq \{0\}, \text{ for some } \Theta \in \mathbb{K}(K_p), \Theta \subset E\right\}. \quad (5.2.8)$$

Proof. By the Proposition 2.2.1 in [139], for $s \in \mathbb{R}, 0 < r < +\infty, 0 < t \leqslant +\infty$, holds
$$B_{r\min(r,t)}^s(K_p) \subset F_{rt}^s(K_p) \subset B_{r\max(r,t)}^s(K_p),$$
thus, by Theorem 5.2.3, we get the proof.

5.2.3 Fourier dimension

In the Section 5.2.2, we discuss the analytic expressions of Hausdorff dimension for non-empty sets in local fields. We now turn to consider the analytic expressions for the Fourier dimension.

1. Analytic definition of Fourier dimension

Definition 5.2.4 (Fourier dimension) *Let $E \subset K_p$ be a non-empty Borel set with $|E| = 0$, the Fourier dimension of E is defined as*

$$\dim_F E = \sup\left\{d : \exists \text{ Radon measure } \mu \neq 0, \text{with (i) compact supp } \mu \subset E;\right.$$
$$\left.\text{(ii) } \mu(K_p) < +\infty; \text{ (iii) } |\mu^\wedge(\xi)| \leqslant |\xi|^{-\frac{d}{2}}, 0 \neq \xi \in K_p\right\}.$$

2. Relationship between Fourier dimension and Hausdorff dimension

Theorem 5.2.5 *For a non-empty Borel set $E \subset K_p$, with $|E| = 0$, it holds*
$$\dim_F E \leqslant \dim_H E.$$

Proof. Let $\dim_F E > 0$. Take $0 < d < \dim_F E$, then there exists a Radon measure $\mu \neq 0$, such that

(i) $\operatorname{supp}\mu \in \mathbb{K}(K_p)$, $\operatorname{supp}\mu \subset E$;

(ii) $\mu(K_p) < +\infty$;

(iii) $|\mu^{\wedge}(\xi)| \leqslant |\xi|^{-\frac{d}{2}}$, $0 \neq \xi \in K_p$.

Since $\mu \in \mathbb{S}^*(K_p)$, we may define a linear fractional as for $\Lambda \in \mathbb{K}(K_p)$, then $\varphi|_\Lambda = 0$ implies $\langle \mu, \varphi \rangle = 0$ for each $\varphi \in \mathbb{S}(K_p)$ that

$$\langle \mu, \varphi \rangle = \int_{K_p} \varphi(x) d\mu(x). \tag{5.2.9}$$

Thus,
$$\|\mu\|_{B_{2\infty}^{-\frac{1-d}{2}}(K_p)} = \sup_j \left(p^{-\frac{1-d}{2}j} \|\mu * \varphi_j\|_{L^2(K_p)} \right)$$
$$= \sup_j \left(p^{-\frac{1-d}{2}j} \|\mu^{\wedge} \varphi_j^{\wedge}\|_{L^2(\Gamma_p)} \right)$$
$$\leqslant \sup_j \left(p^{-\frac{1-d}{2}j} \left\| |\xi|^{-\frac{d}{2}} (\Delta_j^{\wedge} - \Delta_{j-1}^{\wedge}) \right\|_{L^2(\Gamma_p)} \right)$$
$$\leqslant c \sup_j \left(p^{-\frac{1-d}{2}j} p^{\frac{1-d}{2}j} \right) = c,$$

where c is independent of $j \in \mathbb{Z}^+$; $\varphi_j(x) = \Delta_j(x) - \Delta_{j-1}(x)$, $\Delta_j(x) = p^j \Phi_j(x)$; and $\Delta_{-1}(x) \equiv 0$. Thus, $\mu \in B_{2\infty}^{-\frac{1-d}{2}, \Lambda}(K_p)$.

By Theorem 5.2.3, it follows that $\dim_H E > d$. Thus
$$\dim_H E \geqslant \dim_F E.$$
The proof is complete.

3. Example

Compute the Fourier dimension of the Cantor type set $C = D \setminus \left(\bigcup_{j=1}^{+\infty} V_j \right)$.

Take $p = 3$, $q = 2$, and set $C \equiv C_3^2 = D \setminus \left(\bigcup_{j=1}^{+\infty} V_j \right)$ in Example 5.1.1, where V_j are

$$V_1 = 0 \cdot \beta^0 + B^1,$$
$$V_2 = \left(1 \cdot \beta^0 + 0 \cdot \beta^1 + B^2\right) \cup \left(2 \cdot \beta^0 + 0 \cdot \beta^1 + B^2\right),$$
$$\cdots\cdots$$

we have $\dim_H C = \dfrac{\ln 2}{\ln 3}$.

To evaluate $\dim_F C$, for any non-zero Radon measure $\mu \neq 0$ with supp $\mu \subset C$, and $\mu(C) > 0$, we have
$$\mu^\wedge\left(\beta^{-l}\right) = \int_C \overline{\chi(\beta^{-l}x)} d\mu(x) = \int_C e^{-\frac{2\pi i}{3} x_{l-1}} d\mu(x),$$
with $\chi \in \Gamma_{K_3}$ as the non-trivial character on K_3, and take "initial value"
$$\chi\left(\beta^{-j}\right) = \begin{cases} e^{\frac{2\pi i}{3}}, & j = 1, \\ 1, & j \neq 1. \end{cases}$$
Thus, for all $l \in \mathbb{N}$, it follows
$$\mu^\wedge\left(\beta^{-l}\right) = \left(-\frac{1}{2} + \frac{\sqrt{3}}{2}i\right)\mu(I_1) + \left(-\frac{1}{2} - \frac{\sqrt{3}}{2}i\right)\mu(I_2)$$
$$= -\frac{1}{2}\mu(C) + \frac{\sqrt{3}}{2}i\left(\mu(I_1) - \mu(I_2)\right), \tag{5.2.10}$$
where $I_1 = \{x \in C : x_{-1} = 1\}, I_2 = \{x \in C : x_{-1} = 2\}$. By the (5.2.10), we have
$$\left|\mu^\wedge\left(\beta^{-l}\right)\right| \geqslant \frac{1}{2}\mu(C), \quad \forall l \in \mathbb{N}.$$
So that if $|\xi| \to +\infty$, then $\mu^\wedge(\xi)$ does not tend to zero. By the definition of Fourier dimension we have $\dim_F C = 0$.

This example not only evaluate the Fourier dimension of the Cantor type set C, but also show an example in which $\dim_F E < \dim_H E$ holds.

We have the other equivalent definition of Fourier dimension.

Definition 5.2.5 (Fourier dimension) *Let $E \subset K_p$ be a non-empty Borel set with $|E| = 0$. The Fourier dimension is defined as: for all Borel measure $\mu \neq 0$, supp $\mu = E$,*
$$\dim_F E = \sup\left\{\alpha : \mu^\wedge(t) = o\left(t^{-\frac{\alpha}{2}}\right) \to 0, |t| \to +\infty\right\},$$
where $\mu^\wedge$ is the Fourier transformation of μ on local field K_p,
$$\mu^\wedge(\xi) = \int_{K_p} \overline{\chi_\xi(x)} d\mu(x), \quad \xi \in \Gamma_{K_p}.$$
For the classical case, it holds

(i) $\dim_F E \leqslant \dim_H E$;

(ii) $\dim_F C < \dim_H C$ for the Cantor set C;

(iii) $\dim_H E$ is invariant, when the set $E \subset \mathbb{R}^n$ is embedded in $\mathbb{R}^{n+1}$. However, even $\dim_F E|_{\mathbb{R}^n} > 0$, we have $\dim_F E|_{\mathbb{R}^{n+1}} = 0$.

Exercises

1. Prove the Lemma 5.2.1.
2. Prove the Theorem 5.2.1 (i)∼(iii).
3. Study the definitions of Hausdorff dimension and Fourier dimension in K^n.
4. Study the property of Fourier dimension on local fields: if $E \subset (K_p)^n$ is embedded in $(K_p)^{n+1}$, $\dim_H E$ is invariant. However, even $\dim_F E|_{\mathbb{R}^n} > 0$, it holds $\dim_F E|_{\mathbb{R}^{n+1}} = 0$.
5. Study the analytic expressions of the Packing dimensions, box dimensions.

5.3 p-type calculus and fractal dimensions on local fields

We study the relationship between p-type calculus and fractal dimensions.

5.3.1 Structures of K_p, 3-adic Cantor type set, 3-adic Cantor type function

1. Algebraic operations and topological structure on K_p

Recall the sructure of a local field K_p, we list a table.

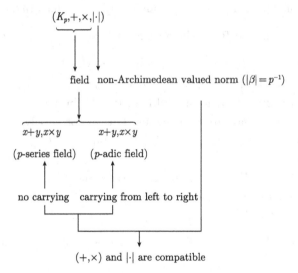

That is, (i) K_p is a field under the algebraic operations $+, \times$; and it is a locally compact topological space under the non-Archimedean valued norm $|\cdot|$; (ii) the algebraic structure $(K_p, +, \times)$ and topological structure $(K_p, |\cdot|)$ are compatible, i.e., $\begin{cases}(x,y) \to x+y \\ (x,y) \to x \times y\end{cases}$ are two continuous mappings under the topology determined by $|\cdot|$. Thus $(K_p, +, \times, |\cdot|)$ becomes a non-trivial, non-discrete, totally disconnected, locally compact topological complete field.

Moreover, we may introduce a distance on $(K_p, +, \times, |\cdot|)$ as

$$d(x,y) = \begin{cases} ||x|-|y||, & |x| \neq |y|, \\ |x-y|, & |x|=|y|, x \neq y, \\ 0, & x=y, \end{cases} \qquad (5.3.1)$$

such that $(K_p, +, \times, |\cdot|, d)$ becomes a complete ultra-metric topological space under the ultra-distance d in (5.3.1).

Denote the set of all compact sets in $(K_p, +, \times, |\cdot|, d)$ by

$$\mathbb{K}(K_p) \equiv \mathbb{K}(K_p, +, \times, |\cdot|, d) = \{\Theta \subset K_p : \Theta \text{ is compact}\}.$$

For a set $B \in \mathbb{K}(K_p)$ and a point $x \in K_p$, the distance from x to B $d(x,B) = \inf\{d(x,y) : y \in B\}$. And for two sets $A, B \in \mathbb{K}(K_p)$, we call

$$d(A,B) = \max\{d(x,B) : x \in A\} \quad \text{and} \quad d(B,A) = \max\{d(x,A) : x \in B\}$$

the distance from A to B and distance from B to A, respectively. Moreover,

$$h(A,B) = \max\{d(A,B), d(B,A)\}$$

is said to be the Hausdorff distance of A and B, and $(\mathbb{K}(K_p), h)$ is said to be the **fractal space on local field** K_p (see Definition 4.3.1 (iii)). A set A in $(\mathbb{K}(K_p), h)$ is said to be a **fractal set**, or for simply, **a fractal**. A complex valued function $f : A \to \mathbb{C}$ defined on $A \in \mathbb{K}(K_p)$ is said to be a **fractal function**.

Remark 1. We emphasize again: the structures of a local field K_p and Euclidean space $\mathbb{R}^n$ have essential difference, for example, for K_3 and $\mathbb{R}^n, n=1$, we take $B^1 \subset K_3$ and $[0,1) \subset \mathbb{R}$, in Fig. 5.3.1.

Let $y \in 0 \cdot \beta^1 + B^2$ and $x \in \left[0, \frac{1}{3}\right)$, a corresponding relationship is

$$y = 0 \cdot \beta^1 + x_2\beta^2 + x_3\beta^3 + \cdots \longleftrightarrow x = 0 \cdot \left(\frac{1}{3}\right)^1 + x_2\left(\frac{1}{3}\right)^2 + x_3\left(\frac{1}{3}\right)^3 + \cdots.$$

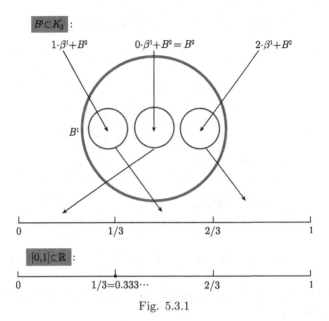

Fig. 5.3.1

In $\mathbb{R}$ case: any rational number has two expressions, finite expression and infinite expression, $x = \frac{1}{3} \in [0,1)$ and $\bar{x} = \frac{1}{3} \in [0,1)$,

$$x = 1 \cdot \left(\frac{1}{3}\right)^1 = 1 \cdot \left(\frac{1}{3}\right)^1 + 0 \cdot \left(\frac{1}{3}\right)^2 + 0 \cdot \left(\frac{1}{3}\right)^3 + 0 \cdot \left(\frac{1}{3}\right)^4 + \cdots = \frac{1}{3},$$

$$\bar{x} = 2 \cdot \left(\frac{1}{3}\right)^2 + 2 \cdot \left(\frac{1}{3}\right)^3 + 2 \cdot \left(\frac{1}{3}\right)^4 + \cdots = \frac{2 \cdot \left(\frac{1}{3}\right)^2}{1 - \frac{1}{3}} = \frac{2 \cdot \frac{1}{9}}{\frac{2}{3}} = \frac{1}{3}$$

are the same element.

In K_3 case: the 3-adic rational number y and 3-adic irrational number $\bar{y}$

$$y = 1 \cdot \beta^1,$$
$$\bar{y} = 0 \cdot \beta^1 + 2 \cdot \beta^2 + 2 \cdot \beta^3 + 2 \cdot \beta^4 + \cdots$$

are two different elements in K_3. Since $y = 1 \cdot \beta^1$ is in the coset $0 \cdot \beta^0 + 1 \cdot \beta^1 + B^2$, and $\bar{y} = 0 \cdot \beta^1 + 2 \cdot \beta^2 + 2 \cdot \beta^3 + 2 \cdot \beta^4 + \cdots$ is in the coset $0 \cdot \beta^0 + 0 \cdot \beta^1 + B^2$, but these two cosets are disjoint each other. This makes the essential difference of the structures of K_3 and $\mathbb{R}$, such that we

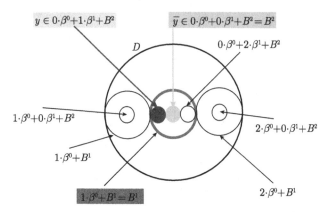

Figure 5.3.2

definitely have no any isomorphic mapping for these two topologies (Fig. 5.3.2).

Remark 2. We may consider the generalization of fractal space $\mathbb{K}(K_p)$ on local field
$$\mathbb{LK}(K_p) \equiv \mathbb{LK}(K_p, +, \times, |\cdot|, h) = \{\Theta \subset K_p : \Theta \text{ is locally compact}\},$$
under the Hausdorff distance, $\mathbb{LK}(K_p)$ is said to be the **generalized fractal space**. However, we focus on $\mathbb{K}(K_p) = \{\Theta \subset K_p : \Theta \text{ is compact}\}$.

Remark 3. The fractal space $(\mathbb{K}(K_p), d)$ on K_p is a complete ultrametric space.

2. 3-adic Cantor type set and 3-adic Cantor type function in K_3

The set $C_3 = D \setminus \bigcup\limits_{j=1}^{+\infty} V_j$ with (see the Section 1.2.8)
$$D = \{x \in K_3 : |x| \leqslant 1\} = \left(0 \cdot \beta^0 + B^1\right) \cup \left(1 \cdot \beta^0 + B^1\right) \cup \left(2 \cdot \beta^0 + B^1\right),$$
$$V_1 = B^1 = \{x \in K_3 : |x| \leqslant 3^{-1}\}$$
$$= \left(0 \cdot \beta^1 + B^2\right) \cup \left(1 \cdot \beta^1 + B^2\right) \cup \left(2 \cdot \beta^1 + B^2\right),$$
$$V_2 = \left(1 \cdot \beta^0 + B^2\right) \cup \left(2 \cdot \beta^0 + B^2\right),$$
$$V_3 = \left(1 \cdot \beta^0 + 1 \cdot \beta^1 + B^3\right) \cup \left(1 \cdot \beta^0 + 2 \cdot \beta^1 + B^3\right)$$
$$\cup \left(2 \cdot \beta^0 + 1 \cdot \beta^1 + B^3\right) \cup \left(2 \cdot \beta^0 + 2 \cdot \beta^1 + B^3\right),$$
$$\cdots$$

is said to be the **3-aidc Cantor type set on** K_3.

Define the so called **3-aidc Cantor type function** (ladder of devil) $\vartheta(x), x \in K_3$, with $x = \sum_{j=-s}^{+\infty} x_j \beta^j$, $x_j \in \{0,1,2\}$, $j = -s, -s+1, \cdots, |\beta| = 3^{-1}$, such that

(i) supp $\vartheta(x) = D$;

(ii) $\forall x \in D$ with $x = \sum_{j=0}^{+\infty} x_j \beta^j$, $x_j \in \{0,1,2\}$ for $j = 0, 1, 2, \cdots$.

① If there exists $k \geqslant 1$, such that $x_j \neq 0$ for $0 \leqslant j \leqslant k-2$, and $x_{k-1} = 0$, then let

$$\vartheta(x) = \sum_{j=0}^{k-2} (x_j - 1) \left(\frac{1}{2}\right)^{j+1} + \left(\frac{1}{2}\right)^k.$$

② If $x_j \neq 0$ for all $0 \leqslant j < +\infty$, then let

$$\vartheta(x) = \sum_{j=0}^{+\infty} (x_j - 1) \left(\frac{1}{2}\right)^{j+1}.$$

③ If $x \notin D$, then let $\vartheta(x) = 0$.

The following Fig. 5.3.3 is a draft of $\vartheta(x)$ on K_3.

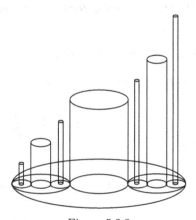

Figure 5.3.3

Readers are familiar with the 3-adic Cantor function on $\mathbb{R}$ with draft in Fig. 5.3.4.

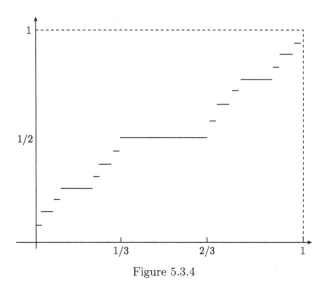

Figure 5.3.4

5.3.2 p-type derivative and p-type integral of $\vartheta(x)$ on K_3

1. Two Lemmas

For studying the p-type derivative and p-type integral of $\vartheta(x)$, we need two lemmas.

Lemma 5.3.1 For the 3-adic Cantor type function $\vartheta(x), \forall x \in B^k, k \in \mathbb{N}$, it holds

$$\vartheta\left(\beta^0 + \cdots + \beta^{k-2} + 2\beta^{k-1} + x\right) = \frac{1}{2^k} + \vartheta\left(\beta^0 + \cdots + \beta^{k-2} + \beta^{k-1} + x\right).$$

Lemma 5.3.2 For the 3-adic Cantor type function $\vartheta(x), \forall x \in D$, it holds

$$\vartheta\left(\beta^0 + \beta^1 + \cdots + \beta^{k-1} + \beta^k x\right) = \frac{1}{2^k}\vartheta(x).$$

The proofs are clear.

2. p-type derivative and p-type integral of $\vartheta(x)$

Let $\omega = e^{\frac{2\pi i}{3}}$, and take $\chi\left(\beta^{-j}\right) = \begin{cases} \omega, & j = 1, \\ 1, & \text{others}. \end{cases}$

Theorem 5.3.1 The 3-adic Cantor type function $\vartheta(x)$ is infinitely p-type integrable; and it is m-order p-type differentiable, with $0 \leqslant m < \dfrac{\ln 2}{\ln 3}$.

Moreover, it holds for $x \in D$,

$$\vartheta^{\langle m \rangle}(x) = \frac{1}{2} + \sum_{l=1}^{-\infty} \frac{3^{lm}}{6^l} \chi\left(\beta^{-l}x\right)\left(\frac{1}{2} + \omega + \frac{1}{2}\chi\left(\beta^{-l}x\right) + \overline{\omega}\chi\left(\beta^{-l}x\right)\right)$$

$$\times \prod_{j=1}^{l-1}\left(2 - \chi\left(\beta^{-j}x\right) - \chi^2\left(\beta^{-j}x\right)\right),$$

and $\vartheta^{\langle m \rangle}(x) = 0$ for $x \notin D$.

Proof. By three steps.

First step. Evaluate $\vartheta^{\wedge}(\xi) = \int_D \vartheta(x)\overline{\chi}_\xi(x)dx$, $|\xi| = 3^l, l \in \mathbb{Z}$.

① If $l \leqslant 0$: since $\forall x \in D$, $|\xi x| \leqslant 1$; then, it follows $\overline{\chi}_\xi(x)|_D = 1$, thus

$$\vartheta^{\wedge}(\xi) = \int_D \vartheta(x)dx = \int_{B^1} \vartheta(x)dx + \int_{\beta^0 + B^1}\vartheta(x)dx + \int_{2\beta^0 + B^1}\vartheta(x)dx$$

$$= \frac{1}{6} + \int_{B^1}\vartheta\left(\beta^0 + x\right)dx + \int_{B^1}\vartheta\left(2\beta^0 + x\right)dx.$$

Then by Lemma 5.3.1, $\vartheta\left(2\beta^0 + x\right) = \frac{1}{2} + \vartheta\left(\beta^0 + x\right)$, thus

$$\vartheta^{\wedge}(\xi) = \frac{1}{3} + 2\int_{B^1}\vartheta\left(\beta^0 + x\right)dx = \frac{1}{3} + 2\int_D \vartheta\left(\beta^0 + \beta x\right)d(\beta x)$$

$$= \frac{1}{3} + \frac{2}{3}\int_D \vartheta\left(\beta^0 + \beta x\right)dx.$$

By Lemma 5.3.2, for $\forall x \in D$, holds $\vartheta\left(\beta^0 + \beta x\right) = \frac{1}{2}\vartheta(x)$, thus

$$\vartheta^{\wedge}(\xi) = \frac{1}{3} + \frac{1}{3}\int_D \vartheta(x)dx.$$

Combining with $\vartheta^{\wedge}(\xi) = \int_D \vartheta(x)dx$, it follows $\vartheta^{\wedge}(\xi) = \frac{1}{2}$ for $l \leqslant 0$.

② If $l \geqslant 1$:

$$\vartheta^{\wedge}(\xi) = \int_D \vartheta(x)\overline{\chi}_\xi(x)dx$$

$$= \int_{B^1}\frac{1}{2}\vartheta(x)dx + \int_{\beta^0 + B^1}\vartheta(x)\overline{\chi}_\xi(x)dx + \int_{2\beta^0 + B^1}\vartheta(x)\overline{\chi}_\xi(x)dx,$$

by the formula $\int_{B^k}\overline{\chi}_\xi(x)dx = \begin{cases}3^{-k}, & l \leqslant k \\ 0, & l > k\end{cases}$ and $\vartheta\left(2\beta^0 + x\right) = \frac{1}{2} + \vartheta\left(\beta^0 + x\right)$, then

$$\vartheta^{\wedge}(\xi) = \frac{1}{6}\begin{cases}1, & l \leqslant 1 \\ 0, & l > 1\end{cases} + \frac{1}{6}\overline{\chi}_\xi^2\left(\beta^0\right)\begin{cases}1, & l \leqslant 1 \\ 0, & l > 1\end{cases} + \left(1 + \overline{\chi}_\xi\left(\beta^0\right)\right)\int_{\beta^0 + B^1}\vartheta(x)\overline{\chi}_\xi(x)dx.$$

To compute the last integral above, when $k \leq l - 1$, by Lemma 5.3.1, we have

$$\int_{\beta^0+\beta^1+\cdots+\beta^{k-1}+B^k} \vartheta(x)\overline{\chi}_\xi(x)dx$$

$$= \int_{\beta^0+\beta^1+\cdots+\beta^{k-1}+B^{k+1}} \frac{1}{2^{k+1}}\overline{\chi}_\xi(x)dx$$

$$+ \int_{\beta^0+\beta^1+\cdots+\beta^{k-1}+\beta^k+B^{k+1}} \vartheta(x)\overline{\chi}_\xi(x)dx$$

$$+ \int_{\beta^0+\beta^1+\cdots+\beta^{k-1}+2\beta^k+B^{k+1}} \vartheta(x)\overline{\chi}_\xi(x)dx$$

$$= \frac{1}{6^{k+1}}\overline{\chi}_\xi\left(\beta^0+\beta^1+\cdots+\beta^{k-1}\right) \begin{cases} 1, & l \leq k+1 \\ 0, & l > k+1 \end{cases}$$

$$+ \frac{1}{6^{k-1}}\overline{\chi}_\xi\left(\beta^0+\beta^1+\cdots\beta^{k-1}+2\beta^k\right) \begin{cases} 1, & l \leq k+1 \\ 0, & l > k+1 \end{cases}$$

$$+ \left(1+\overline{\chi}_\xi\left(\beta^k\right)\right) \int_{\beta^0+\beta^1+\cdots+\beta^k+B^{k+1}} \vartheta(x)\overline{\chi}_\xi(x)dx.$$

When $l \geq 1$, by induction, it follows

$$\vartheta^\wedge(\xi) = \int_D \vartheta(x)\overline{\chi}_\xi(x)dx$$

$$= \left(1+\overline{\chi}_\xi\left(\beta^0\right)\right)\left(1+\overline{\chi}_\xi\left(\beta^1\right)\right)\cdots\left(1+\overline{\chi}_\xi\left(\beta^{l-2}\right)\right)$$

$$\cdot \int_{\beta^0+\beta^1+\cdots+\beta^{l-2}+B^{l-1}} \vartheta(x)\overline{\chi}_\xi(x)dx$$

$$= \left(1+\overline{\chi}_\xi\left(\beta^0\right)\right)\left(1+\overline{\chi}_\xi\left(\beta^1\right)\right)\cdots\left(1+\overline{\chi}_\xi\left(\beta^{l-2}\right)\right)$$

$$\cdot \left\{\frac{1}{6^l}\overline{\chi}_\xi\left(\beta^0+\beta^1+\cdots+\beta^{l-2}\right)+\frac{1}{6^l}\overline{\chi}_\xi\left(\beta^0+\beta^1+\cdots+\beta^{l-2}+2\beta^{l-1}\right)\right.$$

$$\left. + \left(1+\overline{\chi}_\xi\left(\beta^{l-1}\right)\right)\int_{\beta^0+\beta^1+\cdots+\beta^{l-1}+B^l} \vartheta(x)\overline{\chi}_\xi(x)dx\right\}$$

$$= \left(1+\overline{\chi}_\xi\left(\beta^0\right)\right)\left(1+\overline{\chi}_\xi\left(\beta^1\right)\right)\cdots\left(1+\overline{\chi}_\xi\left(\beta^{l-2}\right)\right)$$

$$\cdot \left\{\frac{1}{6^l}\overline{\chi}_\xi\left(\beta^0+\beta^1+\cdots+\beta^{l-2}\right)+\frac{1}{6^l}\overline{\chi}_\xi\left(\beta^0+\beta^1+\cdots+\beta^{l-2}+2\beta^{l-1}\right)\right.$$

$$+ \left(1+\overline{\chi}_\xi\left(\beta^{l-1}\right)\right)\overline{\chi}_\xi\left(\beta^0+\beta^1+\cdots+\beta^{l-1}\right)$$

$$\left. \cdot \int_{B^l} \vartheta\left(\beta^0+\beta^1+\cdots+\beta^{l-1}+x\right)\overline{\chi}_\xi(x)dx\right\}.$$

Since $|\xi| = 3^l$, it follows from Lemma 5.3.2 that
$$\int_{B^l} \vartheta\left(\beta^0 + \beta^1 + \cdots + \beta^{l-1} + x\right) dx = \frac{1}{6^l} \int_D \vartheta(x) dx,$$
then
$$\vartheta^\wedge(\xi) = \left(1 + \overline{\chi}_\xi\left(\beta^0\right)\right)\left(1 + \overline{\chi}_\xi\left(\beta^1\right)\right) \cdots \left(1 + \overline{\chi}_\xi\left(\beta^{l-2}\right)\right)$$
$$\cdot \left\{ \frac{1}{6^l} \overline{\chi}_\xi\left(\beta^0 + \beta^1 + \cdots + \beta^{l-2}\right) + \frac{1}{6^l} \overline{\chi}_\xi\left(\beta^0 + \beta^1 + \cdots + \beta^{l-2} + 2\beta^{l-1}\right) \right.$$
$$\left. + \frac{1}{2} \frac{1}{6^l} \left(1 + \overline{\chi}_\xi\left(\beta^{l-1}\right)\right) \overline{\chi}_\xi\left(\beta^0 + \beta^1 + \cdots + \beta^{l-1}\right) \right\}. \qquad (5.3.2)$$

Second step. Evaluate $\langle \xi \rangle^m \vartheta^\wedge(\xi)$, $0 \leqslant m < +\infty$, $|\xi| = 3^l, l \in \mathbb{Z}$.
By the (5.3.2), we have
① If $l \leqslant 0$, then $\langle \xi \rangle^m \vartheta^\wedge(\xi) = \frac{1}{2}$.
② If $l \geqslant 1$, then
$$\langle \xi \rangle^m \vartheta^\wedge(\xi) = 3^{lm} \left(1 + \overline{\chi}_\xi\left(\beta^0\right)\right)\left(1 + \overline{\chi}_\xi\left(\beta^1\right)\right) \cdots \left(1 + \overline{\chi}_\xi\left(\beta^{l-2}\right)\right)$$
$$\cdot \left\{ \frac{1}{6^l} \overline{\chi}_\xi\left(\beta^0 + \beta^1 + \cdots + \beta^{l-2}\right) \right.$$
$$+ \frac{1}{6^l} \overline{\chi}_\xi\left(\beta^0 + \beta^1 + \cdots + \beta^{l-2} + 2\beta^{l-1}\right)$$
$$\left. + \frac{1}{2} \frac{1}{6^l} \left(1 + \overline{\chi}_\xi\left(\beta^{l-1}\right)\right) \overline{\chi}_\xi\left(\beta^0 + \beta^1 + \cdots + \beta^{l-1}\right) \right\}.$$
$$(5.3.3)$$

Third step. Evaluate $\left(\langle \cdot \rangle^m \vartheta^\wedge(\cdot)\right)^\vee(x)$, $0 \leqslant m < +\infty$, $|x| = 3^k, k \in \mathbb{Z}$.
Since
$$\left(\langle \cdot \rangle^m \vartheta^\wedge(\cdot)\right)^\vee(x) = \frac{1}{2} \int_{|\xi| \leqslant 1} \chi_\xi(x) d\xi + \sum_{l=1}^{+\infty} 3^{lm} J_l = \frac{1}{2} \begin{cases} 1, & k \leqslant 0 \\ 0, & k > 0 \end{cases} + \sum_{l=1}^{+\infty} 3^{lm} J_l,$$
where
$$J_l = \int_{|\xi| = 3^l} \left(1 + \overline{\chi}_\xi\left(\beta^0\right)\right)\left(1 + \overline{\chi}_\xi\left(\beta^1\right)\right) \cdots \left(1 + \overline{\chi}_\xi\left(\beta^{l-2}\right)\right)$$
$$\cdot \left\{ \frac{1}{6^l} \overline{\chi}_\xi\left(\beta^0 + \beta^1 + \cdots + \beta^{l-2}\right) + \frac{1}{6^l} \overline{\chi}_\xi\left(\beta^0 + \beta^1 + \cdots + \beta^{l-2} + 2\beta^{l-1}\right) \right.$$
$$\left. + \frac{1}{2} \frac{1}{6^l} \left(1 + \overline{\chi}_\xi\left(\beta^{l-1}\right)\right) \overline{\chi}_\xi\left(\beta^0 + \beta^1 + \cdots + \beta^{l-1}\right) \right\} \chi_\xi(x) d\xi.$$

By $|\xi| = 3^l, l \in \mathbb{Z}$, it happens two cases:

$$\xi \in \beta^{-l} + \xi_{-l+1}\beta^{-l+1} + \cdots + \xi_{-2}\beta^{-2} + \xi_{-1}\beta^{-1} + D, \quad \xi_{-l+1}, \cdots, \xi_{-1} \in \{0, 1, 2\}$$

and

$$\xi \in 2\beta^{-l} + \xi_{-l+1}\beta^{-l+1} + \cdots + \xi_{-2}\beta^{-2} + \xi_{-1}\beta^{-1} + D, \quad \xi_{-l+1}, \cdots, \xi_{-1} \in \{0, 1, 2\}.$$

Thus,

$$\int_{\beta^{-l}+\xi_{-l+1}\beta^{-l+1}+\cdots+\xi_{-1}\beta^{-1}+D} \left(1+\overline{\chi}_\xi\left(\beta^0\right)\right)\left(1+\overline{\chi}_\xi\left(\beta^1\right)\right)\cdots\left(1+\overline{\chi}_\xi\left(\beta^{l-2}\right)\right)$$

$$\cdot \left\{\frac{1}{6^l}\overline{\chi}_\xi\left(\beta^0+\beta^1+\cdots+\beta^{l-2}\right) + \frac{1}{6^l}\overline{\chi}_\xi\left(\beta^0+\beta^1+\cdots+\beta^{l-2}+2\beta^{l-1}\right)\right.$$

$$\left. +\frac{1}{2}\frac{1}{6^l}\left(1+\overline{\chi}_\xi\left(\beta^{l-1}\right)\right)\overline{\chi}_\xi\left(\beta^0+\beta^1+\cdots+\beta^{l-1}\right)\right\}\chi_\xi(x)d\xi$$

$$=\frac{1}{6^l}\left(\frac{1}{2}+\omega\right)\prod_{j=1}^{l-1}\left(1+\overline{\omega}^{\xi_{-j}}\right)\overline{\omega}^{\xi_{-j}}\chi\cdot\left(\left(\beta^{-l}+\xi_{-l+1}\beta^{-l+1}+\cdots+\xi_{-1}\beta^{-1}\right)x\right)$$

$$\cdot \begin{cases} 1, & k \leqslant 0, \\ 0, & k > 0, \end{cases}$$

moreover,

$$\int_{2\beta^{-l}+\xi_{-l+1}\beta^{-l+1}+\cdots+\xi_{-1}\beta^{-1}+D} \left(1+\overline{\chi}_\xi\left(\beta^0\right)\right)\left(1+\overline{\chi}_\xi\left(\beta^1\right)\right)\cdots\left(1+\overline{\chi}_\xi\left(\beta^{l-2}\right)\right)$$

$$\cdot \left\{\frac{1}{6^l}\overline{\chi}_\xi\left(\beta^0+\beta^1+\cdots+\beta^{l-2}\right) + \frac{1}{6^l}\overline{\chi}_\xi\left(\beta^0+\beta^1+\cdots+\beta^{l-2}+2\beta^{l-1}\right)\right.$$

$$\left. +\frac{1}{2}\frac{1}{6^l}\left(1+\overline{\chi}_\xi\left(\beta^{l-1}\right)\right)\overline{\chi}_\xi\left(\beta^0+\beta^1+\cdots+\beta^{l-1}\right)\right\}\chi_\xi(x)d\xi$$

$$=\frac{1}{6^l}\left(\frac{1}{2}+\overline{\omega}\right)\prod_{j=1}^{l-1}\left(1+\overline{\omega}^{\xi_{-j}}\right)\overline{\omega}^{\xi_{-j}}\chi\cdot\left(\left(2\beta^{-l}+\xi_{-l+1}\beta^{-l+1}+\cdots+\xi_{-1}\beta^{-1}\right)x\right)$$

$$\cdot \begin{cases} 1, & k \leqslant 0, \\ 0, & k > 0. \end{cases}$$

Thus, if $k > 0$, then $J_l = 0$; if $k \leqslant 0$, then

$$J_l = \frac{1}{6^l}\sum_{\xi_{-l+1},\cdots,\xi_{-1}\in\{0,1,2\}}\left(\frac{1}{2}+\omega\right)\prod_{j=1}^{l-1}\left(1+\overline{\omega}^{\xi_{-j}}\right)\overline{\omega}^{\xi_{-j}}$$

$$\cdot \chi\left(\left(\beta^{-l}+\xi_{-l+1}\beta^{-l+1}+\cdots+\xi_{-1}\beta^{-1}\right)x\right)$$

$$+ \frac{1}{6^l} \sum_{\xi_{-l+1},\cdots,\xi_{-1}\in\{0,1,2\}} \left(\frac{1}{2}+\overline{\omega}\right) \prod_{j=1}^{l-1} (1+\overline{\omega}^{\xi_{-j}}) \overline{\omega}^{\xi_{-j}}$$
$$\cdot \chi\left((2\beta^{-l}+\xi_{-l+1}\beta^{-l+1}+\cdots+\xi_{-1}\beta^{-1})x\right)$$
$$= \frac{1}{6^l} \sum_{\xi_{-l+1},\cdots,\xi_{-1}\in\{0,1,2\}} \left(\frac{1}{2}+\omega+\frac{1}{2}\chi(\beta^{-l}x)+\overline{\omega}\chi(\beta^{-l}x)\right)$$
$$\cdot \prod_{j=1}^{l-1} (1+\overline{\omega}^{\xi_{-j}}) \overline{\omega}^{\xi_{-j}} \chi\left((\beta^{-l}+\xi_{-l+1}\beta^{-l+1}+\cdots+\xi_{-1}\beta^{-1})x\right)$$
$$= \frac{1}{6^l} \left(\frac{1}{2}+\omega+\frac{1}{2}\chi(\beta^{-l}x)+\overline{\omega}\chi(\beta^{-l}x)\right)\chi(\beta^{-l}x)$$
$$\cdot \sum_{\xi_{-l+1},\cdots,\xi_{-1}\in\{0,1,2\}} \prod_{j=1}^{l-1} (\overline{\omega}^{\xi_{-j}}+\overline{\omega}^{2\xi_{-j}})\chi(\beta^{-j}x)^{\xi_{-j}}$$
$$= \frac{1}{6^l} \left(\frac{1}{2}+\omega+\frac{1}{2}\chi(\beta^{-l}x)+\overline{\omega}\chi(\beta^{-l}x)\right)\chi(\beta^{-l}x)$$
$$\cdot \prod_{j=1}^{l-1} (2-\chi(\beta^{-j}x)-\chi^2(\beta^{-j}x)).$$

Combining the above evaluates, we have
If $k>0$, then $(\langle\cdot\rangle^m \vartheta^{\wedge}(\cdot))^{\vee}(x)=0$; if $k\leq 0$, then

$$(\langle\cdot\rangle^m \vartheta^{\wedge}(\cdot))^{\vee}(x)=\frac{1}{2}+\sum_{l=1}^{+\infty}\frac{3^{lm}}{6^l}\left(\frac{1}{2}+\omega+\frac{1}{2}\chi(\beta^{-l}x)+\overline{\omega}\chi(\beta^{-l}x)\right)\chi(\beta^{-l}x)$$
$$\cdot \prod_{j=1}^{l-1} (2-\chi(\beta^{-j}x)-\chi^2(\beta^{-j}x)).$$

Thus, we have shown that the p-type derivative of $\vartheta(x)$ is

$$\vartheta^{\langle m\rangle}(x)=\frac{1}{2}+\sum_{l=1}^{+\infty}\frac{3^{lm}}{6^l}\left(\frac{1}{2}+\omega+\frac{1}{2}\chi(\beta^{-l}x)+\overline{\omega}\chi(\beta^{-l}x)\right)\chi(\beta^{-l}x)$$
$$\cdot \prod_{j=1}^{l-1} (2-\chi(\beta^{-j}x)-\chi^2(\beta^{-j}x)).$$

Consider the convergence of the following series

$$\sum_{l=1}^{+\infty}\frac{3^{lm}}{6^l}\left(\frac{1}{2}+\omega+\frac{1}{2}\chi(\beta^{-l}x)+\overline{\omega}\chi(\beta^{-l}x)\right)\chi(\beta^{-l}x)$$

$$\cdot \prod_{j=1}^{l-1} \left(2 - \chi\left(\beta^{-j}x\right) - \chi^2\left(\beta^{-j}x\right)\right) \tag{5.3.4}$$

at any $x \in D$, we see that: when $0 \leqslant m < \dfrac{\ln 2}{\ln 3}$,

$$\sum_{l=1}^{+\infty} \frac{3^{lm}}{6^l} \left| \left(\frac{1}{2} + \omega + \frac{1}{2}\chi\left(\beta^{-l}x\right) + \overline{\omega}\chi\left(\beta^{-l}x\right)\right) \chi\left(\beta^{-l}x\right) \right.$$

$$\left. \cdot \prod_{j=1}^{l-1} \left(2 - \chi\left(\beta^{-j}x\right) - \chi^2\left(\beta^{-j}x\right)\right) \right|$$

$$\leqslant c \sum_{l=1}^{+\infty} \frac{3^{lm}}{6^l} 3^l \leqslant c \sum_{l=1}^{+\infty} \left(\frac{3^m}{2}\right)^l < +\infty;$$

when $m = \dfrac{\ln 2}{\ln 3}$, take $x = \beta^0 + \beta^1 + \cdots$, then

$$\left\{ \sum_{l=1}^{+\infty} \frac{3^{lm}}{6^l} \left(\frac{1}{2} + \omega + \frac{1}{2}\chi\left(\beta^{-l}x\right) + \overline{\omega}\chi\left(\beta^{-l}x\right)\right) \chi\left(\beta^{-l}x\right) \right.$$

$$\left. \cdot \prod_{j=1}^{l-1} \left(2 - \chi\left(\beta^{-j}x\right) - \chi^2\left(\beta^{-j}x\right)\right) \right\} \Bigg|_{x=\beta^0+\beta^1+\cdots}$$

$$= \sum_{l=1}^{+\infty} \frac{2^l}{6^l} \left(\frac{1}{2} + \omega + \frac{1}{2}\omega + \overline{\omega}\omega\right) \omega 3^{l-1} = -\frac{1}{2} \sum_{l=1}^{+\infty} \frac{2^l 3^l}{6^l} = -\infty.$$

These imply that the 3-aidc Cantor type function $\vartheta(x)$ is m-order p-type differentiable with $0 \leqslant m < \dfrac{\ln 2}{\ln 3}$. And for $x \in D$,

$$\vartheta^{\langle m \rangle}(x) = \frac{1}{2} + \sum_{l=1}^{+\infty} \frac{3^{lm}}{6^l} \chi\left(\beta^{-l}x\right) \left(\frac{1}{2} + \omega + \frac{1}{2}\chi\left(\beta^{-l}x\right) + \overline{\omega}\chi\left(\beta^{-l}x\right)\right)$$

$$\cdot \prod_{j=1}^{l-1} \left(2 - \chi\left(\beta^{-j}x\right) - \chi^2\left(\beta^{-j}x\right)\right);$$

for $x \notin D$, $\vartheta^{\langle m \rangle}(x) = 0$.

For the p-type integrability of $\vartheta(x)$, by $\vartheta_{\langle m \rangle}(x) = \vartheta^{\langle -m \rangle}(x)$, $0 \leqslant m < +\infty$, and the series in (5.3.4) converges for $m \leqslant 0$, then we complete the proof of Theorem.

3. Hausdorff dimension of $\vartheta^{\langle m \rangle}(x)$

We evaluate the Hausdorff dimension of $\vartheta^{\langle m \rangle}(x)$.

Lemma 5.3.3 If there exists $L \in \mathbb{N}$, such that $\chi\left(\beta^{-L}x\right) = 1$, then for $\forall l \geqslant L$, it holds

$$P_l \equiv \frac{3^{lm}}{6^l} \chi\left(\beta^{-l}x\right) \left(\frac{1}{2} + \omega + \frac{1}{2}\chi\left(\beta^{-l}x\right) + \overline{\omega}\chi\left(\beta^{-l}x\right)\right)$$

$$\cdot \prod_{j=1}^{l-1} \left(2 - \chi\left(\beta^{-j}x\right) - \chi^2\left(\beta^{-j}x\right)\right) = 0.$$

Proof. Since $\chi\left(\beta^{-L}x\right) = 1$, then $\frac{1}{2} + \omega + \frac{1}{2}\chi\left(\beta^{-l}x\right) + \overline{\omega}\chi\left(\beta^{-l}x\right) = 0$, thus $P_L = 0$.

On the other hand, if $l > L$, then $\left(2 - \chi\left(\beta^{-L}x\right) - \chi^2\left(\beta^{-L}x\right)\right) = 0$, also holds $P_l = 0$.

Definition 5.3.1 (Graph of a fractal function) For a fractal function $f : A \to \mathbb{R}$ defined on set $A \subset K_p$, the set

$$\mathbb{G}(f, A) \equiv \mathbb{G}\left\{\{f(x) \in \mathbb{R} : x \in A\}, A\right\} \subset \mathbb{R} \times K_p$$

is said to be a **graph** of f.

The following theorem is very interesting and important.

Theorem 5.3.2 For $m, 0 \leqslant m < \frac{\ln 2}{\ln 3}$, the m-th p-type derivative of $\vartheta(x)$ at $x \in D$ is

$$\vartheta^{\langle m \rangle}(x) = \frac{1}{2} + \sum_{l=1}^{-\infty} \frac{3^{lm}}{6^j} \chi\left(\beta^{-l}x\right) \left(\frac{1}{2} + \omega + \frac{1}{2}\chi\left(\beta^{-l}x\right) + \overline{\omega}\chi\left(\beta^{-l}x\right)\right)$$

$$\cdot \prod_{j=1}^{l-1} \left(2 - \chi\left(\beta^{-j}x\right) - \chi^2\left(\beta^{-j}x\right)\right),$$

and the Hausdorff dimension of graph $\mathbb{G}\left(\vartheta^{\langle m \rangle}, D\right)$ is 1.

Proof. Denote $D = \bigcup_{j_1,\cdots,j_k \in \{1,2\}} I_{j_1,\cdots,j_k,0}$, where

$I_0 = 0 \cdot \beta^0 + B^1$, $I_{1,0} = 1 \cdot \beta^0 + 0 \cdot \beta^1 + B^2$, $I_{2,0} = 2 \cdot \beta^0 + 0 \cdot \beta^1 + B^2; \cdots$;

$I_{j_1,\cdots,j_k,0} = j_1 \cdot \beta^0 + j_2 \cdot \beta^1 + \cdots + j_k \cdot \beta^{k-1} + 0 \cdot \beta^k + B^{k+1}, \cdots$.

By definition of P_l, it follows that $\vartheta^{\langle m \rangle}(x) = \frac{1}{2} + \sum_{l=1}^{+\infty} P_l$. Thus, if $x \in I_{j_1,\cdots,j_k,0}$, it implies $\chi\left(\beta^{-(k+1)}x\right) = 1$. So that by Lemma 5.3.3, we have

$$\vartheta^{\langle m \rangle}(x) = \frac{1}{2} + \sum_{l=1}^{+\infty} P_l = \frac{1}{2} + \sum_{l=1}^{k} P_l$$

$$= \frac{1}{2} + \sum_{l=1}^{k} \omega^{j_l} \left(\frac{1}{2} + \omega + \frac{1}{2} \omega^{j_l} + \overline{\omega} \omega^{j_l} \right) \prod_{s=1}^{l-1} \left(2 - \omega^{j_s} - \omega^{2j_s} \right).$$

Then, $\vartheta^{\langle m \rangle}(x)$ takes constants on each $I_{j_1,\cdots,j_k,0}$, $j_1,\cdots,j_k \in \{1,2\}, k \geqslant 0$.

Since $\mathbb{G}\left(\vartheta^{\langle m \rangle}, I_{j_1,\cdots,j_k,0}\right) = I_{j_1,\cdots,j_k,0} \times \{c_{j_1,\cdots,j_k}\}$, where $c_{j_1,\cdots,j_k}$ is independent of $j_1,\cdots,j_k$, moreover, the 1-Hausdorff dimension

$$H^1\left(\mathbb{G}\left(\vartheta^{\langle m \rangle}, I_{j_1,\cdots,j_k,0}\right)\right)$$

$$= \lim_{\delta \to 0} H^1_\delta\left(\mathbb{G}\left(\vartheta^{\langle m \rangle}, I_{j_1,\cdots,j_k,0}\right)\right)$$

$$= \lim_{\delta \to 0} \inf \left\{ \sum_{j=1}^{+\infty} |U_j|_d : \{U_j\}_{j=1}^{+\infty} \text{ is } \delta\text{-covering of } \mathbb{G}\left(\vartheta^{\langle m \rangle}, I_{j_1,\cdots,j_k,0}\right) \right\}$$

$$= \lim_{\delta \to 0} \inf \left\{ \sum_{j=1}^{+\infty} |U_j|_d : \{U_j\}_{j=1}^{+\infty} \text{ is } \delta\text{-covering of } I_{j_1,\cdots,j_k,0} \right\} = 3^{-(k+1)},$$

hence, it holds for $j_1,\cdots,j_k \in \{1,2\}, k \geqslant 0$,

$$\dim_H\left(\mathbb{G}\left(\vartheta^{\langle m \rangle}, I_{j_1,\cdots,j_k,0}\right)\right) = 1.$$

Then,

$$\dim_H \mathbb{G}\left(\vartheta^{\langle m \rangle}, D\right) = \dim_H \mathbb{G}\left(\vartheta^{\langle m \rangle}, \bigcup_{j_1,\cdots,j_k \in \{1,2\}, k \geqslant 0} I_{j_1,\cdots,j_k,0}\right)$$

$$= \sup_{j_1,\cdots,j_k \in \{1,2\}, k \geqslant 0} \dim_H \mathbb{G}\left(\vartheta^{\langle m \rangle}, I_{j_1,\cdots,j_k,0}\right) = 1.$$

The proof is complete.

5.3.3 p-type derivative and integral of Weierstrass type function on K_p

1. Weierstrass type function on K_p

The famous Weierstrass function on $\mathbb{R}$ is a typical fractal, and we may define this type function, Weierstrass type function on K_p. For example, on K_2, we define for each $x \in K_2$, $x = \sum_{j=s}^{+\infty} x_j \beta^j$, $x_j \in \{0,1\}$, $j = s, s+1, \cdots, -1, 0, 1, \cdots$, with $|\beta| = 2^{-1}$,

$$W_2(x) = W_2\left(\sum_{j=s}^{+\infty} x_j \beta^j\right) = \begin{cases} \sum_{j=1}^{+\infty} x_j \left(\dfrac{1}{2}\right)^j, & x \in B^1, \\ 0, & x \notin B^1, \end{cases} \quad (5.3.5)$$

this $W_2(x)$ is said to be a **Weierstrass type function**.

2. p-type derivative and p-type integral of $W_2(x)$

Theorem 5.3.3 *The Weierstrass type function $W_2 : K_2 \to \mathbb{R}$ is infinitely p-type integrable; and for $0 \leqslant m < 1$, W_2 is m-order p-type differentiable, with*

$$W_2^{\langle m \rangle}(x) = \begin{cases} \dfrac{1}{4} + \dfrac{2^m}{4} - \dfrac{2^{2m-2}}{1-2^{m-1}} + \sum_{j=1}^{+\infty} x_j \left(\dfrac{1}{2}\right)^{j-(j+1)m}, & x \in B^1, \\ \dfrac{1}{4} - \dfrac{2^m}{4}, & x \in D \setminus B^1, \\ 0, & \text{others.} \end{cases}$$

Moreover, $W_2(x)$ has no $m = 1$ order p-type derivative at any $x \in B^1$.

Proof. By three steps.

First step. Evaluate $W_2^\wedge(\xi) = \int_{B^1} W_2(x) \overline{\chi_\xi}(x) dx$, $|\xi| = 2^l$, $l \in \mathbb{Z}$.

① If $l \leqslant 1$: $\forall x \in B^1$, then $|\xi x| \leqslant 1$, and $\overline{\chi_\xi}(x)\big|_{B^1} = 1$, so that

$$W_2^\wedge(\xi) = \int_{B^1} W_2(x) dx = \int_{B^2} W_2(x) dx + \int_{B^1 \setminus B^2} W_2(x) dx$$

$$= \int_{B^2} W_2(x) dx + \int_{\beta^1 + B^2} W_2(x) dx$$

$$= \int_{B^2} W_2(x) dx + \int_{B^2} W_2(x + \beta) dx.$$

Note that: $\forall x \in B^2$, $W_2(x+\beta) = W_2(x) + W_2(\beta) = W_2(x) + \dfrac{1}{2}$, thus

$$W_2^\wedge(\xi) = 2\int_{B^2} W_2(x) dx + \dfrac{1}{2} \int_{B^2} dx = 2\int_{B^2} W_2(x) dx + \dfrac{1}{8}.$$

On the other hand, $\forall x \in B^2$, holds $W_2(x\beta^{-1}) = 2W_2(x)$, then

$$W_2^\wedge(\xi) = \int_{B^1} W_2(x) dx = \int_{B^2} W_2(x\beta^{-1}) d(x\beta^{-1}) = 4 \int_{B^2} W_2(x) dx.$$

Combining the above evaluations, if $l \leqslant 1$, then
$$W_2^\wedge(\xi) = \frac{1}{4}.$$

② If $l \geqslant 2$:

$$\begin{aligned}
W_2^\wedge(\xi) &= \int_{B^1} W_2(x)\overline{\chi_\xi}(x)dx \\
&= \int_{B^2} W_2(x)\overline{\chi_\xi}(x)dx + \int_{B^2} W_2(x+\beta)\overline{\chi_\xi}(x+\beta)dx \\
&= (1+\overline{\chi_\xi}(\beta))\int_{B^2} W_2(x)\overline{\chi_\xi}(x)dx + \frac{1}{2}\overline{\chi_\xi}(\beta)\int_{B^2}\overline{\chi_\xi}(x)dx.
\end{aligned}$$

By the formula $\int_{B^k} \overline{\chi_\xi}(x)dx = \begin{cases} 2^{-k}, & l \leqslant k, \\ 0, & l > k, \end{cases}$ we get

$$W_2^\wedge(\xi) = (1+\overline{\chi_\xi}(\beta))\int_{B^2} W_2(x)\overline{\chi_\xi}(x)dx + \frac{1}{2}\overline{\chi_\xi}(\beta)\begin{cases} 2^{-2}, & l = 2, \\ 0, & l > 2. \end{cases}$$

To evaluate the integral in above equality, we see that, when $k \leqslant l-1$, then

$$\begin{aligned}
\int_{B^k} W_2(x)\overline{\chi_\xi}(x)dx &= \int_{B^{k+1}} W_2(x)\overline{\chi_\xi}(x)dx \\
&\quad + \int_{B^{k+1}} W_2(x+\beta^k)\overline{\chi_\xi}(x+\beta^k)dx.
\end{aligned}$$

Note that, $\forall x \in B^{k+1}$ implies $W_2(x+\beta^k) = W_2(x) + W_2(\beta^k) = W_2(x) + \left(\frac{1}{2}\right)^k$, so that

$$\begin{aligned}
&\int_{B^k} W_2(x)\overline{\chi_\xi}(x)dx \\
&= (1+\overline{\chi_\xi}(\beta^k))\int_{B^{k+1}} W_2(x)\overline{\chi_\xi}(x)dx + \frac{1}{2^k}\overline{\chi_\xi}(\beta^k)\int_{B^{k+1}}\overline{\chi_\xi}(x)dx \\
&= (1+\overline{\chi_\xi}(\beta^k))\int_{B^{k+1}} W_2(x)\overline{\chi_\xi}(x)dx + \frac{1}{2^k}\overline{\chi_\xi}(\beta^k)\begin{cases} 2^{-k-1}, & l = k+1, \\ 0, & l > k+1 \end{cases} \\
&= (1+\overline{\chi_\xi}(\beta^k))\int_{B^{k+1}} W_2(x)\overline{\chi_\xi}(x)dx + \overline{\chi_\xi}(\beta^k)\begin{cases} 2^{-2k-1}, & l = k+1, \\ 0, & l > k+1. \end{cases}
\end{aligned}$$

Inductively, for $l \geqslant 2$,

$$W_2^\wedge(\xi) = \int_{B^1} W_2(x)\overline{\chi_\xi}(x)dx$$

$$= \left(1+\overline{\chi_\xi}(\beta)\right)\left(1+\overline{\chi_\xi}(\beta^2)\right)\cdots\left(1+\overline{\chi_\xi}(\beta^{l-2})\right)\int_{B^{l-1}} W_2(x)\overline{\chi_\xi}(x)dx$$

$$= \left(1+\overline{\chi_\xi}(\beta)\right)\left(1+\overline{\chi_\xi}(\beta^2)\right)\cdots\left(1+\overline{\chi_\xi}(\beta^{l-1})\right)\int_{B^l} W_2(x)\overline{\chi_\xi}(x)dx$$

$$+ \left(1+\overline{\chi_\xi}(\beta)\right)\left(1+\overline{\chi_\xi}(\beta^2)\right)\cdots\left(1+\overline{\chi_\xi}(\beta^{l-2})\right)\overline{\chi_\xi}(\beta^{l-1})\left(\frac{1}{2}\right)^{2l-1}.$$

And since $|\xi| = 2^l$ implies $\left|\xi\beta^{l-1}\right| = 2$, thus $\xi\beta^{l-1} \in B^{-1}\backslash D$. So $\xi\beta^{l-1} \in \beta^{-1} + D$.

By $\chi_\xi(\beta^{l-1}) = \chi(\xi\beta^{l-1}) = \chi(\beta^{-1}) = -1$, we may deduce $W_2^\wedge(\xi)$ to

$$W_2^\wedge(\xi) = -\left(1+\overline{\chi_\xi}(\beta)\right)\left(1+\overline{\chi_\xi}(\beta^2)\right)\cdots\left(1+\overline{\chi_\xi}(\beta^{l-2})\right)\left(\frac{1}{2}\right)^{2l-1}.$$

Combining the above evaluations, it follows

$$W_2^\wedge(\xi) = \begin{cases} \dfrac{1}{4}, & l \leqslant 1, \\ -\left(1+\overline{\chi_\xi}(\beta)\right)\left(1+\overline{\chi_\xi}(\beta^2)\right)\cdots\left(1+\overline{\chi_\xi}(\beta^{l-2})\right)\left(\dfrac{1}{2}\right)^{2l-1}, & l \geqslant 2. \end{cases}$$

Second step. Evaluate $\langle\xi\rangle^m W_2^\wedge(\xi)$, $m \in \mathbb{R}$,

$$\langle\xi\rangle^m W_2^\wedge(\xi) = \begin{cases} \dfrac{1}{4}, & l \leqslant 0, \\ \dfrac{1}{4}2^m, & l = 1, \\ -\left(1+\overline{\chi_\xi}(\beta)\right)\left(1+\overline{\chi_\xi}(\beta^2)\right)\cdots\left(1+\overline{\chi_\xi}(\beta^{l-2})\right)\left(\dfrac{1}{2}\right)^{2l-1}2^{lm}, & l \geqslant 2. \end{cases}$$

Third step. Evaluate $\left(\langle\cdot\rangle^m W_2^\wedge(\cdot)\right)^\vee(x)$, $m \in \mathbb{R}$.

Let $|x| = 2^k$, $k \in \mathbb{Z}$, then

$$\left(\langle\cdot\rangle^m W_2^\wedge(\cdot)\right)^\vee(x) = \frac{1}{4}\int_{|\xi|\leqslant 1}\chi_\xi(x)d\xi + \frac{1}{4}2^m\int_{|\xi|=2^1}\chi_\xi(x)d\xi$$

$$+ \sum_{l=2}^{+\infty}\int_{|\xi|=2^l}\left\{-\left(1+\overline{\chi_\xi}(\beta)\right)\left(1+\overline{\chi_\xi}(\beta^2)\right)\cdots\right.$$

$$\left.\cdot\left(1+\overline{\chi_\xi}(\beta^{l-2})\right)\right\}\left(\frac{1}{2}\right)^{2l-1}2^{lm}\chi_\xi(x)d\xi$$

$$\equiv J_1 + J_2 + J_3,$$

where

$$J_1 = \frac{1}{4}\int_{|\xi|\leqslant 1} \chi_\xi(x)d\xi = \frac{1}{4}\begin{cases} 1, & k \leqslant 0, \\ 0, & k > 0, \end{cases}$$

$$J_2 = \frac{1}{4}2^m \int_{|\xi|=2^1} \chi_\xi(x)d\xi = \frac{2^m}{4}\begin{cases} 1, & k \leqslant -1, \\ -1, & k = 0, \\ 0, & k > 0, \end{cases}$$

$$J_3 = \sum_{l=2}^{+\infty} I_l,$$

with

$$I_l = \int_{|\xi|=2^l} \left\{ -\left(1+\overline{\chi}_\xi(\beta)\right)\left(1+\overline{\chi}_\xi(\beta^2)\right)\cdots\left(1+\overline{\chi}_\xi(\beta^{l-2})\right)\left(\frac{1}{2}\right)^{2l-1} 2^{lm} \right\} \chi_\xi(x)d\xi$$

$$= -2^{lm-2l+1} \int_{|\xi|=2^l} \left\{ \left(1+\overline{\chi}_\xi(\beta)\right)\left(1+\overline{\chi}_\xi(\beta^2)\right)\cdots\left(1+\overline{\chi}_\xi(\beta^{l-2})\right)\chi_\xi(x) \right\} d\xi.$$

Let

$$E_l = \left\{ \xi \in \Gamma_{K_2} : |\xi| = 2^l, \chi_\xi(\beta) = \chi_\xi(\beta^2) = \cdots = \chi_\xi(\beta^{l-2}) = 1 \right\},$$

and $\chi_\xi(\beta), \chi_\xi(\beta^2), \cdots, \chi_\xi(\beta^{l-2})$ take values only $+1$ or -1, then

$$I_l = -2^{lm-2l+1} \int_{|\xi|=2^l} \left\{ \left(1+\overline{\chi}_\xi(\beta)\right)\left(1+\overline{\chi}_\xi(\beta^2)\right)\cdots\left(1+\overline{\chi}_\xi(\beta^{l-2})\right)\chi_\xi(x) \right\} d\xi$$

$$= -2^{lm-l-1} \int_{E_l} \chi_\xi(x)d\xi.$$

Since $E_l = \beta^{-l} + B_{-1}$, thus

$$I_l = -2^{lm-l-1}\int_{E_l}\chi_\xi(x)d\xi = -2^{lm-l-1}\int_{\beta^{-l}+B_{-1}}\chi_\xi(x)d\xi$$

$$= -2^{lm-l-1}\int_{B_{-1}}\chi_{\beta^{-l}+\eta}(x)d\eta = -2^{lm-l-1}\chi(\beta^{-l}x)\int_{B_{-1}}\chi_\eta(x)d\eta$$

$$= -2^{lm-l-1}\chi(\beta^{-l}x)\begin{cases} 2, & k \leqslant -1 \\ 0, & k > -1 \end{cases} = \begin{cases} -2^{lm-l}\chi(\beta^{-l}x), & k \leqslant -1, \\ 0, & k > -1. \end{cases}$$

Combining the above results, it follows

$$\left(\langle \cdot \rangle^m W_2^\wedge(\cdot)\right)^\vee(x) = J_1 + J_2 + \sum_{l=2}^{+\infty} I_l.$$

Moreover, $\forall x \in B_1$, $x = \sum_{j=1}^{+\infty} x_j \beta^j$, $x_j \in \{0,1\}$, $j \in \mathbb{N}$, and

$$\chi\left(\beta^{-l} x\right) = \chi\left(x_{-1+l}\beta^{-1}\right) = (-1)^{x_{l-1}} = 1 - 2x_{l-1},$$

then, it follows

$$W_2^{\langle m \rangle}(x) = \begin{cases} \dfrac{1}{4} + \dfrac{2^m}{4} + \sum_{l=2}^{+\infty} \left(-2^{lm-l}(1 - 2x_{l-1})\right), & x \in B^1, \\ \dfrac{1}{4} - \dfrac{2^m}{4}, & x \in D \setminus B^1, \\ 0, & x \notin D. \end{cases} \quad (5.3.6)$$

For the series $\sum_{l=2}^{+\infty} \left(-2^{lm-l}(1-2x_{l-1})\right)$ in (5.3.6), if $m \geqslant 1$, since $-2^{lm-l} \cdot (1-2x_{l-1}) \in \{1, -1\}$, thus $\lim_{l \to +\infty} \left\{-2^{lm-l}(1-2x_{l-1})\right\} \neq 0$. This implies that series $\sum_{l=2}^{+\infty}(-2^{lm-l}(1-2x_{l-1}))$ is not convergent for any $m \geqslant 1$, and it is divergent at any point in B^1; if $m < 1$, then

$$\sum_{l=2}^{+\infty} \left|-2^{lm-l}(1-2x_{l-1})\right| \leqslant \sum_{l=2}^{+\infty} 2^{lm-l} = \sum_{l=2}^{+\infty} \left(2^{m-1}\right)^l < +\infty.$$

Thus, the series $\sum_{l=2}^{+\infty} \left(-2^{lm-l}(1-2x_{l-1})\right)$ is absolutely convergent, if and only if $m < 1$.

Finally, if $0 \leqslant m < 1$, one has

$W_2^{\langle m \rangle}(x)$

$$= \begin{cases} \dfrac{1}{4} + \dfrac{2^m}{4} - \dfrac{2^{2m-2}}{1-2^{m-1}} + \sum_{j=1}^{+\infty} x_j \left(\dfrac{1}{2}\right)^{j-(j+1)m}, & |x| \leqslant 2^{-1}, \text{ i.e. } x \in B^1, \\ \dfrac{1}{4} - \dfrac{2^m}{4}, & |x| = 1, \text{ i.e. } x \in D \setminus B^1, \\ 0, & \text{others.} \end{cases}$$

The proof is complete.

Theorem 5.3.4 The Weierstrass type function $W_2 : K_2 \to \mathbb{R}$, $x = \sum_{j=s}^{+\infty} x_j \beta^j$, $s \in \mathbb{Z}$,

$$W_2(x) = W_2\left(\sum_{j=s}^{+\infty} x_j \beta^j\right) = \begin{cases} \sum_{j=1}^{+\infty} x_j \left(\dfrac{1}{2}\right)^j, & x \in B^1, \\ 0, & x \notin B^1 \end{cases}$$

is continuous at any $x \in B^1$.

Proof. $\forall x = \sum_{j=1}^{+\infty} x_j \beta^j \in B^1$, $\forall \varepsilon > 0$, take $K = \left[\log_{\frac{1}{2}} \dfrac{\varepsilon}{4}\right]$, then $\forall k > K$, $\forall y \in B^k$, $|W_2(x+y) - W_2(x)| \leqslant \sum_{j=k}^{+\infty} 2\left(\dfrac{1}{2}\right)^k = 4\left(\dfrac{1}{2}\right)^k < \varepsilon$, this implies the continuity of $W_2(x)$.

We can conclude that: the Weierstrass type function $W_2 : K_2 \to \mathbb{R}$ is continuous, m-order ($0 \leqslant m < 1$) p-type differentiable, however, it does not have 1 order p-type derivative; moreover, it is infinitely p-type integralble.

We now may generalize the definition of $W_2 : K_2 \to \mathbb{R}$ in (5.3.5) to that on K_p, denoted by $W_p : K_p \to \mathbb{C}$,

$$W_p(x) = W_p\left(\sum_{j=s}^{+\infty} x_j \beta^j\right) = \sum_{j=s}^{+\infty} x_j \left(\dfrac{1}{p}\right)^j,$$

with $x = \sum_{j=s}^{+\infty} x_j \beta^j$, $x_j \in \{0, 1, \cdots, p-1\}$, $j = s, s+1, \cdots$, $s \in \mathbb{Z}$, $|\beta| = p^{-1}$.

The Theorem 5.3.3 and Theorem 5.3.4 hold for $W_p(x)$. But the expression of p-type derivative $W_p^{\langle m \rangle}(x)$ is more complex[54].

3. Examples of the Weierstrass type functions and p-type calculus of them on K_p

(i) $W_2^{\langle 0 \rangle}(x) = W_2(x)$.

(ii) $W_2^{\langle \frac{1}{2} \rangle}(x) = \begin{cases} -\dfrac{3}{4} - \dfrac{\sqrt{2}}{4} + \sqrt{2} \sum_{j=1}^{+\infty} x_j \left(\dfrac{1}{\sqrt{2}}\right)^j, & x \in B^1, \\ \dfrac{1 - \sqrt{2}}{4}, & x \in D \backslash B^1, \\ 0, & \text{others.} \end{cases}$

(iii) $(W_2)_{\langle 1 \rangle}(x) = \begin{cases} \dfrac{7}{24} + \dfrac{1}{2}\sum\limits_{j=1}^{+\infty} x_j \left(\dfrac{1}{4}\right)^j, & x \in B^1, \\ \dfrac{1}{8}, & x \in D\backslash B^1, \\ 0, & \text{others.} \end{cases}$

The graphs in Fig. 5.3.5 are those of $W_2^{\langle \frac{1}{2} \rangle}(x)$ and $(W_2)_{\langle 1 \rangle}(x)$.

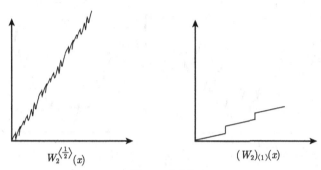

Figure 5.3.5

5.3.4 p-type derivative and integral of second Weierstrass type function on K_p

1. The second Weierstrass type function on K_p

There is the other form of Weierstrass type function defined on local field K_p, motivated by some classical forms. Recall one Weierstrass function on $\mathbb{R}^1$,

$$W(x) = \sum_{k=0}^{+\infty} \alpha^k \cos\left(\beta^k \pi x\right), \quad 0 < \alpha < 1, \quad \alpha\beta > 1 + \frac{\pi}{2}(1-\alpha),$$

since $\cos\left(\beta^k \pi x\right) = \operatorname{Re}(e^{\beta^k \pi x \mathrm{i}})$ motivates us to consider real part $\operatorname{Re}\chi_\xi(x)$ of character function $\chi_\xi(x)$, and define the second Weierstrass type function as

$$W(x) = \sum_{k=1}^{+\infty} p^{(s-2)k} \operatorname{Re}\chi\left(\beta^{-k} x\right), \quad 1 \leqslant s < 2, \quad x \in D. \tag{5.3.7}$$

Fig. 5.3.6 consists the graphs of the second Weierstrass type function $W(x)$ for $p = 3$, $s = 1.5$; $p = 5$, $s = 1.55$; $p = 7$, $s = 1.55$; $p = 11$, $s = 1.45$, respectively.

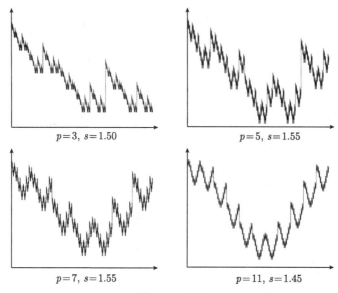

Figure. 5.3.6

2. The p-type differentiability and integrability of second Weierstrass type function

We study the second Weierstrass type function $W(x)$ in (5.3.7) in this subsection.

Lemma 5.3.4 *For the function* $\text{Re}\chi\left(\beta^{-k}x\right)$ *defined on* $D \subset K_p$, *if* $k \geqslant 1$ *and* $m \in \mathbb{R}$, *then*

$$\left\{\text{Re}\chi\left(\beta^{-k}x\right)\right\}^{\langle m \rangle} = p^{km}\text{Re}\chi\left(\beta^{-k}x\right), \quad x \in D.$$

Proof. ① Evaluate the Fourier transformation of $\text{Re}\chi\left(\beta^{-k}x\right)$
Consider $\forall x \in D$, $x = x_0\beta^0 + x_1\beta^1 + \cdots$, and $|\xi| = p^l$, $l \in \mathbb{Z}$, then

$$\left[\text{Re}\chi\left(\beta^{-k}\cdot\right)\right]^{\wedge}(\xi)$$

$$= \int_D \text{Re}\chi\left(\beta^{-k}x\right)\overline{\chi}_\xi(x)dx = \int_D \cos\frac{2\pi x_{k-1}}{p}\overline{\chi}_\xi(x)dx$$

$$= \sum_{x_{k-1}=0}^{p-1}\left(\cos\frac{2\pi x_{k-1}}{p}\sum_{0\leqslant x_i<p,0\leqslant i\leqslant k-2}\int_{x_0\beta^0+x_1\beta^1+\cdots+x_{k-1}\beta^{k-1}+B^k}\overline{\chi}_\xi(x)dx\right)$$

$$= \sum_{x_{k-1}=0}^{p-1}\left(\cos\frac{2\pi x_{k-1}}{p}\sum_{0\leqslant x_i<p,0\leqslant i\leqslant k-2}\overline{\chi}_\xi\left(x_0\beta^0+x_1\beta^1+\cdots+x_{k-1}\beta^{k-1}\right)\int_{B^k}\overline{\chi}_\xi(x)dx\right)$$

$$= \sum_{x_{k-1}=0}^{p-1} \left(\cos \frac{2\pi x_{k-1}}{p} \sum_{0 \leqslant x_i < p, 0 \leqslant i \leqslant k-2} \overline{\chi_\xi} \left(x_0 \beta^0 + x_1 \beta^1 + \cdots + x_{k-1} \beta^{k-1} \right) p^{-k} \right)$$

$$= \sum_{x_{k-1}=0}^{p-1} \left(\cos \frac{2\pi x_{k-1}}{p} \sum_{0 \leqslant x_i < p, 0 \leqslant i \leqslant k-2} \left(\overline{\chi_\xi} \left(\beta^0 \right) \right)^{x_0} \left(\overline{\chi_\xi} \left(\beta^1 \right) \right)^{x_1} \cdots \left(\overline{\chi_\xi} \left(\beta^{k-1} \right) \right)^{x_{k-1}} p^{-k} \right),$$

for $l \leqslant k$, here we use the integral $\int_{B^k} \overline{\chi_\xi}(x)dx = \begin{cases} p^{-k}, & l \leqslant k, \\ 0, & l > k. \end{cases}$ Thus, the Fourier transformation of $\text{Re}\chi\left(\beta^{-k}x\right)$ is zero when $l > k$. When $l \leqslant k$, we evaluate: If $0 \leqslant i \leqslant k-2$, then

$$\xi_{-(i+1)} \neq 0 \Rightarrow \chi_\xi\left(\beta^i\right) \neq 1 \Rightarrow \sum_{0 \leqslant x_i < p} \left(\overline{\chi_\xi}\left(\beta^i\right)\right)^{x_i} = 0,$$

$$\xi_{-(i+1)} = 0 \Rightarrow \chi_\xi\left(\beta^i\right) = 1 \Rightarrow \sum_{0 \leqslant x_i < p} \left(\overline{\chi_\xi}\left(\beta^i\right)\right)^{x_i} = p,$$

these imply that for $l \leqslant k$, $\xi_{-1} = \cdots = \xi_{-(k-1)} = 0$,

$$\int_D \text{Re}\chi\left(\beta^{-k}x\right) \overline{\chi_\xi}(x)dx = \frac{1}{p} \sum_{x_{k-1}=0}^{p-1} \cos \frac{2\pi x_{k-1}}{p} \left(\overline{\chi_\xi}\left(\beta^{k-1}\right)\right)^{x_{k-1}},$$

otherwise, $\int_D \text{Re}\chi\left(\beta^{-k}x\right) \overline{\chi_\xi}(x)dx = 0$. So that, we have three cases.

case 1. If $1 \leqslant l < k$, then $\xi_{-1} = \cdots = \xi_{-(k-1)} = 0$ can not be true, thus

$$\int_D \text{Re}\chi\left(\beta^{-k}x\right) \overline{\chi_\xi}(x)dx = 0.$$

case 2. If $l < 1$, then $\xi_{-1} = \cdots = \xi_{-(k-1)} = \xi_{-k} = 0$ can be true, certainly, thus

$$\int_D \text{Re}\chi\left(\beta^{-k}x\right) \overline{\chi_\xi}(x)dx = \frac{1}{p} \sum_{x_{k-1}=0}^{p-1} \cos \frac{2\pi x_{k-1}}{p} \left(\overline{\chi_\xi}\left(\beta^{k-1}\right)\right)^{x_{k-1}}$$

$$= \frac{1}{p} \sum_{x_{k-1}=0}^{p-1} \cos \frac{2\pi x_{k-1}}{p} = 0.$$

case 3. If $l = k$, and $\xi_{-1} = \cdots = \xi_{-(k-1)} = 0$, $\xi_{-k} \neq 0$, thus $\xi = \xi_{-k}\beta^{-k} + \xi_0\beta^0 + \cdots$,

$$\int_D \text{Re}\chi\left(\beta^{-k}x\right) \overline{\chi_\xi}(x)dx = \frac{1}{p} \sum_{x_{k-1}=0}^{p-1} \cos \frac{2\pi x_{k-1}}{p} \left(\overline{\chi_\xi}\left(\beta^{k-1}\right)\right)^{x_{k-1}}$$

$$= \frac{1}{p} \sum_{j=0}^{p-1} \cos \frac{2\pi j}{p} (\overline{\omega})^{j\xi_{-k}}.$$

Combining the above cases, for $\xi = \xi_{-k}\beta^{-k} + \xi_0\beta^0 + \cdots$, $\xi_{-k} \neq 0$, it follows

$$\left[\text{Re}\chi\left(\beta^{-k}\cdot\right)\right]^{\wedge}(\xi) = \int_D \text{Re}\chi\left(\beta^{-k}x\right) \overline{\chi}_\xi(x)dx = \frac{1}{p} \sum_{j=0}^{p-1} \cos \frac{2\pi j}{p} (\overline{\omega})^{j\xi_{-k}}.$$

② Evaluate the p-type derivative and p-type integral of $\text{Re}\chi\left(\beta^{-k}x\right)$

$$\left(\text{Re}\chi\left(\beta^{-k}x\right)\right)^{\langle m \rangle} = \int_{\Gamma_p} \langle\xi\rangle^m \left\{\int_D \text{Re}\chi\left(\beta^{-k}y\right) \overline{\chi}_\xi(y)dy\right\} \chi_x(\xi)d\xi$$

$$= \sum_{t=1}^{p-1} p^{km-1} \left\{\int_{t\beta^{-k}+D} \sum_{j=0}^{p-1} \cos \frac{2\pi j}{p} (\overline{\omega})^{tj} \chi_x(\xi)d\xi\right\}$$

$$= \sum_{t=1}^{p-1} p^{km-1} \left\{\sum_{j=0}^{p-1} \cos \frac{2\pi j}{p} (\overline{\omega})^{tj} \chi_x\left(t\beta^{-k}\right)\right\} \int_D \chi_x(\xi)d\xi$$

$$= \sum_{t=1}^{p-1} p^{km-1} \left\{\sum_{j=0}^{p-1} \cos \frac{2\pi j}{p} (\overline{\omega})^{tj} \chi_x\left(t\beta^{-k}\right)\right\} \quad (x \in D)$$

$$= \sum_{t=1}^{p-1} p^{km-1} \left\{\sum_{j=0}^{p-1} \cos \frac{2\pi j}{p} (\overline{\omega})^{tj} \left(\chi\left(\beta^{-k}x\right)\right)^t\right\}$$

$$= p^{km-1} \sum_{t=1}^{p-1} \left\{\sum_{j=0}^{p-1} \cos \frac{2\pi j}{p} (\overline{\omega})^{tj} \left(\chi\left(\beta^{-k}x\right)\right)^t\right\}$$

$$= p^{km-1} \sum_{j=0}^{p-1} \cos \frac{2\pi j}{p} \sum_{t=1}^{p-1} (\overline{\omega})^{tj} \left(\chi\left(\beta^{-k}x\right)\right)^t.$$

Note that, for $0 \leqslant j < p$, if $\chi\left(\beta^{-k}x\right) = \omega^j$, then

$$\sum_{t=1}^{p-1} (\overline{\omega})^{tj} \left(\chi\left(\beta^{-k}x\right)\right)^t = p - 1;$$

otherwise, $\sum_{t=1}^{p-1} (\overline{\omega})^{tj} \left(\chi\left(\beta^{-k}x\right)\right)^t = -1.$

Thus, for $x \in D$ and $k \geqslant 1, m \in \mathbb{R}$, it follows

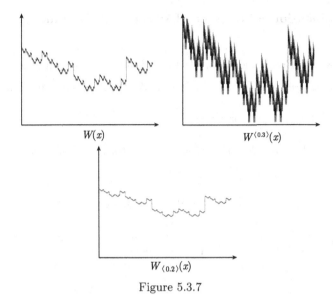

Figure 5.3.7

$$\left(\operatorname{Re}\chi\left(\beta^{-k}x\right)\right)^{\langle m \rangle} = p^{km-1}\left(p\operatorname{Re}\chi\left(\beta^{-k}x\right) - \sum_{j=0}^{p-1}\cos\frac{2\pi j}{p}\right) = p^{km}\operatorname{Re}\chi\left(\beta^{-k}x\right).$$

The proof is complete.

Theorem 5.3.5 *The second Weierstrass type function at* $x = \sum_{j=0}^{+\infty} x_j \beta^j$,

$$W(x) = \sum_{k=1}^{+\infty} p^{(s-2)k}\operatorname{Re}\chi\left(\beta^{-k}x\right), \quad 1 \leqslant s < 2, \quad x \in D$$

is infinitely p-type integrable; and is m-order p-type differentiable with $0 \leqslant m < 2 - s$; *and*

$$W^{\langle m \rangle}(x) = \sum_{k=1}^{+\infty} p^{(s+m-2)k}\operatorname{Re}\chi\left(\beta^{-k}x\right), \quad x \in D.$$

One may complete the proof by the Lemma 5.3.4 and the convergence of the series.

It is clear that the Weiestrasstype function $W_2(x)$ is a special case of $W(x)$, when $p = 2, s = 1$. Fig. 5.3.7 depicts the graphs of $W(x)$, $W^{\langle 0.3 \rangle}(x)$, $W_{\langle 0.2 \rangle}(x)$ for $p = 5$ and $s = 1.45$.

3. Dimensions of the second Weierstrass type function $W(x)$

We study the Box dimension, Packing dimension and Hausdorff dimension of the graph $\mathbb{G}(W(x), D)$ of second Weierstrass type function $W(x) = \sum_{k=1}^{+\infty} p^{(s-2)k} \operatorname{Re}\chi(\beta^{-k}x)$, for $1 \leqslant s < 2$.

Lemma 5.3.5 For the second Weierstrass type function $W(x)$, we have for $p = 2$,

$$\sup_{x \in D} W(x) = W(0) = \frac{2^{s-2}}{1 - 2^{s-2}},$$

$$\inf_{x \in D} W(x) = W(\beta^0 + \beta^1 + \cdots) = \frac{-2^{s-2}}{1 - 2^{s-2}};$$

for $p > 2$,

$$\sup_{x \in D} W(x) = W(0) = \frac{p^{s-2}}{1 - p^{s-2}},$$

$$\inf_{x \in D} W(x) = W\left(\frac{p-1}{2}\beta^0 + \frac{p-1}{2}\beta^1 + \cdots\right) = \frac{p^{s-2}}{1 - p^{s-2}} \cos\frac{(p-1)\pi}{p}.$$

Proof. The proof can be obtained by the definition

$$W(x) = \sum_{k=1}^{+\infty} p^{(s-2)k} \operatorname{Re}\chi(\beta^{-k}x) = \sum_{k=1}^{+\infty} p^{(s-2)k} \cos\left(\frac{2\pi}{p} x_{k-1}\right).$$

Lemma 5.3.6 For $W(x)$, $\forall x \in y + B^n$, $\forall y = y_0\beta^0 + y_1\beta^1 + \cdots + y_{n-1}\beta^{n-1}$, with $n \in \mathbb{N}$, it follows that

$$W(x) - W(y) = p^{(s-2)n}\left(W\left(\beta^{-n}(x-y)\right) - W(0)\right).$$

Proof. $\forall x \in y + B^n$, it follows

$$W(x) = \sum_{k=1}^{n} p^{(s-2)k} \cos\left(\frac{2\pi}{p} y_{k-1}\right) + \sum_{k=n+1}^{+\infty} p^{(s-2)k} \cos\left(\frac{2\pi}{p} x_{k-1}\right)$$

$$= W(y) - \sum_{k=n+1}^{+\infty} p^{(s-2)k} + \sum_{k=n+1}^{+\infty} p^{(s-2)k} \cos\left(\frac{2\pi}{p} x_{k-1}\right)$$

$$= W(y) - \frac{p^{(s-2)(n+1)}}{1 - p^{s-2}} + p^{(s-2)n} \sum_{k=1}^{+\infty} p^{(s-2)k} \cos\left(\frac{2\pi}{p} x_{k+n-1}\right)$$

$$= W(y) + p^{(s-2)n}\left(W\left(\beta^{-n}(x-y)\right) - \frac{p^{s-2}}{1-p^{s-2}}\right),$$

then by Lemma 5.3.5, the lemma is proved.

Lemma 5.3.7 *For the second Weierstrass type function $W(x)$, if $n \in \mathbb{N}$, and $\forall x \in y + B^n$, $y = y_0\beta^0 + y_1\beta^1 + \cdots + y_{n-1}\beta^{n-1}$, we have*

$$\sup\{|W(x) - W(x')| : x, x' \in y + B^n\} = 2^{(s-2)n+1}\frac{2^{s-2}}{1-2^{s-2}}$$

when $p = 2$; and

$$\sup\{|W(x) - W(x')| : x, x' \in y + B^n\} = p^{(s-2)n}\left(1 - \cos\frac{(p-1)\pi}{p}\right)\frac{p^{s-2}}{1-p^{s-2}}$$

when $p > 2$.

For the Box dimension and Packing dimension of the graph of $W(x)$, we have

Theorem 5.3.6 *For the second Weierstrass type function*

$$W(x) = \sum_{k=1}^{+\infty} p^{(s-2)k}\operatorname{Re}\chi\left(\beta^{-k}x\right), \quad 1 \leqslant s < 2, \quad x \in D,$$

the Box dimension and Packing dimension of graph $\mathbb{G}(W(x), D)$ are

$$\dim_B \mathbb{G}(W(x), D) = \dim_P \mathbb{G}(W(x), D) = s.$$

Proof. We prove for $p > 2$. By the Lemma 5.3.7, it has

$$p^n N_n(\mathbb{G}(W(x), D)) = p^n \frac{p^{(s-2)n}\left(1 - \cos\frac{(p-1)\pi}{p}\right)\frac{p^{s-2}}{1-p^{s-2}}}{p^{-n}} = p^{sn}.$$

Since $\forall n \in \mathbb{N}$,

$$\frac{\ln(p^n N_n(\mathbb{G}))}{\ln(p^n)} = \frac{\ln(p^{sn})}{\ln(p^n)} = s,$$

then $\dim_B \mathbb{G}(W(x), D)$ exists, and $\dim_B \mathbb{G}(W(x), D) = s$. Thus, by Theorem 5.1.9,

$$\dim_B \mathbb{G}(W(x), D) = \dim_P \mathbb{G}(W(x), D) = s.$$

Next, to discuss the Hausdorff dimension of the graph $\mathbb{G}(W(x), D)$ of second Weiestrass type function $W(x)$, we define the following Borel probability measure and image measure of Borel sets.

Definition 5.3.2 (Borel probability measure) *For a real function $f : A \to \mathbb{R}$ defined on a set $A \subset K_p$ in local field K_p, and for a Borel set $U \subset K_p \times \mathbb{R}$ in the product space $K_p \times \mathbb{R}$, denoted by*

$\nu(U) = |\{x \in A : (x, f(x)) \in U\}|$, then ν is said to be a **Borel probability measure of** U.

Definition 5.3.3 (Image measure) For a Borel measurable function $f : D \to \mathbb{R}$ defined on $D \subset K_p$, then the **image measure** μ_f **with respect to the Haar measure on** D **under** f is defined as

$$\mu_f(E) = |f^{-1}(E)|, \quad \forall E \subset \mathbb{R}.$$

If μ_f is absolutely continuous with respect to the Lebesgue measure on $\mathbb{R}$, then by the Radon–Nidodym Theorem[30], there exists a Borel measurable function $\alpha_f : E \to \mathbb{R}$, such that $\mu_f(E) = \int_E \alpha_f(x)dx$.

Theorem 5.3.7 For the second Weierstrass type function

$$W(x) = \sum_{k=1}^{+\infty} p^{(s-2)k} \mathrm{Re}\chi\left(\beta^{-k}x\right), \quad 1 \leqslant s < 2, \quad x \in D,$$

if its image measure μ_W is absolutely continuous, and thus $\alpha_W(x)$ exists, $\alpha_W \in L^\infty(D)$, then the Hausdorff dimension of the graph $\mathbb{G}(W(x), D)$ is $\dim_H \mathbb{G}(W(x), D) = s$.

Proof. Fix $y = y_0\beta^0 + \cdots + y_{n-1}\beta^{n-1} + y_n\beta^n + \cdots \in D, n \in \mathbb{N}$. For a ball in $D \times \mathbb{R}$, $B((y, W(y)), p^{-n})$ with the center $(y, W(y)) \in D \times \mathbb{R}$ and radius p^{-n}, we evaluate

$$\nu\left(B\left((y, W(y)), p^{-n}\right)\right) = \left|\{x \in y + B^n : W(x) \in (W(y) - p^{-n}, W(y) + p^{-n})\}\right|.$$

Denote $y' = y_0\beta^0 + \cdots + y_{n-1}\beta^{n-1}$, and

$$J = \left(W(y) - W(y') - p^{-n}, W(y) - W(y') + p^{-n}\right).$$

By Lemma 5.3.6,

$$\begin{aligned}
\nu\left(B\left((y, W(y)), p^{-n}\right)\right) &= \left|\{x \in y + B^n : W(x) \in (W(y) - p^{-n}, W(y) + p^{-n})\}\right| \\
&= \left|\{x \in y' + B^n : W(x) - W(y') \in J\}\right| \\
&= \left|\{x \in B^n : p^{(s-2)n}\left(W(\beta^{-n}x) - W(0)\right) \in J\}\right| \\
&= p^{-n}\left|\{x \in D : p^{(s-2)n}(W(x) - W(0)) \in J\}\right| \\
&= p^{-n}\left|\{x \in D : p^{(s-2)n}W(x) \in p^{(s-2)n}W(0) + J\}\right| \\
&= p^{-n}\left|\{x \in D : W(x) \in p^{(2-s)n}\left(p^{(s-2)n}W(0) + J\right)\}\right| \\
&= p^{-n}\mu_W\left(p^{(2-s)n}\left(p^{(s-2)n}W(0) + J\right)\right) \\
&\leqslant 2cp^{-n}p^{(2-s)n}p^{-n} = 2c\left(p^{-n}\right)^s.
\end{aligned}$$

By the mass distribution principle[5], it follows $\dim_H \mathbb{G}(W(x), D) \geqslant s$. On the other hand, by Theorem 5.1.7 and Theorem 5.3.6, we have

$$s \leqslant \dim_H \mathbb{G}(W(x), D) \leqslant \dim_B \mathbb{G}(W(x), D) = s.$$

This shows the result of Theorem.

If we try to weaken the condition in Theorem 5.3.7 of μ_w to "$\exists \alpha_W(x) \in L^r(K_p), r > 1$", then the other two Lemmas are needed, and the proofs of the two lemmas are complex and technical. However, if one tries to remove the condition "$\exists \alpha_W(x) \in L^r(K_p), r > 1$", it is more difficult, we refer to [57], and state the result without proof.

Theorem 5.3.8 *For the second Weierstrass type function*

$$W(x) = \sum_{k=1}^{+\infty} p^{(s-2)k} \mathrm{Re}\chi\left(\beta^{-k}x\right), \quad 1 \leqslant s < 2, \quad x \in D.$$

(i) *If $p = 2$, then*

$$\dim_H \mathbb{G}(W(x), D) = s, \quad a.e. \quad s \in (1, 2).$$

(ii) *If $p > 2$, then*

$$\dim_H \mathbb{G}(W(x), D) = s, \quad a.e. \quad s \in \left(\log_p(2p-1), 2 + \log_p y(b_p)\right),$$

with $b_p = \dfrac{1 - \cos\dfrac{(p-1)\pi}{p}}{1 - \cos\dfrac{2\pi}{p}}$, $y(b_p)$ is the smallest positive number satisfying the equations

$$g(y(b_p)) = g'(y(b_p)) = 0,$$

where $g(x) = 1 + \sum_{j=1}^{+\infty} g_j x^j, g_j \in [-b, b]$, is some power series.

4. Dimensions of the graph of p-type derivative and integral of $W(x)$

Theorem 5.3.9 *The Box dimension, Packing dimension, Hausdorff dimension of the Graph $\mathbb{G}\left(W^{\langle m \rangle}(x), D\right)$ of p-type derivative and p-type integral of second Weierstrass type function $W(x) = \sum_{k=1}^{+\infty} p^{(s-2)k} \mathrm{Re}\chi\left(\beta^{-k}x\right)$, $1 \leqslant s < 2$, have the following relationships*

(i) $\dim_B \mathbb{G}\left(W^{\langle m \rangle}(x), D\right) = s + m$, $\forall m \in [1-s, 2-s)$;

$\dim_P \mathbb{G}\left(W^{\langle m \rangle}(x), D\right) = s + m$, $\forall m \in [1-s, 2-s)$.

(ii) If $p = 2$, then

$\dim_H \mathbb{G}\left(W^{\langle m \rangle}(x), D\right) = s + m$, a.e. $m \in (1-s, 2-s)$;

if $p > 2$, then

$\dim_H \mathbb{G}\left(W^{\langle m \rangle}(x), D\right) = s + m$, a.e. $m \in (a, b)$

with $(a, b) \equiv \left(\log_p(2p-1) - s, 2 + \log_p y(b_p) - s\right)$.

This Theorem has deep and important sense: it describes the linear relationships between the dimensions of a fractal function with its p-type derivatives. The topic is the most important one in the area of fractal analysis, and lots of open problems are waiting to be studied.

Exercises

1. Give the expression of the Cantor type set

$$C_3 = D \Big\backslash \bigcup_{j=1}^{+\infty} V_j = D \Big\backslash \left(\bigcap_{k=1}^{\infty} \left(\bigcup_{j=1}^{2^k} V_{k,j} \right) \right)$$

of $V_{k,j}$.

2. By using the Walsh-Fourier series to show the Cantor type function can be expressed as

$$\vartheta(x) = \frac{1}{2} - \frac{1}{12} \sum_{j=1}^{+\infty} \frac{1}{6^r} \Lambda(j_0) \cdots \Lambda(j_{-r+1})(1 - \omega^{j_r}) w_j(x),$$

where $\omega = \exp \dfrac{2\pi i}{3}$, and $\Lambda(j) = \begin{cases} 2, & j = 0, \\ \exp \dfrac{-2\pi i}{3} j, & j = 1, 2. \end{cases}$ Give the expressions of $j_0, j_{-1}, \cdots, j_{-r+1}$.

3. For $W_p(x)$, prove the Theorems 5.3.3 and 5.3.4.
4. For the second Weierstrass type function $W(x)$, study its Hausdorff dimension and the Hausdorff dimension of its p-type derivative $W^{\langle m \rangle}(x)$.
5. For the Cantor type function $C(x)$, Weierstrass type function $W_p(x)$, the second Weierstrass type function $W(x)$, study the Fourier dimension, the Fourier dimension of its p-type derivative $W^{\langle m \rangle}(x)$, and the relationships between other dimensions of them.

Chapter 6

Fractal PDE on Local Fields

The topics of ordinary and partial differential equations based upon local fields as underlying spaces are quite new, and so are those topics of fractal ODE and PDE on local fields.

As we know, the realm of ODE and PDE based upon $\mathbb{R}^n$ is so important in all classical and modern scientific area, the achievements in the realm have been applied widely, since every object in universe is in status of motion, so that it has rate of change, that is, velocity; and has acceleration. Moreover, majority of motions can be described by ODE or PDE.

However, lots of objects in nature do not have the Newton derivative which indicates velocity of an object, a typical and familiar example is the famous Weierstrass function, it does not have Newton derivative at any points. The fractals, regarded as objects, have no Newton derivative too. But these objects would have "velocity" since they still indicate certain motion. Thus a challenging task presents to mathematicians and physicists: What is rate of change, or "velocity", of a fractal? Find and define new calculus to fit new object—fractal. After introducing the fractal space $(\mathbb{K}(K_p), h)$ and p-type calculus, it is reasonable to develop a very new mathematical area — fractal differential equations.

In this chapter, we start from special examples, then discuss a general theory for fractal partial differential equations based upon a local field K_p as an underlying space.

6.1 Special examples

6.1.1 Classical 2-dimension wave equation with fractal boundary

1. The problem of classical 2-dimension wave equation with fractal boundary

Wave equation is one of three classical partial differential equations, its study in the area of PDE has the typical sense because it is to describe the membrane vibration. Recall that in classical case, the boundaries of Dirichlet problem of 2-dimension wave equations are smooth curves, or piecewise smooth curves, and the results are more accurate, complete, and deep. However, if the boundaries are not smooth but a fractal, for example, the **dyadic von Koch type curve** γ as a boundary of domain D, the Dirichlet type problem proposed as[35],[99]

$$\begin{cases} \dfrac{\partial^2 u(t,x,y)}{\partial t^2} = \dfrac{\partial^2 u(t,x,y)}{\partial x^2} + \dfrac{\partial^2 u(t,x,y)}{\partial y^2}, & t>0, (x,y) \in D, \\ u(t,x,y)|_{t=0} = \varphi(x,y), & (x,y) \in D, \\ \dfrac{\partial u(t,x,y)}{\partial t}\bigg|_{t=0} = \psi(x,y), & (x,y) \in D, \\ u(t,x,y)|_\gamma = 0, & t>0, \end{cases} \quad (6.1.1)$$

where $D \subset \mathbb{R}^2$ is a 2-dimension domain with boundary γ, the dyadic von Koch, $\gamma = \partial D$. Denote the approximation curves of fractal γ by $\gamma_1, \gamma_2, \cdots, \gamma_k, \cdots$, and corresponding domains are $D_1, D_2, \cdots, D_k, \cdots$, see Fig. 6.1.1. The initial functions $\varphi(x,y) = \lim\limits_{k \to +\infty} \varphi_k(x,y)$ and $\psi(x,y) = \lim\limits_{k \to +\infty} \psi_k(x,y)$ converge in D, where φ_k, ψ_k are defined in $D_k, k = 1, 2, \cdots$.

2. Solution of the 2-dimension wave equation with fractal boundary

Suppose that the solution of problem (6.1.1) is $u(t,x,y)$, we consider the kth-**approximation problem** given by

$$\begin{cases} \dfrac{\partial^2 u_k(t,x,y)}{\partial t^2} = \dfrac{\partial^2 u_k(t,x,y)}{\partial x^2} + \dfrac{\partial^2 u_k(t,x,y)}{\partial y^2}, & t>0, (x,y) \in D_k, \\ u_k(t,x,y)|_{t=0} = \varphi_k(x,y), & (x,y) \in D_k, \\ \dfrac{\partial u_k(t,x,y)}{\partial t}\bigg|_{t=0} = \psi_k(x,y), & (x,y) \in D_k, \\ u_k(t,x,y)|_{\gamma_k} = 0, & t>0, \end{cases} \qquad (6.1.2)$$

with k^{th}-**approximation boundary** γ_k.

To solve the equation $\dfrac{\partial^2 u_k(t,x,y)}{\partial t^2} = \dfrac{\partial^2 u_k(t,x,y)}{\partial x^2} + \dfrac{\partial^2 u_k(t,x,y)}{\partial y^2}$, we suppose that

$$u_k(t,x,y) = T_k(t) v_k(x,y).$$

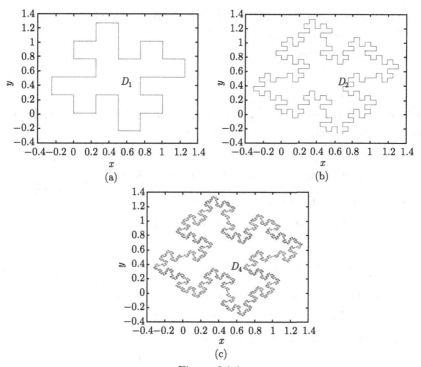

Figure 6.1.1

Then, we have
$$\dfrac{d^2 T_k(t)}{dt^2} v_k(x,y) = T_k(t) \left\{ \dfrac{\partial^2 v_k(x,y)}{\partial x^2} + \dfrac{\partial^2 v_k(x,y)}{\partial y^2} \right\}. \qquad (6.1.3)$$

If $T_k(t) \neq 0$, $v_k(x,y) \neq 0$, then by the method of separation of variables, it follows

$$\frac{1}{v_k(x,y)} \left\{ \frac{\partial^2 v_k(x,y)}{\partial x^2} + \frac{\partial^2 v_k(x,y)}{\partial y^2} \right\} = \frac{1}{T_k(t)} \frac{d^2 T_k(t)}{dt^2}. \qquad (6.1.4)$$

Let the right side be $-(\lambda^2 + \mu^2)$, with positive constants λ, μ. Then,

$$\frac{\partial^2 v_k(x,y)}{\partial x^2} + \frac{\partial^2 v_k(x,y)}{\partial y^2} = -(\lambda^2 + \mu^2) v_k(x,y), \qquad (6.1.5)$$

and

$$\frac{d^2 T_k(t)}{dt^2} + (\lambda^2 + \mu^2) T_k(t) = 0. \qquad (6.1.6)$$

For equation (6.1.5), we have a special solution

$$v_k(x,y) = \sin \lambda x \sin \mu y. \qquad (6.1.7)$$

Determine constants λ and μ by the boundary condition $v_k(x,y)|_{\gamma_k} = 0$.
For $k \in \mathbb{N}$, consider the problem

$$\begin{cases} \dfrac{\partial^2 v_k(x,y)}{\partial x^2} + \dfrac{\partial^2 v_k(x,y)}{\partial y^2} = -(\lambda^2 + \mu^2) v_k(x,y), & (x,y) \in D_k, \\ v_k(x,y)|_{\gamma_k} = 0, \end{cases} \qquad (6.1.8)$$

where γ_k is the kth-approximation of von Koch type curve γ, and the domain of γ_k is D_k, it is the kth-approximation of domain D.

① When $k = 1$.

By Fig. 6.1.1(a), the values of x and y in D_1 are in $\left[-\dfrac{1}{4}, \dfrac{5}{4}\right]$. Since $\sin \lambda x \sin \mu y|_{\gamma_1} = 0$, we have the eigen-values of (6.1.7),

$$\lambda = 4^1 m\pi, \quad m = 1, 2, \cdots \quad \text{and} \quad \mu = 4^1 n\pi, \quad n = 1, 2, \cdots.$$

② When $k = 2$.

By Fig. 6.1.1(b), the values of x and y in D_2 are in $\left[-\dfrac{5}{16}, \dfrac{21}{16}\right]$. Since $\sin \lambda x \sin \mu y|_{\gamma_2} = 0$, we have the eigen-values of (6.1.7)

$$\lambda = 4^2 m\pi, \quad m = 1, 2, \cdots \quad \text{and} \quad \mu = 4^2 n\pi, \quad n = 1, 2, \cdots.$$

③ In general, when $k \in \mathbb{N}$, we have the eigen-values of (6.1.7),

$$\lambda = 4^k m\pi, \quad m = 1, 2, \cdots, \quad k \in \mathbb{N}$$

and

$$\mu = 4^k n\pi, \quad n = 1, 2, \cdots, \quad k \in \mathbb{N}.$$

Thus, for $k \in \mathbb{N}$, the problem (6.1.8) has solution

$$v_{k,m,n}(x,y) = \sin 4^k m\pi x \sin 4^k n\pi y.$$

Correspondingly, the solution of equation (6.1.6) is

$$T_{k,m,n}(t) = A_{k,m,n} \cos 4^k \sqrt{m^2+n^2} \pi t + B_{k,m,n} \sin 4^k \sqrt{m^2+n^2} \pi t,$$

where $A_{k,m,n}$ and $B_{k,m,n}$ are constants depending on k, m, n.

Summarize the above results, the problem

$$\begin{cases} \dfrac{\partial^2 u_k(t,x,y)}{\partial t^2} = \dfrac{\partial^2 u_k(t,x,y)}{\partial x^2} + \dfrac{\partial^2 u_k(t,x,y)}{\partial y^2}, & t>0, (x,y)\in D_k, \\ u_k(t,x,y)|_{\gamma_k} = 0, & t>0 \end{cases}$$

has a formal solution

$$u_k(t,x,y) = \sum_{m=1}^{+\infty}\sum_{n=1}^{+\infty} T_{k,m,n}(t) v_{k,m,n}(x,y)$$

$$= \sum_{m=1}^{+\infty}\sum_{n=1}^{+\infty} \left\{ A_{k,m,n} \cos 4^k \sqrt{m^2+n^2}\pi t + B_{k,m,n} \sin 4^k \sqrt{m^2+n^2}\pi t \right\}$$

$$\cdot \sin 4^k m\pi x \cdot \sin 4^k n\pi y.$$

This is a lacunary series, to determine $A_{k,m,n}$ and $B_{k,m,n}$, we suppose that $\varphi_k(x,y)$ and $\psi_k(x,y)$ can be expressed as lacunary series

$$\varphi_k(x,y) = \sum_{m=1}^{+\infty}\sum_{n=1}^{+\infty} a_{m,n}^k \sin 4^k m\pi x \sin 4^k n\pi y,$$

$$\psi_k(x,y) = \sum_{m=1}^{+\infty}\sum_{n=1}^{+\infty} b_{m,n}^k \sin 4^k m\pi x \sin 4^k n\pi y,$$

and $\sum_{m=1}^{+\infty}\sum_{n=1}^{+\infty} |a_{m,n}^k| < +\infty$, $\sum_{m=1}^{+\infty}\sum_{n=1}^{+\infty} |b_{m,n}^k| < +\infty$. By the method of Fourier series, we have

$$A_{k,m,n} = 4 \int_{D_k} \varphi_k(x,y) \sin 4^k m\pi x 4^k n\pi y dx dy, \qquad (6.1.9)$$

$$B_{k,m,n} = \dfrac{4}{4^k \pi \sqrt{m^2+n^2}} \int_{D_k} \psi_k(x,y) \sin 4^k m\pi x \sin 4^k n\pi y dx dy. \qquad (6.1.10)$$

Hence, a formal solution of (6.1.1) is obtained

$$u(t,x,y) = \lim_{k\to+\infty} u_k(t,x,y)$$

$$= \lim_{k \to +\infty} \sum_{m=1}^{+\infty} \sum_{n=1}^{+\infty} \left\{ A_{k,m,n} \cos 4^k \sqrt{m^2+n^2}\pi t + B_{k,m,n} \sin 4^k \sqrt{m^2+n^2}\pi t \right\}$$
$$\cdot \sin 4^k m\pi x \cdot \sin 4^k n\pi y, \qquad (6.1.11)$$

the coefficients $A_{k,m,n}$ and $B_{k,m,n}$ of (6.1.11) are determined by (6.1.9) and (6.1.10).

How about differentiability of the function $u(t,x,y) = \lim_{k \to +\infty} u_k(t,x,y)$ in (6.1.11)?

Lemma 6.1.1 *Let function $f(x)$ have derivative $f'(x_0)$ at x_0; and $\eta > 0$ be a constant. Suppose that $\{x_n\}$ and $\{\bar{x}_n\}$ are two real sequences satisfying $x_n \leqslant x_0 \leqslant \bar{x}_n$ and $\bar{x}_n - x_n \geqslant \eta(\bar{x}_n - x_0)$; and $\lim_{n \to +\infty} \bar{x}_n = x_0 = \lim_{n \to +\infty} x_n$. Then we have*

$$\lim_{n \to +\infty} \frac{f(\bar{x}_n) - f(x_n)}{\bar{x}_n - x_n} = f'(x_0).$$

Theorem 6.1.1 *Let $\varphi(x)$ be a bounded function on $\mathbb{R}$ satisfying Lipschitz condition with constant $A > 0$. If for an integer $b \geqslant 2$, the series $\sum_{k=1}^{+\infty} \varphi(b^k x)$ is absolutely uniformly convergent in $x \in (-\infty, +\infty) = \mathbb{R}$, and there exist constants $l > 0$, $\eta > 0$, $a \geqslant 0$, such that*

(i) *For any integers $k \in \mathbb{Z}$,*

$$\varphi(kl) = a, \qquad (6.1.12)$$

$$\left| \varphi\left(kl + \frac{l}{b}\right) - \varphi(kl) \right| \geqslant \eta. \qquad (6.1.13)$$

(ii) *The following inequality holds*

$$\frac{\eta}{l} - \frac{A}{b(b-1)} > 0. \qquad (6.1.14)$$

Then, the function $f(x) = \sum_{k=1}^{+\infty} \varphi(b^k x)$ is continuous at $x \in \mathbb{R}$, and it does not have finite right derivatives and finite left derivatives at any $x \in \mathbb{R}$.

Proof. By assumption, the series $f(x) = \sum_{k=1}^{+\infty} \varphi(b^k x)$ is absolutely uniformly convergent in $x \in (-\infty, +\infty) = \mathbb{R}$, so $f(x)$ is continuous at $x \in \mathbb{R}$.

We prove that the finite right derivative of $f(x)$ at any $x \in (-\infty, +\infty)$ does not exist.

Take $x \in \mathbb{R}$, without loss of generality, take $x > 0$. Let $\eta_n = \dfrac{l}{b^n}$, then for any $n \in \mathbb{N}$, there exists unique $N_n \in \mathbb{N}$ satisfing

$$(N_n - 1)\eta_n \leqslant x < N_n \eta_n, \qquad (6.1.15)$$

that is, $(N_n - 1)lb^{-n} \leqslant x < N_n lb^{-n}$. Let

$$x_n = N_n l b^{-n}, \quad \bar{x}_n = \left(N_n + \frac{1}{b}\right) lb^{-n},$$

then

$$\bar{x}_n - x_n = \frac{\eta_n}{b} = \frac{l}{b^{n+1}}. \qquad (6.1.16)$$

Estimate

$$|f(\bar{x}_n) - f(x_n)| = \left| \sum_{k=1}^{+\infty} \varphi(b^k \bar{x}_n) - \sum_{k=1}^{+\infty} \varphi(b^k x_n) \right|$$

$$= \left| \sum_{k=1}^{n-1} [\varphi(b^k \bar{x}_n) - \varphi(b^k x_n)] + [\varphi(b^n \bar{x}_n) - \varphi(b^n x_n)] \right.$$

$$\left. + \sum_{k=n+1}^{+\infty} [\varphi(b^k \bar{x}_n) - \varphi(b^k x_n)] \right|$$

$$\geqslant |\varphi(b^n \bar{x}_n) - \varphi(b^n x_n)| - \left| \sum_{k=1}^{n-1} [\varphi(b^k \bar{x}_n) - \varphi(b^k x_n)] \right.$$

$$\left. + \sum_{k=n+1}^{+\infty} [\varphi(b^k \bar{x}_n) - \varphi(b^k x_n)] \right|, \qquad (6.1.17)$$

by the Lipschitz condition and (6.1.16), if $k < n$, then

$$|\varphi(b^k \bar{x}_n) - \varphi(b^k x_n)| \leqslant \frac{Alb^k}{b^{n+1}}; \qquad (6.1.18)$$

if $k = n$, by (6.1.13), then

$$|\varphi(b^n \bar{x}_n) - \varphi(b^n x_n)| = \left| \varphi\left(b^n \left(N_n + \frac{1}{b}\right) lb^{-n}\right) - \varphi(b^n N_n l b^{-n}) \right|$$

$$= \left| \varphi\left(N_n l + \frac{l}{b}\right) - \varphi(N_n l) \right| \geqslant \eta; \qquad (6.1.19)$$

if $k \geqslant n+1$, by $b^k x_n = (b^{k-n} N_n) l$ and $b^k \bar{x}_n = b^k \left(N_n + \dfrac{1}{b}\right) lb^{-n}$, with the integers $b^{k-n} N_n \in \mathbb{N}$, $b^{k-n-1}(bN_n + 1) \in \mathbb{N}$, as well as by (6.1.12), then

$$\varphi(b^n \bar{x}_n) - \varphi(b^n x_n) = a - a = 0. \qquad (6.1.20)$$

Combine (6.1.17)~(6.1.20), for $n \in \mathbb{N}$, it follows

$$|f(\bar{x}_n) - f(x_n)| \geqslant \eta - \frac{Al}{b^{n+1}} \sum_{k=1}^{+\infty} b^k = \eta - \frac{Al}{b^{n+1}} \frac{b(b^{n-1} - 1)}{b - 1} > \eta - \frac{Al}{b(b-1)}. \tag{6.1.21}$$

Again by (6.1.16), (6.1.21), and for $n \in \mathbb{N}$ large enough, we get

$$\left| \frac{f(\bar{x}_n) - f(x_n)}{\bar{x}_n - x_n} \right| > \frac{\eta}{\dfrac{l}{b^{n+1}}} - \frac{Al}{b(b-1)} \frac{1}{\dfrac{l}{b^{n+1}}} \geqslant \left(\frac{\eta}{l} - \frac{A}{b(b-1)} \right) b^{n+1}. \tag{6.1.22}$$

By assumption (6.1.14) and (6.1.22), it follows that

$$\lim_{n \to +\infty} \left| \frac{f(\bar{x}_n) - f(x_n)}{\bar{x}_n - x_n} \right| = +\infty. \tag{6.1.23}$$

On the other hand, (6.1.15) and (6.1.16) show that $x < x_n < \bar{x}_n$, this implies $\bar{x}_n - x_n \to 0$, as $n \to +\infty$, and $\bar{x}_n - x_n \geqslant \dfrac{\eta_n}{b} \geqslant \dfrac{\bar{x}_n - x_n}{b+1}$. So, by the Lemma 6.1.1, and by (6.1.23), $f(x)$ does not have the finite right derivative at x.

$f(x)$ dose not have the finite left derivative, similarly.

Now, we consider the differentiability of the formal solution

$$u(t, x, y) = \lim_{k \to +\infty} u_k(t, x, y).$$

Suppose that $(t, x, y) \in [0, +\infty) \times D_k$, where t and y are fixed at moment, and x is the variable of functions. Then, the following two sums

$$\lambda_k(t, x, y) = \sum_{m=1}^{+\infty} \sum_{n=1}^{+\infty} \left\{ A_{k,m,n} \cos 4^k \sqrt{m^2 + n^2} \pi t \right\} \cdot \sin 4^k m\pi x \cdot \sin 4^k n\pi y,$$

$$\mu_k(t, x, y) = \sum_{m=1}^{+\infty} \sum_{n=1}^{+\infty} \left\{ B_{k,m,n} \sin 4^k \sqrt{m^2 + n^2} \pi t \right\} \cdot \sin 4^k m\pi x \cdot \sin 4^k n\pi y$$

have same differentiability. We suppose that the above sums and limit can be written as the form

$$\lambda_k(t, x, y) + \mu_k(t, x, y) = \Theta\left(4^k t, 4^k x, 4^k y\right), \tag{6.1.24}$$

$$u(x, y, t) = \lim_{k \to +\infty} \Theta\left(4^k t, 4^k x, 4^k y\right). \tag{6.1.25}$$

Theorem 6.1.2 *Suppose that the series* $\sum_{k=1}^{+\infty} \psi\left(4^k t, 4^k x, 4^k y\right)$ *is absolutely and uniformly convergent in its domain. For* (t_0, x, y_0), $0 \leqslant t_0 < +\infty$, $(x, y_0) \in D_k$, $k \in \mathbb{N}$, *denote*

$$\varphi(x) = \psi\left(4^k t_0, x, 4^4 y_0\right), \quad x \in \mathbb{R}.$$

Let $\varphi(x)$ be a bounded function on $\mathbb{R}$ satisfying Lipschitz condition with Lip constant A, and there exist constant $\eta > 0$, such that

(i) For all integers $k \in \mathbb{Z}$,

$$\left|\varphi\left(k + \frac{l}{4}\right) - \varphi(k)\right| \geq \eta. \tag{6.1.26}$$

(ii) The following inequality holds

$$\eta - \frac{A}{12} > 0. \tag{6.1.27}$$

Then the sum function $\sum_{k=1}^{+\infty} \varphi\left(4^k x\right)$ is continuous, but it does not have the finite right and left derivatives at any point.

Proof. By assumption, the series $\sum_{k=1}^{+\infty} \psi\left(4^k t, 4^k x, 4^k y\right)$ is absolutely and uniformly convergent in $x \in \mathbb{R}$, thus the sum is continuous.

On the other hand, $\varphi(x)$ is a bounded function satisfying Lipschitz condition with constant $A > 0$ on $\mathbb{R}$, and satisfying Theorem 6.1.1 with constants $b = 4$, $l = 1$, then the sum function $\sum_{k=1}^{+\infty} \varphi\left(4^k x\right)$ is continuous, but it does not have the finite right and left derivatives at any points x. The proof is complete.

Since the problem (6.1.1) has a formal solution (6.1.25), we take ψ in Theorem 6.1.2

$$\psi\left(4^k t, 4^k x, 4^k y\right) = \Theta\left(4^k t, 4^k x, 4^k y\right) - \Theta\left(4^{k-1} t, 4^{k-1} x, 4^{k-1} y\right),$$

then the conditions of Theorem 6.1.2 are satisfied. Hence, we conclude that the formal solution (6.1.25) of problem (6.1.1) can be a fractal function, that is, it is continuous at $x \in D$, but is not differentiable at any x.

The following are some numerical examples. Take $\varphi_k(x, y) = 0$, and $\psi_k(x, y) \neq 0$. Then

$$A_{k,m,n} = 0, \quad B_{k,m,n} = \frac{4}{4^k \pi \sqrt{m^2 + n^2}} \alpha_{k,m,n},$$

with $\alpha_{k,m,n} = \int_{D_k} \psi_k(x, y) \sin 4^k m\pi x \sin 4^k n\pi y \, dx \, dy$.

Example 6.1.1 Let $\psi_k(x, y) = \begin{cases} \dfrac{1}{k^2}, & (x, y) \in \left[\dfrac{2}{5}, \dfrac{3}{5}; \dfrac{2}{5}, \dfrac{3}{5}\right], \\ 0, & \text{otherwise}. \end{cases}$

Solution. Evaluate

$$\alpha_{k,m,n} = \int_{D_k} \psi_k(x,y) \sin 4^k m\pi x \sin 4^k n\pi y\, dxdy$$

$$= \frac{1}{k^2} \int_{\frac{2}{5}}^{\frac{3}{5}} \sin 4^k m\pi x\, dx \int_{\frac{2}{5}}^{\frac{3}{5}} \sin 4^k n\pi y\, dy$$

$$= \frac{1}{k^2} \left(-\frac{1}{4^k m\pi} \cos 4^k m\pi x \right) \bigg|_{\frac{2}{5}}^{\frac{3}{5}} \left(-\frac{1}{4^k n\pi} \cos 4^k n\pi y \right) \bigg|_{\frac{2}{5}}^{\frac{3}{5}}$$

$$= \frac{1}{k^2} \left(\frac{1}{4^k m\pi} 2 \sin \frac{4^k m\pi}{2} \sin \frac{4^k m\pi}{10} \right) \left(\frac{1}{4^k n\pi} 2 \sin \frac{4^k n\pi}{2} \sin \frac{4^k n\pi}{10} \right)$$

$$= \frac{1}{k^2} \frac{4}{(4^k \pi)^2 mn} \left(\sin \frac{4^k m\pi}{2} \sin \frac{4^k m\pi}{10} \right) \left(\sin \frac{4^k n\pi}{2} \sin \frac{4^k n\pi}{10} \right).$$

Thus,

$$u_k(t,x,y)$$

$$= \sum_{m=1}^{+\infty} \sum_{n=1}^{+\infty} \left\{ B_{k,m,n} \sin 4^k \sqrt{m^2+n^2} \pi t \right\} \cdot \sin 4^k m\pi x \cdot \sin 4^k n\pi y$$

$$= \sum_{m=1}^{+\infty} \sum_{n=1}^{+\infty} \left\{ \frac{16}{k^2 (4^k \pi)^3 \sqrt{m^2+n^2} mn} \sin \frac{4^k m\pi}{2} \sin \frac{4^k m\pi}{10} \sin \frac{4^k n\pi}{2} \sin \frac{4^k n\pi}{10} \right\}$$

$$\cdot \sin 4^k \pi \sqrt{m^2+n^2}\, t \sin 4^k m\pi x \cdot \sin 4^k n\pi y.$$

For solution $u_k(t,x,y)$, the coefficients $B_{k,m,n} = \dfrac{4\alpha_{k,m,n}}{4^k \pi \sqrt{m^2+n^2}}$. It is easy to check the conditions of Theorem 6.1.2 are satisfied, so that $u(t,x,y)$ is continuous, but it does not have the differentiability, it is a fractal solution of the problem (6.1.1).

See Fig. 6.1.2.

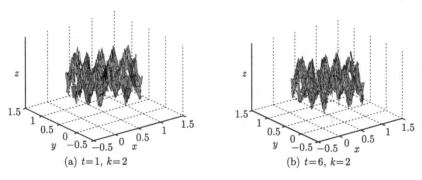

(a) $t=1$, $k=2$ (b) $t=6$, $k=2$

Figure 6.1.2

Example 6.1.2 Let $\psi_k(x,y) = \begin{cases} x^k y^k, & (x,y) \in \left[\dfrac{2}{5}, \dfrac{3}{5}; \dfrac{2}{5}, \dfrac{3}{5}\right], \\ 0, & \text{otherwise.} \end{cases}$

Solution. Evaluate

$$\alpha_{k,m,n} = \int_{D_k} \psi_k(x,y) \sin 4^k m\pi x \sin 4^k n\pi y \, dx dy$$

$$= \int_{\frac{2}{5}}^{\frac{3}{5}} x^k \sin 4^k m\pi x \, dx \int_{\frac{2}{5}}^{\frac{3}{5}} y^k \sin 4^k n\pi y \, dy$$

$$= \left\{ \cos 4^k m\pi x \sum_{j=0}^{\left[\frac{k}{2}\right]} \frac{(-1)^{j+1} k!}{(k-2j)!} \frac{x^{k-2j}}{(4^k m\pi)^{2j+1}} \right.$$

$$\left. + \sin 4^k m\pi x \sum_{j=0}^{\left[\frac{k-1}{2}\right]} \frac{(-1)^j k!}{(k-2j-1)!} \frac{x^{k-2j-1}}{(4^k m\pi)^{2j+2}} \right\} \Bigg|_{x=\frac{2}{5}}^{\frac{3}{5}}$$

$$\cdot \left\{ \cos 4^k n\pi y \sum_{j=0}^{\left[\frac{k}{2}\right]} \frac{(-1)^{j+1} k!}{(k-2j)!} \frac{y^{k-2j}}{(4^k n\pi)^{2j+1}} \right.$$

$$\left. + \sin 4^k n\pi y \sum_{j=0}^{\left[\frac{k-1}{2}\right]} \frac{(-1)^j k!}{(k-2j-1)!} \frac{y^{k-2j-1}}{(4^k n\pi)^{2j+2}} \right\} \Bigg|_{y=\frac{2}{5}}^{\frac{3}{5}}.$$

See Fig. 6.1.3.

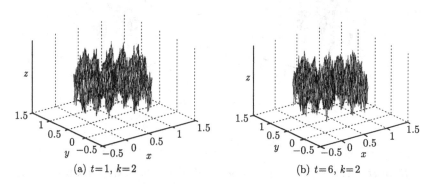

(a) $t=1$, $k=2$ (b) $t=6$, $k=2$

Figure 6.1.3

It is easy to check $|\alpha_{k,m,n}| < +\infty$, thus the coefficients $B_{k,m,n} =$

$\dfrac{4\alpha_{k,m,n}}{4^k\pi\sqrt{m^2+n^2}}$ of solution $u_k(t,x,y)$ have similar properties with Example 6.1.1, then the formal solution $u(t,x,y)$ is continuous, but it does not have the differentiability, it is a fractal solution of the problem (6.1.1).

Example 6.1.3 Let $\psi_k(x,y) = \begin{cases} \sin 4^k x \sin 4^k y, & (x,y) \in \left[\dfrac{2}{5},\dfrac{3}{5};\dfrac{2}{5},\dfrac{3}{5}\right], \\ 0, & \text{otherwise.} \end{cases}$

Solution. Evaluate

$$\alpha_{k,m,n} = \int_{D_k} \psi_k(x,y)\sin 4^k m\pi x \sin 4^k n\pi y\,dxdy$$

$$= \int_{\frac{2}{5}}^{\frac{3}{5}} \sin 4^k x \sin 4^k m\pi x\,dx \int_{\frac{2}{5}}^{\frac{3}{5}} \sin 4^k y \sin 4^k n\pi y\,dy,$$

$$= \dfrac{1}{4^k(m^2\pi^2-1)}\left(\sin 4^k m\pi x \cos 4^k x - m\pi \cos 4^k m\pi x \sin 4^k x\right)\Big|_{x=\frac{2}{5}}^{\frac{3}{5}}$$

$$\cdot \dfrac{1}{4^k(n^2\pi^2-1)}\left(\sin 4^k n\pi y \cos 4^k y - n\pi \cos 4^k n\pi y \sin 4^k y\right)\Big|_{y=\frac{2}{5}}^{\frac{3}{5}},$$

where $B_{k,m,n} = \dfrac{4\alpha_{k,m,n}}{4^k\pi\sqrt{m^2+n^2}}$ is not like those of $B_{k,m,n}$ in the Examples 6.1.1 and 6.1.2, but the formal solution $u(t,x,y)$ of problem (6.1.1) is still a fractal, see Fig. 6.1.4.

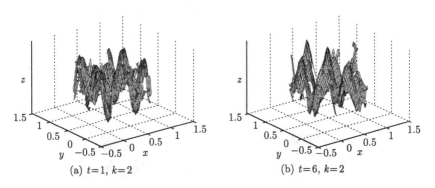

(a) $t=1,\ k=2$ \qquad (b) $t=6,\ k=2$

Figure 6.1.4

Remark 1. In the physics point of view, the Examples 6.1.1~6.1.3 show: a membrane has a little velocity on the square

$$\left[\dfrac{2}{5},\dfrac{3}{5};\dfrac{2}{5},\dfrac{3}{5}\right],$$

then it has a small vibration, as a fractal.

Remark 2. In general, if the initial conditions of problem (6.1.1) have the order as

$$a^k \Phi\left(\frac{1}{4^k m^s n^t}\right), \quad 0 < a < 1, \quad s > 1, \quad t > 1,$$

then problem (6.1.1) may have fractal solutions.

6.1.2 p-type 2-dimension wave equation with fractal boundary

1. p-type 2-dimension wave equation

p-type 2-dimension wave partial differential equation has the physics sense[76], it is a vibration of a membrane defined on a 2-dimendion local field $K_p \times K_p$, the equation of motion is

$$\frac{\partial^{\langle 2 \rangle} u(t,x,y)}{\partial t^{\langle 2 \rangle}} = \frac{\partial^{\langle 2 \rangle} u(t,x,y)}{\partial x^{\langle 2 \rangle}} + \frac{\partial^{\langle 2 \rangle} u(t,x,y)}{\partial y^{\langle 2 \rangle}},$$

where $\dfrac{\partial^{\langle 2 \rangle} u(t,x,y)}{\partial t^{\langle 2 \rangle}}, \dfrac{\partial^{\langle 2 \rangle} u(t,x,y)}{\partial x^{\langle 2 \rangle}}, \dfrac{\partial^{\langle 2 \rangle} u(t,x,y)}{\partial y^{\langle 2 \rangle}}$ are 2-order p-type partial derivatives of $u(t,x,y)$; and $\dfrac{\partial^{\langle 1 \rangle} u(t,x,y)}{\partial t^{\langle 1 \rangle}}$ is the p-type "velocity", i.e. the rate about change of time $t \in \mathbb{R}^+$, and $(x,y) \in K_p \times K_p$; $u(t,x,y)$ is the position of the membrane.

We consider the problem

$$\begin{cases} \dfrac{\partial^{\langle 2 \rangle} u(t,x,y)}{\partial t^{\langle 2 \rangle}} = \dfrac{\partial^{\langle 2 \rangle} u(t,x,y)}{\partial x^{\langle 2 \rangle}} + \dfrac{\partial^{\langle 2 \rangle} u(t,x,y)}{\partial y^{\langle 2 \rangle}}, & t > 0, (x,y) \in \Omega, \\ u(t,x,y)|_{t=0} = \varphi(x,y), & (x,y) \in \Omega, \\ \dfrac{\partial^{\langle 1 \rangle} u(t,x,y)}{\partial t^{\langle 1 \rangle}}\bigg|_{t=0} = \psi(x,y), & (x,y) \in \Omega, \\ u(t,x,y)|_\gamma = 0, & t > 0, \end{cases}$$

(6.1.28)

where the domain $\Omega \subset K_p \times K_p$ with the boundary γ which is the p-**adic von Koch type curve**, and the approximation curves of γ are $\gamma_1, \gamma_2, \cdots$, their domains are $\Omega_1, \Omega_2, \cdots$, respectively. The Fig. 6.1.5 is the $\gamma_1, \gamma_2, \gamma_4$.

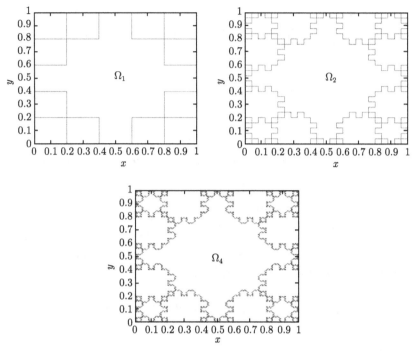

Figure 6.1.5

We suppose that the initial conditions $\varphi(x,y) = \lim_{k \to +\infty} \varphi_k(x,y)$ and $\psi(x,y) = \lim_{k \to +\infty} \psi_k(x,y)$ are convergent in Ω, and the domains of $\varphi_k(x,y)$ and $\psi_k(x,y)$ are Ω_k, $k = 1, 2, \cdots$.

To solve the problem (6.1.28), we need the knowledge of p-series field.

Denote $K_p = [0, +\infty)$, with a prime $p \geqslant 2$, then $\forall x \in K_p$, it has a unique expression
$$x = x_{-s}\beta^{-s} + \cdots + x_{-1}\beta^{-1} + x_0\beta^0 + x_1\beta^1 + \cdots,$$
where $x_j \in \{0, 1, \cdots, p-1\}$, $j \geqslant -s, s \in \mathbb{P}$, $|\beta| = p^{-1}$, the operation "+" is term by term, mod p, no carrying, i.e, $x, y \in K_p \Rightarrow x + y = (x_j + y_j \bmod p)$.

The character group of K_p is $\Gamma_p = \{w_y \in \mathbb{C} : y \in K_p\}$, the well-know **Walsh function system**. For each character w_y, its action on $x \in K_p$ is $w_y(x) = \exp\dfrac{2\pi i}{p} y \otimes x$, with
$$y \otimes x = \sum_k x_k y_{-1-k} \,(\bmod\, p) = x \otimes y = \sum_k y_k x_{-1-k} \,(\bmod\, p).$$

D is the compact subgroup of K_p, and the character group of D is Γ_D:

$$\Gamma_D = \{w_j(x) : D \to \mathbb{C}, x \in D, j \in \mathbb{P}\}, \quad w_j(x) = \exp\frac{2\pi\mathrm{i}}{p}j \otimes x.$$

In the 2-dimension case, compact subgroup $D \times D$ has the character group $\Gamma_D \times \Gamma_D$:

$$\Gamma_D \times \Gamma_D = \{w_{js}(x,y) = w_j(x)w_s(y) : D \times D \to \mathbb{C}, (x,y) \in D \times D, j, s \in \mathbb{P}\}.$$

A function $f(x,y) : D \times D \to \mathbb{C}$ can be expanded as a Walsh-Fourier series

$$f(x,y) = \sum_{j,s=0}^{+\infty} a_{js} w_j(x) w_s(y),$$

where $w_j(x) = \exp\dfrac{2\pi\mathrm{i}}{p}j \otimes x$, $w_s(y) = \exp\dfrac{2\pi\mathrm{i}}{p}s \otimes y$. The coefficients are

$$a_{js} = \int_{D \times D} f(x,y)\overline{w}_j(x)\overline{w}_s(y)dxdy.$$

Let the p-adic expressions of x, y, j, s be

$$x = (x_0, x_1, \cdots), \quad x_0, x_1, \cdots \in \{0, 1, \cdots, p-1\},$$
$$y = (y_0, y_1, \cdots), \quad y_0, y_1, \cdots \in \{0, 1, \cdots, p-1\},$$
$$j = (j_{-t}, j_{-t+1}, \cdots, j_0), \quad j_{-t}, j_{-t+1}, \cdots, j_0 \in \{0, 1, \cdots, p-1\}, \quad t \in \mathbb{P},$$
$$s = (s_{-r}, s_{-r+1}, \cdots, s_0), \quad s_{-r}, s_{-r+1}, \cdots, s_0 \in \{0, 1, \cdots, p-1\}, \quad t \in \mathbb{P},$$

then,

$$w_j(x) = \exp\frac{2\pi\mathrm{i}}{p}j \otimes x = \cos\frac{2\pi}{p}\left(\sum_k x_k j_{-1-k}\right) + \mathrm{i}\sin\frac{2\pi}{p}\left(\sum_k x_k j_{-1-k}\right),$$

$$w_s(y) = \exp\frac{2\pi\mathrm{i}}{p}s \otimes y = \cos\frac{2\pi}{p}\left(\sum_k y_k s_{-1-k}\right) + \mathrm{i}\sin\frac{2\pi}{p}\left(\sum_k y_k s_{-1-k}\right).$$

The definitions of p-type derivatives $\dfrac{\partial^{\langle 2\rangle}u}{\partial t^{\langle 2\rangle}}, \dfrac{\partial^{\langle 2\rangle}u}{\partial x^{\langle 2\rangle}}, \dfrac{\partial^{\langle 2\rangle}u}{\partial y^{\langle 2\rangle}}, \dfrac{\partial^{\langle 1\rangle}u}{\partial t^{\langle 1\rangle}}$ are given in Chapter 3, and we use the equivalent definition[62]: for a Haar measurable function $f : K_p \to \mathbb{C}$, if

$$\lim_{N \to +\infty} \sum_{k=0}^{N} p^k \left\{\sum_{j=0}^{p-1} A_j f\left(x + jp^{-k-1}\right)\right\}$$

exists and finite, where

$$A_0 = \frac{p-1}{2}, \quad A_k = \frac{\omega^k}{1-\omega^k}, \quad k = 1, 2, \cdots, p-1,$$

then the limit is said to be the p-type derivative at x, denoted by $f^{\langle 1 \rangle}(x)$.

We need the following properties of the p-type derivatives[7],[62],[76]:

$$\frac{\partial^{\langle 1 \rangle} w_{js}(x,y)}{\partial x^{\langle 1 \rangle}} = jw_j(x)w_s(y), \quad \frac{\partial^{\langle 1 \rangle} w_{js}(x,y)}{\partial y^{\langle 1 \rangle}} = sw_j(x)w_s(y),$$

$$\frac{\partial^{\langle 2 \rangle} w_{js}(x,y)}{\partial x^{\langle 2 \rangle}} = j^2 w_j(x)w_s(y), \quad \frac{\partial^{\langle 2 \rangle} w_{js}(x,y)}{\partial y^{\langle 2 \rangle}} = s^2 w_j(x)w_s(y).$$

2. Solutions of p-type 2-dimension wave equation with fractal boundary

We solve for the kth-approximation problem of 2-dimension wave equation in the p-type sense 2-order partial differential equation

$$\begin{cases} \dfrac{\partial^{\langle 2 \rangle} u_k(t,x,y)}{\partial t^{\langle 2 \rangle}} = \dfrac{\partial^{\langle 2 \rangle} u_k(t,x,y)}{\partial x^{\langle 2 \rangle}} + \dfrac{\partial^{\langle 2 \rangle} u_k(t,x,y)}{\partial y^{\langle 2 \rangle}}, & t > 0, (x,y) \in \Omega_k, \\ u_k(t,x,y)|_{t=0} = \varphi_k(x,y), & (x,y) \in \Omega_k, \\ \dfrac{\partial^{\langle 1 \rangle} u_k(t,x,y)}{\partial t^{\langle 1 \rangle}}\bigg|_{t=0} = \psi_k(x,y), & (x,y) \in \Omega_k, \\ u(t,x,y)|_{\gamma_k} = 0, & t > 0, \end{cases}$$

(6.1.29)

where $\gamma_1, \gamma_2, \cdots, \gamma_k, \cdots$ are the first, second, $\cdots$, k^{th} approximation of the p-adic von Koch type curve. The initial functions $\varphi(x,y) = \lim\limits_{k \to +\infty} \varphi_k(x,y)$, $\psi(x,y) = \lim\limits_{k \to +\infty} \psi_k(x,y)$ are convergent in domain Ω, and the domains of $\varphi_k(x,y), \psi_k(x,y)$ are Ω_k, $k = 1, 2, \cdots$.

Suppose that $u_k(t,x,y) = T_k(t)v_k(x,y)$ is the formal solution of (6.1.29), then for $k = 1, 2, 3, \cdots$, we have

$$\frac{\partial^{\langle 2 \rangle} u_k(t,x,y)}{\partial t^{\langle 2 \rangle}} = \frac{d^{\langle 2 \rangle} T_k(t)}{dt^{\langle 2 \rangle}} v_k(x,y),$$

$$\frac{\partial^{\langle 2 \rangle} u_k(t,x,y)}{\partial x^{\langle 2 \rangle}} = T_k(t) \frac{\partial^{\langle 2 \rangle} v_k(x,y)}{\partial x^{\langle 2 \rangle}}, \quad \frac{\partial^{\langle 2 \rangle} u_k(t,x,y)}{\partial y^{\langle 2 \rangle}} = T_k(t) \frac{\partial^{\langle 2 \rangle} v_k(x,y)}{\partial y^{\langle 2 \rangle}}.$$

Substitute into $\dfrac{\partial^{\langle 2 \rangle} u_k(t,x,y)}{\partial t^{\langle 2 \rangle}} = \dfrac{\partial^{\langle 2 \rangle} u_k(t,x,y)}{\partial x^{\langle 2 \rangle}} + \dfrac{\partial^{\langle 2 \rangle} u_k(t,x,y)}{\partial y^{\langle 2 \rangle}}$, then we get

$$\frac{d^{\langle 2 \rangle} T_k(t)}{dt^{\langle 2 \rangle}} v_k(x,y) = T_k(t) \left(\frac{\partial^{\langle 2 \rangle} v_k(x,y)}{\partial x^{\langle 2 \rangle}} + \frac{\partial^{\langle 2 \rangle} v_k(x,y)}{\partial y^{\langle 2 \rangle}} \right).$$

If $T_k(t) \neq 0, v_k(x,y) \neq 0$, it follows that

$$\frac{1}{T_k(t)} \frac{d^{\langle 2 \rangle} T_k(t)}{dt^{\langle 2 \rangle}} = \frac{1}{v_k(x,y)} \left(\frac{\partial^{\langle 2 \rangle} v_k(x,y)}{\partial x^{\langle 2 \rangle}} + \frac{\partial^{\langle 2 \rangle} v_k(x,y)}{\partial y^{\langle 2 \rangle}} \right).$$

By the separation of variables, let the right side of above equation be $\lambda^2 + \mu^2$, $\lambda > 0, \mu > 0$, then

$$\frac{1}{T_k(t)} \frac{d^{\langle 2 \rangle} T_k(t)}{dt^{\langle 2 \rangle}} = \frac{1}{v_k(x,y)} \left(\frac{\partial^{\langle 2 \rangle} v_k(x,y)}{\partial x^{\langle 2 \rangle}} + \frac{\partial^{\langle 2 \rangle} v_k(x,y)}{\partial y^{\langle 2 \rangle}} \right)$$
$$= \lambda^2 + \mu^2, \quad \lambda > 0, \mu > 0.$$

So, we have

$$\frac{\partial^{\langle 2 \rangle} v_k}{\partial x^{\langle 2 \rangle}} + \frac{\partial^{\langle 2 \rangle} v_k}{\partial y^{\langle 2 \rangle}} = \left(\lambda^2 + \mu^2 \right) v_k, \tag{6.1.30}$$

$$\frac{d^{\langle 2 \rangle} T_k}{dt^{\langle 2 \rangle}} = \left(\lambda^2 + \mu^2 \right) T_k. \tag{6.1.31}$$

It is easy to verify that the function

$$v_{k,\lambda,\mu}(x,y) = \sin \left(\frac{2\pi}{p} (\lambda \otimes x) \right) \sin \left(\frac{2\pi}{p} (\mu \otimes y) \right)$$

is a solution of (6.1.30). To determine λ and μ, we use the condition $v_{k,\lambda,\mu}|_{\gamma_k} = 0$, thus

$$\lambda_{k,m,p} \equiv \lambda(k,m,p), \quad \mu_{k,n,p} = \mu(k,n,p), \quad m,n = 1,2,\cdots.$$

Moreover, the solution can be written as

$$v_{k,m,n}(x,y) \equiv \sin \left(\frac{2\pi}{p} (\lambda_{k,m,p} \otimes x) \right) \sin \left(\frac{2\pi}{p} (\mu_{k,n,p} \otimes y) \right), \ k,m,n=1,2,\cdots. \tag{6.1.32}$$

Correspondingly, a solution of (6.1.31) is

$$T_{k,m,n}(t) = A_{k,m,n} \cos \left(\frac{2\pi}{p} (\lambda_{k,m,p} \otimes t) \right) + B_{k,m,n} \sin \left(\frac{2\pi}{p} (\mu_{k,n,p} \otimes t) \right),$$
$$k,m,n = 1,2,\cdots.$$

Combine the above, we have

$$u_{k,m,n}(t,x,y) = T_{k,m,n}(t) v_{k,m,n}(x,y)$$
$$= \left\{ A_{k,m,n} \cos \left(\frac{2\pi}{p} \lambda_{k,m,p} \otimes t \right) + B_{k,m,n,p} \sin \left(\frac{2\pi}{p} \mu_{k,n,p} \otimes t \right) \right\}$$

$$\cdot \sin\left(\frac{2\pi}{p}\lambda_{k,m,p}\otimes x\right)\sin\left(\frac{2\pi}{p}\mu_{k,n,p}\otimes y\right),$$

$$k,m,n = 1,2,\cdots \qquad (6.1.33)$$

with coefficients

$$A_{k,m,n} = \frac{\int_{\Omega_k}\varphi_k(x,y)\cdot\sin\left(\frac{2\pi}{p}\lambda_{k,m,p}\otimes x\right)\cdot\sin\left(\frac{2\pi}{p}\mu_{k,n,p}\otimes y\right)dxdy}{\int_{\Omega_k}\left\{\sin\left(\frac{2\pi}{p}\lambda_{k,m,p}\otimes x\right)\right\}^2\cdot\left\{\sin\left(\frac{2\pi}{p}\mu_{k,n,p}\otimes y\right)\right\}^2 dxdy},$$

$$(6.1.34)$$

$$B_{k,m,n} = \frac{\int_{\Omega_k}\psi_k(x,y)\cdot\sin\left(\frac{2\pi}{p}\lambda_{k,m,p}\otimes x\right)\cdot\sin\left(\frac{2\pi}{p}\mu_{k,m,p}\otimes y\right)dxdy}{p^k\sqrt{m^2+n^2}\int_{\Omega_k}\left\{\sin\left(\frac{2\pi}{p}\lambda_{k,m,p}\otimes x\right)\right\}^2\cdot\left\{\sin\left(\frac{2\pi}{p}\mu_{k,m,p}\otimes y\right)\right\}^2 dxdy}.$$

$$(6.1.35)$$

So that the problem (6.1.29) has a formal solution

$$u_k(t,x,y) = \sum_{m=1}^{+\infty}\sum_{n=1}^{+\infty}T_{k,m,n}(t)v_{k,m,n}(x,y), \qquad (6.1.36)$$

and the problem (6.1.28) has a formal solution

$$u(t,x,y) = \lim_{k\to+\infty}u_k(t,x,y) = \lim_{k\to+\infty}\sum_{m=1}^{+\infty}\sum_{n=1}^{+\infty}T_{k,m,n}(t)v_{k,m,n}(x,y).$$

$$(6.1.37)$$

3. Numerical examples

Example 6.1.4 Take $p = 3$, the first, second, fourth approximation curves of the **3-adic von Koch type curve** are shown in Fig. 6.1.6.

For $p = 3$, take the generator $\beta \in K_3$ with $|\beta| = 3^{-1}$, and denote

$$x = x_0\beta^0 + x_1\beta^1 + x_2\beta^2 + \cdots + x_s\beta^s + \cdots, \quad x_0, x_1, \cdots, x_s, \cdots \in \{0,1,2\},$$

$$\lambda = \lambda_{-s}\beta^{-s} + \lambda_{-s-1}\beta^{-s-1} + \cdots + \lambda_{-1}\beta^{-1} + \lambda_0\beta^0 + \lambda_1\beta^1 + \cdots,$$

$$\lambda_{-s}, \cdots \in \{0,1,2\}, s \in \mathbb{P},$$

$$y = y_0\beta^0 + y_1\beta^1 + y_2\beta^2 + \cdots + y_s\beta^s + \cdots, \quad y_0, y_1, \cdots, y_s, \cdots \in \{0,1,2\},$$

$$\mu = \mu_{-s}\beta^{-s} + \mu_{-s-1}\beta^{-s-1} + \cdots + \mu_{-1}\beta^{-1} + \mu_0\beta^0 + \mu_1\beta^1 + \cdots,$$

$$\mu_{-s}, \cdots \in \{0,1,2\}, s \in \mathbb{P}.$$

Since $\lambda \otimes x = \sum_k x_k \lambda_{-1-k}$ and $x \in \Omega$, we have $x_{-1} = x_{-2} = \cdots = 0$, then, the value $\lambda \otimes x$ is independent of $\lambda_0, \lambda_1, \lambda_2, \cdots$, we may take $\lambda_0 = \lambda_1 = \lambda_2 = \cdots = 0$, and

$$\lambda \otimes x = \sum_k x_k \lambda_{-1-k} = \lambda_{-1} x_0 + \lambda_{-2} x_1 + \cdots + \lambda_{-s+1} x_{s-2} + \lambda_{-s} x_{s-1},$$

$$\sin \frac{2\pi}{3} (\lambda \otimes x) = \sin \frac{2\pi}{3} (\lambda_{-1} x_0 + \lambda_{-2} x_1 + \cdots + \lambda_{-s+1} x_{s-2} + \lambda_{-s} x_{s-1}).$$

Similarly,

$$\mu \otimes y = \sum_k y_k \mu_{-1-k} = \mu_{-1} y_0 + \mu_{-2} y_1 + \cdots + \mu_{-s+1} y_{s-2} + \mu_{-s} y_{s-1},$$

$$\sin \frac{2\pi}{3} (\mu \otimes y) = \sin \frac{2\pi}{3} (\mu_{-1} y_0 + \mu_{-2} y_1 + \cdots + \mu_{-s+1} y_{s-2} + \mu_{-s} y_{s-1}).$$

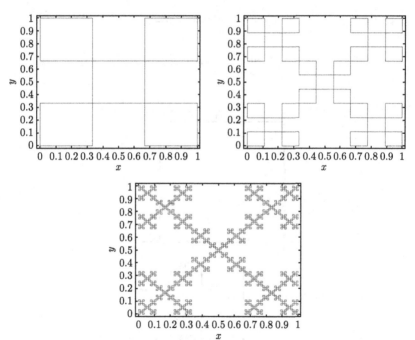

Figure 6.1.6

Now we determine the values λ, μ by the boundary conditions $v_k|_{\gamma_k} = 0$, $k = 1, 2, \cdots$.

For the 0-th approximation and 1-st approximation of the 3-adic von koch type curve, we draft them in the Fig. 6.1.7.

The 0-th approximation $\gamma_0 = \gamma_{0,1} \cup \gamma_{0,2} \cup \gamma_{0,3} \cup \gamma_{0,4}$ with

$\gamma_{0,1} = \{(x,y) : x \in [0,1), y = 0\}, \quad \gamma_{0,2} = \{(x,y) : x = 1, y \in [0,1)\},$
$\gamma_{0,3} = \{(x,y) : x \in [0,1), y = 1\}, \quad \gamma_{0,4} = \{(x,y) : x = 0, y \in [0,1)\}.$

On $\gamma_{0,1} = \{(x,y) : x \in [0,1), y = 0\}$, we determine λ and μ —

$\gamma_{0,1}\colon x \in [0,1) \leftrightarrow x = x_0 \left(\frac{1}{3}\right)^0 + x_1 \left(\frac{1}{3}\right)^1 + \cdots, \quad x_0 = 0, \ x_1, x_2, \cdots \in \{0, 1, 2\};$

$y = 0 \leftrightarrow y = y_0 \left(\frac{1}{3}\right)^0 + y_1 \left(\frac{1}{3}\right)^1 + \cdots, \quad y_0 = y_1 = \cdots = 0;$

① $\sin\left(\frac{2\pi}{3}\lambda \otimes x\right)\bigg|_{\gamma_{0,1}} = \sin\frac{2\pi}{3}\left(\lambda_{-1}x_0 + \lambda_{-2}x_1 + \cdots + \lambda_{-s}x_{s-1}\right)\bigg|_{\gamma_{0,1}}$

$= \sin\frac{2\pi}{3}\left(\lambda_{-2}x_1 + \lambda_{-3}x_2 + \cdots + \lambda_{-s}x_{s-1}\right)\bigg|_{\gamma_{0,1}};$

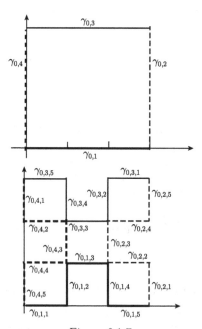

Figure 6.1.7

② $\sin\left(\dfrac{2\pi}{3}\mu\otimes y\right)\bigg|_{\gamma_{0,1}} = \sin\dfrac{2\pi}{3}(\mu_{-1}y_0 + \mu_{-2}y_1 + \cdots + \mu_{-s}y_{s-1})\bigg|_{\gamma_{0,1}}$

$= \sin\dfrac{2\pi}{3}(\mu_{-1}0 + \mu_{-2}0 + \cdots + \mu_{-s}0)\bigg|_{\gamma_{0,1}} = 0.$

Thus, for any λ and μ, we have $\sin\dfrac{2\pi}{3}(\lambda\otimes x)\sin\dfrac{2\pi}{3}(\mu\otimes y)\bigg|_{\gamma_{0,1}} = 0$;

On $\gamma_{0,2} = \{(x,y) : x = 1, y \in [0,1)\}$, we determine λ and μ —

$\gamma_{0,2}: x = 1 \leftrightarrow x = 1\left(\dfrac{1}{3}\right)^0$, $x_0 = 1, x_1 = x_2 = \cdots = 0$;

$y = [0,1) \leftrightarrow y = y_0\left(\dfrac{1}{3}\right)^0 + y_1\left(\dfrac{1}{3}\right)^1 + \cdots$, $y_0 = 0, y_1, y_2, \cdots \in \{0,1,2\}$;

① $\sin\left(\dfrac{2\pi}{3}\lambda\otimes x\right)\bigg|_{\gamma_{0,2}} = \sin\dfrac{2\pi}{3}(\lambda_{-1}x_0 + \lambda_{-2}x_1 + \cdots + \lambda_{-s}x_{s-1})\bigg|_{\gamma_{0,2}}$

$= \sin\dfrac{2\pi}{3}(\lambda_{-1}x_0)\bigg|_{\gamma_{0,2}} = \sin\dfrac{2\pi}{3}(\lambda_{-1}\cdot 1).$

Thus, $\lambda_{-1} = 0$ implies $\sin\dfrac{2\pi}{3}(\lambda_{-1}\cdot 1) = 0$; so $\lambda_{-1} = 0$ implies that $\sin\dfrac{2\pi}{3}(\lambda\otimes x)\bigg|_{\gamma_{0,2}} = 0.$

② $\sin\left(\dfrac{2\pi}{3}\mu\otimes y\right)\bigg|_{\gamma_{0,2}} = \sin\dfrac{2\pi}{3}(\mu_{-1}y_0 + y_{-2}y_1 + \cdots + \mu_{-s}y_{s-1})\bigg|_{\gamma_{0,2}}$

$= \sin\dfrac{2\pi}{3}(\mu_{-2}y_1 + \mu_{-3}y_2 + \cdots + \mu_{-s}y_{s-1})\bigg|_{\gamma_{0,2}}.$

Thus, $\lambda = m\cdot 3^2, m \in \mathbb{N}$, implies $\sin\dfrac{2\pi}{3}(\lambda\otimes x)\sin\dfrac{2\pi}{3}(\mu\otimes y)\bigg|_{\gamma_{0,2}} = 0.$

On $\gamma_{0,3} = \{(x,y) : x \in [0,1), y = 1\}$, we determine λ and μ —

$\gamma_{0,3}: x = [0,1) \leftrightarrow x = x_0\left(\dfrac{1}{3}\right)^0 + x_1\left(\dfrac{1}{3}\right)^1 + \cdots$, $x_0 = 0, x_1, x_2, \cdots \in \{0,1,2\}$;

$y = 1 \leftrightarrow y = 1\left(\dfrac{1}{3}\right)^0$, $y_0 = 1, y_1 = y_2 = \cdots = 0$;

① $\sin\left(\dfrac{2\pi}{3}\lambda\otimes x\right)\bigg|_{\gamma_{0,3}} = \sin\dfrac{2\pi}{3}(\lambda_{-1}x_0 + \lambda_{-2}x_1 + \cdots + \lambda_{-s}x_{s-1})\bigg|_{\gamma_{0,3}}$

$= \sin\dfrac{2\pi}{3}(\lambda_{-2}x_1 + \lambda_{-3}x_2 + \cdots + \lambda_{-s}x_{s-1})\bigg|_{\gamma_{0,3}};$

② $\left.\sin\left(\dfrac{2\pi}{3}\mu\otimes y\right)\right|_{\gamma_{0,3}} = \left.\sin\dfrac{2\pi}{3}\left(\mu_{-1}\cdot y_0 + \mu_{-2}y_1 + \cdots + \mu_{-s}y_{s-1}\right)\right|_{\gamma_{0,3}}$

$\qquad = \left.\sin\dfrac{2\pi}{3}(\mu_{-1}y_0)\right|_{\gamma_{0,3}} = \sin\dfrac{2\pi}{3}(\mu_{-1}\cdot 1).$

Hence $\mu_{-1} = 0$ implies $\sin\dfrac{2\pi}{3}(\mu_{-1}) = 0$, so $\mu_{-1} = 0$ implies that $\left.\sin\dfrac{2\pi}{3}(\mu\otimes y)\right|_{\gamma_{0,3}} = 0$. Thus, $\mu = n\cdot 3^2$, $n\in\mathbb{N}$, implies that
$\left.\sin\dfrac{2\pi}{3}(\lambda\otimes x)\sin\dfrac{2\pi}{3}(\mu\otimes y)\right|_{\gamma_{0,3}} = 0.$

On $\gamma_{0,4} = \{(x,y) : x = 0, y \in [0,1)\}$, we determine λ and μ ——

$\gamma_{0,4}$: $x = 0 \leftrightarrow x = 0\left(\dfrac{1}{3}\right)^0 + 0\left(\dfrac{1}{3}\right)^1$, $x_0 = x_1 = x_2 = \cdots = 0$;

$y = [0,1) \leftrightarrow y = y_0\left(\dfrac{1}{3}\right)^0 + y_1\left(\dfrac{1}{3}\right)^1 + \cdots$, $y_0 = 0, y_1, y_2, \cdots \in \{0,1,2\}$;

① $\left.\sin\left(\dfrac{2\pi}{3}\lambda\otimes x\right)\right|_{\gamma_{0,4}} = \left.\sin\dfrac{2\pi}{3}\left(\lambda_{-1}x_0 + \lambda_{-2}x_1 + \cdots + \lambda_{-s}x_{s-1}\right)\right|_{\gamma_{0,4}}$

$\qquad = \left.\sin\dfrac{2\pi}{3}(\lambda_{-1}0 + \lambda_{-2}0)\right|_{\gamma_{0,4}} = 0;$

② $\left.\sin\left(\dfrac{2\pi}{3}\mu\otimes y\right)\right|_{\gamma_{0,4}} = \left.\sin\dfrac{2\pi}{3}\left(\mu_{-1}y_0 + \mu_{-2}y_1 + \cdots + \mu_{-s}y_{s-1}\right)\right|_{\gamma_{0,4}}$

$\qquad = \left.\sin\dfrac{2\pi}{3}\left(\mu_{-2}y_1 + \mu_{-3}y_2 + \cdots + \mu_{-s}y_{s-1}\right)\right|_{\gamma_{0,4}}.$

Thus, any λ and μ imply that $\left.\sin\dfrac{2\pi}{3}(\lambda\otimes x)\sin\dfrac{2\pi}{3}(\mu\otimes y)\right|_{\gamma_{0,4}} = 0.$

Summing up the above, for the 0-th approximation, we take $\lambda = m\cdot 3^2$ and $\mu = n\cdot 3^2$, $m, n \in \mathbb{N}$, then the boundary condition $v_k|_{\gamma_0} = 0$ is satisfied.

By induction, we may determine that: for the k-th approximation γ_k,

$$\gamma_k = \bigcup_{j_1=1}^{4}\bigcup_{j_2=1}^{5}\cdots\bigcup_{j_{k+1}=1}^{5} \gamma_{0,j_1,j_2,\cdots,j_{k+1}}, \quad k\in\mathbb{N},$$

the eigen-value sequence is $\lambda = 3^{k+2}m$, $\mu = 3^{k+2}n$, $m, n \in \mathbb{N}$, $k\in\mathbb{P}$. Then, the boundary condition $v_k|_{\gamma_k} = 0$ is satisfied.

Take the eigen-value sequence of the problem (6.1.30)

$$\lambda_{k,m,3} = 3^{k+2}m, \quad \mu_{k,n,3} = 3^{k+2}n, \quad m,n = 1,2,\cdots,$$

the solution can be written as

$$v_{k,m,n}(x,y) \equiv \sin\frac{2\pi}{3}\left(3^{k+2}m\otimes x\right)\sin\frac{2\pi}{3}\left(3^{k+2}n\otimes y\right),$$
$$k=0,1,2,\cdots,m,n=1,2,\cdots.$$

Correspondingly, the equation (6.1.31) has a solution

$$T_{k,m,n}(t) \equiv A_{k,m,n}\cos\frac{2\pi}{3}\left(3^{k+2}m\otimes t\right) + B_{k,m,n}\sin\frac{2\pi}{3}\left(3^{k+2}n\otimes t\right),$$
$$k=0,1,2,\cdots,m,n=1,2,\cdots.$$

Combine the above, we have

$$u_{k,m,n}(t,x,y) = T_{k,m,n}(t)v_{k,m,n}(x,y)$$
$$= \left\{A_{k,m,n}\cos\frac{2\pi}{3}\left(3^{k+2}m\otimes t\right) + B_{k,m,n}\sin\frac{2\pi}{3}\left(3^{k+2}n\otimes t\right)\right\}$$
$$\cdot \sin\frac{2\pi}{3}\left(3^{k+2}m\otimes x\right)\sin\frac{2\pi}{3}\left(3^{k+2}n\otimes y\right), \quad k\in\mathbb{P}, m,n\in\mathbb{N},$$

where

$$A_{k,m,n} = \frac{\int_{\Omega_k}\varphi_k(x,y)\cdot\sin\frac{2\pi}{3}\left(3^{k+2}m\otimes x\right)\cdot\sin\frac{2\pi}{3}\left(3^{k+2}n\otimes y\right)dxdy}{\int_{\Omega_k}\left\{\sin\frac{2\pi}{3}\left(3^{k+2}m\otimes x\right)\right\}^2\cdot\left\{\sin\frac{2\pi}{3}\left(3^{k+2}n\otimes y\right)\right\}^2 dxdy},$$

$$B_{k,m,n} = \frac{\int_{\Omega_k}\psi_k(x,y)\cdot\sin\frac{2\pi}{3}\left(3^{k+2}m\otimes x\right)\cdot\sin\frac{2\pi}{3}\left(3^{k+2}n\otimes y\right)dxdy}{3^{k+2}\sqrt{m^2+n^2}\int_{\Omega_k}\left\{\sin\frac{2\pi}{3}\left(3^{k+2}m\otimes x\right)\right\}^2\cdot\left\{\sin\frac{2\pi}{3}\left(3^{k+2}\otimes y\right)\right\}^2 dxdy}.$$

The problem (6.1.29) has a formal solution

$$u_k(t,x,y) = \sum_{m=1}^{+\infty}\sum_{n=1}^{+\infty}T_{k,m,n}(t)v_{k,m,n}(x,y),$$

and the formal solution of problem (6.1.28) is

$$u(t,x,y) = \lim_{k\to+\infty}u_k(t,x,y) = \lim_{k\to+\infty}\sum_{m=1}^{+\infty}\sum_{n=1}^{+\infty}u_{k,m,n}(t,x,y).$$

Take $p=5$, $\varphi(x,y)=0$, and $\psi_k(x,y)$ is taken as that of in Example 6.1.1, then, we may show that the numerical approximation solution in Fig. 6.1.8.

We may also show other numerical examples, for example, $p=5$, $p=7$, and so on, and may draw the approximation pictures of the solutions.

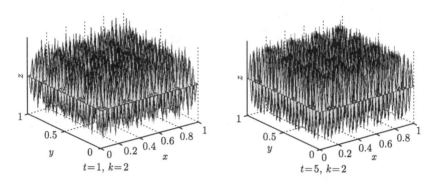

Figure 6.1.8

6.2 Further study on fractal analysis over local fields

Example 6.1.4 shows a special p-type 2-order partial differential equation and its solution over local field K_p, this motivates us to establish a general theory of fractal PDE on local fields.

6.2.1 *Pseudo-differential operator T_α*

The theory of pseudo-differential operators plays a significant role in the fractal PDE on local fields. As a groundwork of fractal analysis and fractal PDE, we consider the underlying space as a p-series field, since p-series field has simple algebraic operations, $+$, $\times$, mod p, no carrying; and a simple topological structure, as well as a simple character group. We concentrate on the studying of theory of fractal analysis and fractal PDE on p-series as a model, such that the general theory for local fields, including p-series fields, p-adic fields, and algebraic extensions of these two fields, can be developed. We refer to the current bibliographies of this section, [35],[60],[72],[99].

1. Pseudo-differential operator on a local field K_p

Preliminaries: readers are familiar with knowledge of Fourier analysis on local field K_p, such as, test function space $\mathbb{S}(K_p)$, distribution space $\mathbb{S}^*(K_p)$, symbol class $S^\alpha_{\rho\delta}(K_p) \equiv S^\alpha_{\rho\delta}(K_p \times \Gamma_p)$, Fourier transformations and inverse Fourier transformations of $\varphi \in \mathbb{S}(K_p)$ and $f \in \mathbb{S}^*(K_p)$, convolution of a function with a distribution, $\cdots$, as well as various properties of them, refer to Chapter 3.

A pseudo-differential operator T_α on local field K_p is defined as follows. Let $\xi \in \Gamma_{K_p}$, $\langle \xi \rangle = \max\{1, |\xi|\}$, then $\langle \xi \rangle^\alpha \in S^\alpha_{\rho\sigma}(K_p)$, $\alpha \in \mathbb{R}$, $\rho \geqslant 0$, $\sigma \geqslant 0$.

Denoted by T_α the pseudo-differential operator with symbol $\langle \xi \rangle^\alpha$, then (see(3.2.2))

$$T_\alpha \varphi(x) = (\langle \cdot \rangle^\alpha \varphi^\wedge(\cdot))^\vee (x), \quad \forall \varphi \in \mathbb{S}(K_p), x \in K_p, \qquad (6.2.1)$$

and

$$\langle T_\alpha f, \varphi \rangle = \langle f, T_\alpha \varphi \rangle, \quad \forall f \in \mathbb{S}^*(K_p). \qquad (6.2.2)$$

For $\alpha > 0$, then T_α is said to be an **α-order p-type derivative operator**; and for $\alpha < 0$, T_α is said to be an **$(-\alpha)$-order p-type integral operator**; If $\alpha = 0$, we agree with $T_0 : T_0 f = f = If$ is an **identity operator**.

To determine the kernel of pseudo-differential operator T_α on local field K_p, we define a distribution $\pi_\alpha \in \mathbb{S}^*(K_p)$.

Definition 6.2.1 (Distribution π_α) Let $\alpha \in \mathbb{C}$. For $\mathrm{Re}\,\alpha > 0$, a distribution $\pi_\alpha \in \mathbb{S}^*(K_p)$ is defined as

$$\langle \pi_\alpha, \varphi \rangle = \int_{K_p} |x|^{\alpha-1} \varphi(x) dx, \quad \forall \varphi \in \mathbb{S}(K_p). \qquad (6.2.3)$$

The above integral is convergent absolutely, thus it guarantees (6.2.3) is well defined. Further, note that π_α is holomorphic in $\mathrm{Re}\,\alpha > 0$, then we may extend π_α as an analytic function to the complex field $\mathbb{C}$, such that $\forall \varphi \in \mathbb{S}(K_p)$, it holds

$$\langle \pi_\alpha, \varphi \rangle = \int_{B^0} |x|^{\alpha-1} (\varphi(x) - \varphi(0)) dx + \int_{K_p \setminus B^0} |x|^{\alpha-1} \varphi(x) dx$$

$$+ \frac{1-p^{-1}}{1-p^{-\alpha}} \varphi(0). \qquad (6.2.4)$$

Clearly, on the complex field $\mathbb{C}$, π_α is analytic except $\alpha_k = \dfrac{2k\pi i}{\ln p}$, $k \in \mathbb{Z}$, which are simple poles with the residue $\dfrac{p-1}{p \ln p} \delta$, hence for any $\alpha \in \mathbb{R}, \alpha \neq 0$, the distribution π_α is well defined.

(6.2.3) and (6.2.4) can be simplified[104]: for $\alpha \in \mathbb{R}, \alpha \neq 0$,

$$\langle \pi_\alpha, \varphi \rangle = \int_{K_p} |x|^{\alpha-1} (\varphi(x) - \varphi(0)) dx, \quad \forall \varphi \in \mathbb{S}(K_p). \qquad (6.2.5)$$

To find the kernel of T_α, we need two lemmas.

Lemma 6.2.1 Let $\alpha \in \mathbb{R}$, if $\alpha \neq 0$, then
$$\int_{B^0} |x|^{-\alpha-1} \left(\chi(-\xi x)-1\right) dx = \left(\frac{p^{-\alpha} - p^{-\alpha-1}}{1-p^{-\alpha}} + \frac{1-p^{-\alpha-1}}{1-p^\alpha} |\xi|^\alpha\right)(1-\Delta_0),$$
where $\Delta_0(x) = \begin{cases} 1, & x \in B^0 \\ 0, & x \notin B^0 \end{cases}$, $B^0 = \{x \in K_p : |x| \leq 1\}$, and $\xi \in \Gamma_p$.

Proof. Take $t = \xi x$, then $dt = |\xi| dx$, we get
$$\int_{B^0} |x|^{-\alpha-1} \left(\chi(-\xi x) - 1\right) dx = |\xi|^{-1} \int_{|t| \leq |\xi|} |\xi^{-1} t|^{-\alpha-1} \left(\chi(-t) - 1\right) dt$$
$$= |\xi|^\alpha \int_{|t| \leq |\xi|} |t|^{-\alpha-1} \left(\chi(-t) - 1\right) dt.$$

If $|\xi| \leq 1$, then $\chi(-t) = 1$, and $\int_{|t| \leq |\xi|} |t|^{-\alpha-1} \left(\chi(-t) - 1\right) dt = 0$. If $|\xi| = p^N > 1$, that is, $N > 0$, then
$$\int_{|t| \leq |\xi|} |t|^{-\alpha-1} \left(\chi(-t) - 1\right) dt = \int_{p \leq |t| \leq |\xi|} |t|^{-\alpha-1} \left(\chi(-t) - 1\right) dt$$
$$= \sum_{r=1}^N p^{-\alpha r - r} \left(\int_{|t|=p^r} \chi(-t) dt - p^r \left(1 - \frac{1}{p}\right)\right)$$
$$= -p^{-\alpha-1} - \left(1 - \frac{1}{p}\right) p^{-\alpha} \frac{1-p^{-\alpha N}}{1-p^{-\alpha}}$$
$$= \frac{p^{-\alpha} - p^{-\alpha-1}}{1-p^{-\alpha}} |\xi|^{-\alpha} + \frac{1-p^{-\alpha-1}}{1-p^\alpha}.$$

Thus,
$$\int_{B^0} |x|^{-\alpha-1} \left(\chi(-\xi x)-1\right) dx = \left(\frac{p^{-\alpha}-p^{-\alpha-1}}{1-p^{-\alpha}} + \frac{1-p^{-\alpha-1}}{1-p^\alpha} |\xi|^\alpha\right)(1-\Delta_0).$$

Definition 6.2.2 (Locally constant function) $\psi : K_p \to \mathbb{C}$ is said to be **a locally constant function**, if $\forall x \in K_p$, $\exists l(x) \in \mathbb{Z}$, s.t. $\psi(x+y) = \psi(x)$ for $y \in B^{l(x)}$.

Denote by $\mathbb{H}(K_p)$ the set of all locally constant functions.

Lemma 6.2.2 Let $\alpha \in \mathbb{R}$, and let
$$\kappa_\alpha = \begin{cases} \left(\dfrac{1-p^\alpha}{1-p^{-\alpha-1}} \pi_{-\alpha} + \dfrac{1-p^\alpha}{1-p^{\alpha+1}}\right) \Delta_0, & \alpha \neq 0, -1, \\ \delta, & \alpha = 0, \\ \left(1 - \dfrac{1}{p}\right)(1 - \log_p |x|) \Delta_0, & \alpha = -1, \end{cases}$$

then $(\kappa_\alpha)^\wedge = \langle \xi \rangle^\alpha$.

Proof. By definition of κ_α, it is a distribution with compact support, supp $\kappa_\alpha \subset B^0$, and it is easy to check $(\kappa_\alpha)^\wedge$ is a locally constant function.

(a) If $\alpha \neq 0, -1$, by (6.2.4), the Fubini Theorem and Lemma 6.2.1, it follows that for $\varphi \in \mathbb{S}(K_p)$,

$$\left\langle \left(\frac{1-p^\alpha}{1-p^{-\alpha-1}} \pi_{-\alpha} \Delta_0 \right)^\wedge, \varphi \right\rangle$$

$$= \left\langle \frac{1-p^\alpha}{1-p^{-\alpha-1}} \pi_{-\alpha} \Delta_0, \varphi^\wedge \right\rangle$$

$$= \int_{B^0} \frac{1-p^\alpha}{1-p^{-\alpha-1}} |x|^{-\alpha-1} (\varphi^\wedge(x) - \varphi^\wedge(0)) \, dx + \frac{1-p^{-1}}{1-p^{-\alpha-1}} \varphi^\wedge(0)$$

$$= \frac{1-p^\alpha}{1-p^{-\alpha-1}} \int_{B^0} |x|^{-\alpha-1} \int_{\Gamma_p} \varphi(\xi)(\chi(-\xi x) - 1) \, d\xi dx + \frac{1-p^{-1}}{1-p^{-\alpha-1}} \langle 1, \varphi \rangle$$

$$= \frac{1-p^\alpha}{1-p^{-\alpha-1}} \int_{\Gamma_p} \varphi(\xi) \int_{B^0} |x|^{-\alpha-1} (\chi(-\xi x) - 1) \, dxd\xi + \frac{1-p^{-1}}{1-p^{-\alpha-1}} \langle 1, \varphi \rangle$$

$$= \frac{1-p^\alpha}{1-p^{-\alpha-1}} \left\langle \left(\frac{p^{-\alpha} - p^{-\alpha-1}}{1-p^{-\alpha}} + \frac{1-p^{-\alpha-1}}{1-p^\alpha} |\xi|^\alpha \right) (1 - \Delta_0), \varphi \right\rangle$$

$$+ \frac{1-p^{-1}}{1-p^{-\alpha-1}} \langle 1, \varphi \rangle$$

$$= \left\langle \left(|\xi|^\alpha - \frac{1-p^{-1}}{1-p^{-\alpha-1}} \right) (1 - \Delta_0), \varphi \right\rangle + \frac{1-p^{-1}}{1-p^{-\alpha-1}} \langle 1, \varphi \rangle$$

$$= \left\langle |\xi|^\alpha (1 - \Delta_0) + \frac{1-p^{-1}}{1-p^{-\alpha-1}} \Delta_0, \varphi \right\rangle.$$

Hence,

$$(\kappa_\alpha)^\wedge = |\xi|^\alpha (1 - \Delta_0) + \frac{1-p^{-1}}{1-p^{-\alpha-1}} \Delta_0 + \frac{1-p^\alpha}{1-p^{\alpha+1}} \Delta_0$$

$$= |\xi|^\alpha (1 - \Delta_0) + \Delta_0 = \langle \xi \rangle^\alpha.$$

(b) If $\alpha = -1$,

$$(\kappa_{-1})^\wedge = \left(1 - \frac{1}{p}\right) \int_{B^0} (1 - \log_p |x|) \chi(-\xi x) \, dx$$

$$= \left(1 - \frac{1}{p}\right) \left(\Delta_0 - \int_{B^0} \log_p |x| \, \chi(-\xi x) \, dx \right).$$

Evaluate $\int_{B^0} \log_p |x| \chi(-\xi x) \, dx$:

For $|\xi| \leq 1$, then

$$\int_{B^0} \log_p |x| \, \chi(-\xi x) \, dx = \sum_{r=0}^{+\infty} \int_{|x|=p^{-r}} \log_p |x| dx$$

$$= \sum_{r=0}^{+\infty} (-r) \, p^{-r} \left(1 - \frac{1}{p}\right) = \frac{1}{1-p}.$$

Thus,

$$(\kappa_{-1})^{\wedge}(\xi) = \left(1 - \frac{1}{p}\right) \left(1 - \frac{1}{1-p}\right) = 1, \quad |\xi| \leqslant 1.$$

For $|\xi| = p^N > 1$, then

$$\int_{|x| \leqslant |\xi|^{-1}} \log_p |x| \, \chi(-\xi x) \, dx + \int_{|\xi|^{-1} < |x| \leqslant 1} \log_p |x| \, \chi(-\xi x) \, dx$$

$$= \sum_{r=N}^{+\infty} (-r) \, p^{-r} \left(1 - \frac{1}{p}\right) + \sum_{r=0}^{N-1} (-r) \int_{|x|=p^{-r}} \chi(-\xi x) \, dx$$

$$= -\left(1 - \frac{1}{p}\right) \sum_{r=N}^{+\infty} r p^{-r} + (-N+1)(-p^{-N}) = \frac{p^{-N}}{p^{-1}-1} = \frac{|\xi|^{-1}}{p^{-1}-1}.$$

Thus,

$$(\kappa_{-1})^{\wedge}(\xi) = \left(1 - \frac{1}{p}\right) \left(-\frac{|\xi|^{-1}}{p^{-1}-1}\right) = |\xi|^{-1}, \quad |\xi| > 1;$$

Combining the above, we have

$$(\kappa_{-1})^{\wedge}(\xi) = \langle \xi \rangle^{-1}.$$

(c) If $\alpha = 0$, it follows that

$$(\kappa_\alpha)^{\wedge} = \delta^{\wedge} = 1 = \langle \xi \rangle^0.$$

The proof is complete.

Theorem 6.2.1 κ_α has the semi-group property

$$\kappa_\alpha * \kappa_\beta = \kappa_{\alpha+\beta}, \quad \alpha, \beta \in \mathbb{R}.$$

Proof. For $\alpha, \beta \in \mathbb{R}$, supp $\kappa_\alpha \subset B^0$, supp $\kappa_\beta \subset B^0$, thus $\kappa_\alpha * \kappa_\beta$ exists, then

$$(\kappa_{\alpha+\beta})^{\wedge} = \langle \xi \rangle^{\alpha+\beta} = \langle \xi \rangle^{\alpha} \cdot \langle \xi \rangle^{\beta} = (\kappa_\alpha)^{\wedge} \cdot (\kappa_\beta)^{\wedge} = (\kappa_\alpha * \kappa_\beta)^{\wedge},$$

we have $\kappa_\alpha * \kappa_\beta = \kappa_{\alpha+\beta}$.

Now we turn to prove the properties of the operator T_α.

Theorem 6.2.2 Let $\alpha \in \mathbb{R}$. Then

(i) $\forall f \in \mathbb{S}^*(K_p) \Rightarrow T_\alpha f = \kappa_\alpha * f$, that is, T_α has the convolution kernel κ_α;

(ii) $\forall f \in \mathbb{S}^*(K_p)$, $\alpha, \beta \in \mathbb{R} \Rightarrow T_{\alpha+\beta} f = T_\alpha T_\beta f = T_{\beta+\alpha} f$.

Hence, $T_\alpha T_{-\alpha} f = T_0 f = f$, that is, $(T_\alpha)^{-1} = T_{-\alpha}$.

Proof. (i) For $\alpha \in \mathbb{R}$ and $f \in \mathbb{S}^*(K_p)$, since supp $\kappa_\alpha \subset B^0$, so that $\kappa_\alpha * f$ exists. By Lemma 6.2.2, it follows that

$$T_\alpha f = (\langle \xi \rangle^\alpha f^\wedge)^\vee = ((\kappa_\alpha)^\wedge \cdot f^\wedge)^\vee = ((\kappa_\alpha * f)^\wedge)^\vee = \kappa_\alpha * f.$$

(ii) $T_\alpha T_\beta f = T_\alpha(T_\beta f) = \kappa_\alpha * (\kappa_\beta * f) = (\kappa_\alpha * \kappa_\beta) * f = \kappa_{\alpha+\beta} * f = T_{\alpha+\beta} f$.

We find the fixed point set of T_α in the space $\mathbb{S}^*(K_p)$ then.

Since $T_0 = I$ is the identity operator, so that $\forall f \in \mathbb{S}^*(K_p) \Rightarrow T_0 f = f$. Thus, we may suppose $\alpha \neq 0$.

Theorem 6.2.3 *For $\alpha \in \mathbb{R}$, the spaces $\mathbb{S}(K_p)$, $\mathbb{S}^*(K_p)$ and $\mathbb{H}(K_p)$ are invariant spaces under the operator T_α.*

It is clear by the properties of T_α.

Next, we see that the set of fixed points of T_α is depending on the support of distribution $f \in \mathbb{S}^*(K_p)$.

Definition 6.2.3 (Space $\mathbb{E}(K_p)$) *For $f \in \mathbb{S}^*(K_p)$, we define the space*

$$\mathbb{E}(K_p) = \{f \in \mathbb{S}^*(K_p) : \text{supp } f^\wedge \subset \Gamma^0\}.$$

It is the set of all distributions with compact support supp$f^\wedge$ of Fourier transformation $f^\wedge$ of $f \in \mathbb{S}^*(K_p)$, and supp$f^\wedge \subset \Gamma^0$.

Theorem 6.2.4 *Let $\alpha \in \mathbb{R}$, $\alpha \neq 0$. Then $T_\alpha g = g$ if and only if $g \in \mathbb{E}(K_p)$. i.e., $\mathbb{E}(K_p)$ is the set of all fixed points of T_α.*

Proof. Sufficiency. Take $g \in \mathbb{E}(K_p)$, by supp $g^\wedge \subset \Gamma^0$, then

$$(\kappa_\alpha * g)^\wedge = (\kappa_\alpha)^\wedge \cdot g^\wedge = \langle \xi \rangle^\alpha f^\wedge = g^\wedge,$$

the last equality $\langle \xi \rangle^\alpha f^\wedge = g^\wedge$ holds because $\xi \in B^0 \Rightarrow \langle \xi \rangle = 1$. Thus by Theorem 6.2.1 and the uniqueness of Fourier transformation, it follows $T_\alpha g = \kappa_\alpha * g = g$.

Necessity. Let $T_\alpha g = g$, $g \in \mathbb{S}^*(K_p)$. If supp $g^\wedge \not\subset \Gamma^0$, then there exists a test function $\varphi \in \mathbb{S}(K_p)$ with supp $\varphi \subset K_p \backslash B^0$, such that

$$\langle g^\wedge, \varphi \rangle \neq 0. \tag{6.2.6}$$

By $\varphi \in \mathbb{S}(K_p)$, there exists an integer $N \in \mathbb{N}$, such that supp $\varphi \subset B^{-N}$, and
$$\varphi = \sum_{r=1}^{N} \varphi \cdot \Phi_{B^{-r}\setminus B^{-r+1}} = \sum_{r=1}^{N} \varphi_r,$$
where $\varphi_r = \varphi \cdot \Phi_{B^{-r}\setminus B^{-r+1}} \in \mathbb{S}(K_p)$.

Thus, (6.2.6) implies that $\exists r_0, 1 \leqslant r_0 \leqslant N$, s.t. $\langle g^\wedge, \varphi_{r_0}\rangle \neq 0$.

On the other hand, by $T_\alpha g = \kappa_\alpha * g = g \Rightarrow (\kappa_\alpha * g)^\wedge = g^\wedge \Rightarrow \langle \xi \rangle^\alpha g^\wedge = g^\wedge$, we get

$\langle g^\wedge, \varphi_{r_0}\rangle = \langle \langle \xi \rangle^\alpha g^\wedge, \varphi_{r_0}\rangle = \langle g^\wedge, \langle \xi \rangle^\alpha \varphi_{r_0}\rangle = \langle g^\wedge, p^{\alpha r_0} \varphi_{r_0}\rangle = p^{\alpha r_0} \langle g^\wedge, \varphi_{r_0}\rangle$,

then, $p^{\alpha r_0} = 1$. However, this contradicts with $\alpha \neq 0$ and $r_0 \neq 0$, so we conclude that supp $g^\wedge \subset \Gamma^0$ holds. This implies $g \in \mathbb{E}(K_p)$. The proof is complete.

Theorem 6.2.5 *The fixed point set $\mathbb{E}(K_p)$ of T_α is the subset of $\mathbb{H}(K_p)$, i.e.,*
$$\mathbb{E}(K_p) = \{f \in \mathbb{H}(K_p) : f \text{ is constant on coset of } B^0\}.$$

The continuity of pseudo-differential operator T_α on $\alpha \in \mathbb{R}$ is described as follows.

Theorem 6.2.6 *Let $\alpha \in \mathbb{R}$. Then κ_α is continuous with respect to α in the space $\mathbb{S}^*(K_p)$; specially, we have $\lim\limits_{\alpha \to 0} \kappa_\alpha = \delta$ and $\lim\limits_{\alpha \to -1} \kappa_\alpha = \kappa_{-1}$ in $\mathbb{S}^*(K_p)$.*

Proof. Clearly, for $\alpha \in \mathbb{R}$, $\langle \xi \rangle^\alpha$ is continuous with respect to the parameter α in $\mathbb{S}^*(K_p)$. Moreover, by the continuity of inverse Fourier transformation in $\mathbb{S}^*(K_p)$, it follows that: $(\langle \xi \rangle^\alpha)^\vee$ is continuous with respect to α in $\mathbb{S}^*(K_p)$. Hence, the kernel κ_α is continuous with respect to α in $\mathbb{S}^*(K_p)$.

Theorem 6.2.7 *Let $\alpha \in \mathbb{R}$. For $f \in \mathbb{S}^*(K_p)$, we have $\lim\limits_{\beta \to \alpha} T^\beta f = T^\alpha f$.*

Proof. Firstly, we prove: $\forall \varphi \in \mathbb{S}(K_p)$ implies $\lim\limits_{\beta \to \alpha} T^\beta \varphi = T^\alpha \varphi$.

In fact, let $l \leqslant N$, denoted by
$$D_l^N(K_p) = \{\varphi \in \mathbb{S}(K_p) : \text{supp } \varphi \subset B^l, \varphi \text{ is constant on cosets of } B^N\}.$$

By Theorem 3.1.5, $\forall \varphi \in \mathbb{S}(K_p)$, there exists the index pair (N, l), such that $\varphi \in D_l^N(K_p)$, and $\varphi^\wedge \in D_N^l(\Gamma_p)$. Since $\langle \xi \rangle^\beta - \langle \xi \rangle^\alpha \in \mathbb{H}(K_p)$, and $\langle \xi \rangle^\beta - \langle \xi \rangle^\alpha$ is constant on the coset of Γ^0, thus, we have $\left(\langle \xi \rangle^\beta - \langle \xi \rangle^\alpha\right) \varphi^\wedge(\xi) \in D_{\max\{0,N\}}^l(K_p)$. This implies that

$$T^\beta\varphi - T^\alpha\varphi = \left(\left(\langle\cdot\rangle^\beta - \langle\cdot\rangle^\alpha\right)\varphi^\wedge(\cdot)\right)^\vee \in D_l^{\max\{0,N\}}.$$

On the other hand, by $\varphi^\wedge \in D_N^l\Gamma_p$, we have

$$T^\beta\varphi - T^\alpha\varphi = \int_{\Gamma_p}\left(\left(\langle\xi\rangle^\beta - \langle\xi\rangle^\alpha\right)\varphi^\wedge(\xi)\right)\chi_\xi(x)d\xi$$

$$= \int_{\Gamma^l\backslash\Gamma^0}\left(\left(\langle\xi\rangle^\beta - \langle\xi\rangle^\alpha\right)\varphi^\wedge(\xi)\right)\chi_\xi(x)d\xi.$$

Hence, $|T^\beta\varphi - T^\alpha\varphi| \leq M\int_{\Gamma^l\backslash\Gamma^0}\left|\langle\xi\rangle^\beta - \langle\xi\rangle^\alpha\right|d\xi$ with constant M depending on φ. By Lebesgue Dominated Theorem, the limit $T^\beta\varphi - T^\alpha\varphi \to 0$ holds uniformly as $\beta \to \alpha$.

Hence, it holds as $\beta \to \alpha$,

$$T^\beta\varphi \xrightarrow{S} T^\alpha\varphi, \quad \forall \varphi \in \mathbb{S}(K_p).$$

Secondly, for $f \in \mathbb{S}^*(K_p)$ and $\varphi \in \mathbb{S}(K_p)$,

$$\langle T^\beta f - T^\alpha f, \varphi\rangle = \langle f, T^\beta\varphi - T^\alpha\varphi\rangle.$$

As $\beta \to \alpha$, it has $T^\beta\varphi \xrightarrow{S} T^\alpha\varphi$. Then by the continuity of distribution f, it holds

$$\langle f, T^\beta\varphi - T^\alpha\varphi\rangle \to 0,$$

this implies that $T^\beta f \xrightarrow{S^*} T^\alpha f$, $\forall f \in \mathbb{S}^*(K_p)$. The proof is complete.

We give examples to evaluate p-type derivatives.

Example 6.2.1 Let $\alpha > 0$. Evaluate $T_\alpha\varphi$, $\varphi \in \mathbb{S}(K_p)$.

Solution. For $\alpha > 0$, by Lemma 6.2.2, for $\varphi \in \mathbb{S}(K_p)$,

$$T_\alpha\varphi(x) = \kappa_\alpha * \varphi(x) = \frac{1-p^\alpha}{1-p^{-\alpha-1}}\pi_{-\alpha}\Delta_0 * \varphi(x) + \frac{1-p^\alpha}{1-p^{\alpha+1}}\Delta_0 * \varphi(x).$$

Evaluate (by Theorem 3.1.26),

$$\pi_{-\alpha}\Delta_0 * \varphi(x) = \langle \pi_{-\alpha}\Delta_0, \varphi(x-\cdot)\rangle = \langle \pi_{-\alpha}, \Delta_0\varphi(x-\cdot)\rangle$$

$$= \int_{B^0}|y|^{-\alpha-1}\left(\varphi(x-y) - \varphi(x)\right)dy - \varphi(x)\int_{K_p\backslash B^0}|y|^{-\alpha-1}dy$$

$$= \int_{x+B^0}\frac{\varphi(y)-\varphi(x)}{|y-x|^{\alpha+1}}dy - \varphi(x)\left(1-p^{-1}\right)\frac{p^{-\alpha}}{1-p^{-\alpha}},$$

then for $\alpha > 0$, $\varphi \in \mathbb{S}(K_p)$,

$$T_\alpha\varphi(x) = \frac{1-p^\alpha}{1-p^{-\alpha-1}}\int_{x+B^0}\frac{\varphi(y)-\varphi(x)}{|y-x|^{\alpha+1}}dy$$

$$+ \frac{1-p^\alpha}{1-p^{\alpha+1}} \int_{x+B^0} \varphi(y) dy + \frac{p-1}{p-p^{-\alpha}} \varphi(x), \quad x \in K_p.$$

Example 6.2.2 Let $\alpha < 0$, $\alpha \neq -1$. Evaluate $T_\alpha \varphi$, $\varphi \in \mathbb{S}(K_p)$.

Solution. For $\alpha < 0$, $\alpha \neq -1$, by Lemma 6.2.2,

$$T_\alpha \varphi(x) = \kappa_\alpha * \varphi(x) = \frac{1-p^\alpha}{1-p^{-\alpha-1}} \pi_{-\alpha} * \varphi(x) + \frac{1-p^\alpha}{1-p^{\alpha+1}} \Delta_0 * \varphi(x).$$

Evaluate (by Theorem 3.1.26),

$$\pi_{-\alpha} \Delta_0 * \varphi(x) = \langle \pi_{-\alpha} \Delta_0, \varphi(x - \cdot) \rangle = \langle \pi_{-\alpha}, \Delta_0 \varphi(x - \cdot) \rangle$$

$$= \int_{B^0} |y|^{-\alpha-1} \varphi(x-y) dy = \int_{x+B^0} |y-x|^{-\alpha-1} \varphi(y) dy.$$

Hence, for $\alpha < 0, \alpha \neq -1$, $\varphi \in \mathbb{S}(K_p)$,

$$T_\alpha \varphi(x) = \frac{1-p^\alpha}{1-p^{-\alpha-1}} \int_{x+B^0} |y-x|^{-\alpha-1} \varphi(y) dy + \frac{1-p^\alpha}{1-p^{\alpha+1}} \int_{x+B^0} \varphi(y) dy.$$

Example 6.2.3 Let $\alpha = -1$. Evaluate $T_{-1}\varphi$, $\varphi \in \mathbb{S}(K_p)$.

Solution. By $\alpha = -1$ in Lemma 6.2.2,

$$T_{-1}\varphi(x) = \kappa_{-1} * \varphi(x) = \left(1 - \frac{1}{p}\right)(1 - \log_p |x|) \Delta_0 * \varphi(x)$$

$$= \left(1 - \frac{1}{p}\right) \langle 1 - \log_p |y|, \Delta_0 \varphi(x-y) \rangle$$

$$= \left(1 - \frac{1}{p}\right) \int_{B^0} (1 - \log_p |y|) \varphi(x-y) dy$$

$$= \left(1 - \frac{1}{p}\right) \int_{x+B^0} (1 - \log_p |y-x|) \varphi(y) dy,$$

hence, for $\varphi \in \mathbb{S}(K_p)$

$$T_{-1}\varphi(x) = \left(1 - \frac{1}{p}\right) \int_{x+B^0} (1 - \log_p |y-x|) \varphi(y) dy, \quad x \in K_p.$$

Example 6.2.4 Evaluate $T_\alpha \chi_\eta(x)$, $\chi_\eta \in \Gamma_p$, $\eta \in K_p$.

Solution. Let $\alpha \in \mathbb{R}$, $\eta \in K_p$, then by $(\chi_\eta(\cdot))^\wedge = \delta_\eta$ (Example 3.3.1), for any $\varphi \in \mathbb{S}(K_p)$, we have

$$\langle T_\alpha \chi_\eta(\cdot), \varphi \rangle = \left\langle (\langle \xi \rangle^\alpha \delta_\eta)^\vee, \varphi \right\rangle = \langle \delta_\eta, \langle \xi \rangle^\alpha \varphi^\vee(\xi) \rangle$$

$$= \langle \eta \rangle^\alpha \varphi^\vee(\eta) = \langle \eta \rangle^\alpha \langle \chi_\eta(\cdot), \varphi \rangle = \langle \langle \eta \rangle^\alpha \chi_\eta(\cdot), \varphi \rangle,$$

thus, for $\alpha \in \mathbb{R}$, $x_\eta \in \Gamma_p, \eta \in K_p$,

$$T_\alpha \chi_\eta(x) = \langle \eta \rangle^\alpha \chi_\eta(x), \quad x \in K_p.$$

Example 6.2.5 Evaluate $T_\alpha 1$, $\alpha \in \mathbb{R}$.
Solution. Since $1^\wedge = \delta$, then $\forall \varphi \in \mathbb{S}(K_p)$,

$$\langle T_\alpha 1, \varphi \rangle = \left\langle (\langle \xi \rangle^\alpha \delta)^\vee, \varphi \right\rangle = \langle \langle \xi \rangle^\alpha \delta, \varphi^\vee(\xi) \rangle = \langle \delta, \langle \xi \rangle^\alpha \varphi^\vee(\xi) \rangle$$
$$= \langle 0 \rangle^\alpha \varphi^\vee(0) = \varphi^\vee(0) = \int_{\Gamma_p} \varphi(\xi) \chi_0(\xi) d\xi = \langle 1, \varphi \rangle;$$

thus, for $\alpha \in \mathbb{R}$,

$$T_\alpha 1 = 1, \quad \text{in } \mathbb{S}^*(K_p).$$

Example 6.2.6 Evaluate $T_\alpha \delta$, $\alpha \in \mathbb{R}$.
Solution. By $T_\alpha \delta = \kappa_\alpha * \delta = \kappa_\alpha$, we get immediately

$$T_\alpha \delta = \kappa_\alpha, \quad \text{in } \mathbb{S}^*(K_p).$$

2. The spectrum theory of pseudo-differential operator T_α on local fields

In order to establish the fractal PDE on local fields, we study the spectrum theory of pseudo-differential operator T_α on K_p.

Firstly, the properties of T_α on the Hilbert space $L^2(K_p)$ are discussed.

Definition 6.2.4 (Domain of T_α) Let $\alpha \in \mathbb{R}$, we denote

$$\mathbb{D}(T_\alpha) = \{ f \in L^2(K_p) : \langle \xi \rangle^\alpha f^\wedge(\xi) \in L^2(\Gamma_p) \}.$$

Then, $\mathbb{D}(T_\alpha)$ is the domain of operator T_α in space $L^2(K_p)$, that is,

$$\forall f \in \mathbb{D}(T_\alpha) \Rightarrow T_\alpha f = (\langle \cdot \rangle^\alpha f^\wedge(\cdot))^\vee.$$

Lemma 6.2.3 $\langle \xi \rangle^\alpha \in L^2(\Gamma_p)$ if and only if $\alpha < -\dfrac{1}{2}$.
Proof. By

$$\int_{\Gamma_p} \langle \xi \rangle^{2\alpha} d\xi = \int_{\Gamma^0} 1 \cdot d\xi + \int_{\Gamma_p \backslash \Gamma^0} |\xi|^{2\alpha} d\xi = 1 + \sum_{r=1}^{+\infty} p^{2\alpha r} p^r (1 - p^{-1})$$
$$= 1 + (1 - p^{-1}) \sum_{r=1}^{+\infty} p^{(1+2\alpha)r} = \frac{p^{2\alpha} - 1}{p^{2\alpha+1} - 1},$$

the series $\displaystyle\sum_{r=1}^{+\infty} p^{(1+2\alpha)r}$ converges if and only if $\alpha < -\dfrac{1}{2}$.

Theorem 6.2.8 For the domain $\mathbb{D}(T_\alpha)$ of T_α, if $\alpha \leqslant 0$, then $\mathbb{D}(T_\alpha) = L^2(K_p)$; if $\alpha > 0$, then $\mathbb{D}(T_\alpha) \subsetneqq L^2(K_p)$. Moreover, $\mathbb{D}(T_\alpha)$ is dense in $L^2(K_p)$, that is, $\overline{\mathbb{D}(T_\alpha)} = L^2(K_p)$.

Proof. If $\alpha \leqslant 0$, then $\langle\xi\rangle^\alpha \leqslant 1$, thus
$$f \in L^2(K_p) \Rightarrow f^\wedge \in L^2(\Gamma_p) \Rightarrow |\langle\xi\rangle^\alpha f^\wedge(\xi)| \leqslant |f^\wedge(\xi)| \in L^2(\Gamma_p),$$
this implies $\mathbb{D}(T_\alpha) = L^2(K_p)$.

If $\alpha > 0$, by Lemma 6.2.3 and the property of Fourier transformation, $F : L^2(K_p) \to L^2(\Gamma_p)$ is equimetric and isomorphic on $L^2(K_p)$, then there exists $g \in L^2(K_p)$, such that
$$g^\wedge(\xi) = \langle\xi\rangle^{-\alpha-\frac{1}{2}} \in L^2(\Gamma_p).$$
Then, by Lemma 6.2.3 again, $\langle\xi\rangle^\alpha g^\wedge = \langle\xi\rangle^{-\frac{1}{2}} \notin L^2(\Gamma_p)$, this implies that: $g \in L^2(K_p)$, but $g \notin \mathbb{D}(T_\alpha)$, so $\mathbb{D}(T_\alpha) \subsetneq L^2(K_p)$.

Thus, we have $\mathbb{D}(T_\alpha) \begin{cases} \subsetneq L^2(K_p), & \alpha > 0, \\ = L^2(K_p), & \alpha \leqslant 0. \end{cases}$

About density, $\mathbb{S}(K_p) \subset \mathbb{D}(T_\alpha)$ and $\overline{\mathbb{S}(K_p)} = L^2(K_p)$ imply $\overline{\mathbb{D}(T_\alpha)} = L^2(K_p)$.

The proof is complete.

Theorem 6.2.9 *For the range $T_\alpha(\mathbb{D}(T_\alpha))$ of T_α, we have*
$$T_\alpha(\mathbb{D}(T_\alpha)) \begin{cases} = L^2(K_p), & \alpha > 0, \\ \subsetneq L^2(K_p), & \alpha \leqslant 0. \end{cases}$$
Moreover, $T_\alpha(\mathbb{D}(T_\alpha))$ is dense in $L^2(K_p)$, $\overline{T_\alpha(\mathbb{D}(T_\alpha))} = L^2(K_p)$.

Proof. If $\alpha \geqslant 0$, then $\langle\xi\rangle^{-\alpha} \leqslant 1$, thus take $g \in L^2(K_p)$, and consider $T_\alpha f = g$, then
$$f = T_{-\alpha} g = \left(\langle\cdot\rangle^{-\alpha} g^\wedge(\cdot)\right)^\vee.$$
Thus,
$$g \in L^2(K_p) \Rightarrow g^\wedge \in L^2(\Gamma_p) \Rightarrow \langle\xi\rangle^{-\alpha} g^\wedge \in L^2(\Gamma_p)$$
$$\Rightarrow f = \left(\langle\cdot\rangle^{-\alpha} g^\wedge\right)^\vee \in L^2(K_p) \Rightarrow T_\alpha(\mathbb{D}(T_\alpha)) = L^2(K_p).$$

If $\alpha < 0$, by Lemma 6.2.3, and the property of Fourier transformation, $F : L^2(K_p) \to L^2(\Gamma_p)$ is equimetric and isomorphic on $L^2(K_p)$, there exists $g \in L^2(K_p)$, s.t.
$$g^\wedge(\xi) = \langle\xi\rangle^{\alpha-\frac{1}{2}} \in L^2(\Gamma_p).$$
Then the equality $T_\alpha f = g$ gives
$$f = T_{-\alpha} g = \left(\langle\cdot\rangle^{-\alpha} g^\wedge\right)^\vee = \left(\langle\cdot\rangle^{-\frac{1}{2}}\right)^\vee \notin L^2(K_p).$$

So there is no any $g \in L^2(K_p)$, s.t. $T_\alpha f = g$, this implies $T_\alpha(\mathbb{D}(T_\alpha)) \subsetneq L^2(K_p)$.

About density, we take $\varphi \in \mathbb{S}(K_p)$, then the equation $T_\alpha f = \varphi$ has an unique solution in $\mathbb{S}(K_p) \subset \mathbb{D}(T_\alpha)$, so that $T_\alpha(\mathbb{D}(T_\alpha)) \supset \mathbb{S}(K_p)$. This implies $\overline{T_\alpha(\mathbb{D}(T_\alpha))} = L^2(K_p)$.

Theorem 6.2.10 *If $\alpha \in \mathbb{R}$, then T_α is a non-negative self-adjoint operator on the space $L^2(K_p)$.*

Proof. By Parseval equality, it is easy to get that for $\varphi, \psi \in \mathbb{D}(T_\alpha)$, it holds

$$\langle T_\alpha \psi, \varphi \rangle = \langle T_{\frac{\alpha}{2}} \psi, T_{\frac{\alpha}{2}} \varphi \rangle = \int_{\Gamma_p} \langle \xi \rangle^\alpha \psi^\wedge(\xi) \overline{\varphi^\wedge(\xi)} d\xi.$$

Then, $\|T_\alpha \psi\|_2^2 = (T_\alpha \psi, T_\alpha \psi) = \int_{\Gamma_p} \langle \xi \rangle^{2\alpha} |\psi^\wedge(\xi)|^2 d\xi$, where $(T_\alpha \psi, T_\alpha \psi)$ is the inner product in $L^2(\Gamma_p)$. Thus, it follows

$$(T_\alpha \psi, T_\alpha \psi) = \|T_\alpha \psi\|_{L^2(K_p)}^2 > 0, \quad \forall \psi \in \mathbb{D}(T_\alpha), \psi \neq 0.$$

By theory of non-negative self-adjoint operators[104], we have $T_{\frac{\alpha}{2}} = (T_\alpha)^{\frac{1}{2}}$, and

$$\mathbb{D}(T_\alpha) = \{\psi \in \mathbb{D}(T_{\frac{\alpha}{2}}) : T_{\frac{\alpha}{2}} \psi \in \mathbb{D}(T_{\frac{\alpha}{2}})\}.$$

Moreover, there exists a non-negative quadratic form $Q^\alpha(\cdot, \cdot)$ on $L^2(K_p)$ with domain $\mathbb{D}(T_{\frac{\alpha}{2}}) \times \mathbb{D}(T_{\frac{\alpha}{2}})$, such that

$$Q^\alpha(\varphi, \psi) = (T_{\frac{\alpha}{2}} \varphi, T_{\frac{\alpha}{2}} \psi), \quad (\varphi, \psi) \in \mathbb{D}(T_{\frac{\alpha}{2}}) \times \mathbb{D}(T_{\frac{\alpha}{2}}).$$

Thus, we may introduce a new inner product $\overline{Q}^\alpha(\varphi, \psi) = Q^\alpha(\varphi, \psi) + (\varphi, \psi)$, and the set $\left(\mathbb{D}(T_{\frac{\alpha}{2}}), \overline{Q}^\alpha\right)$ becomes a Hilbert space. The proof is complete.

How about the eigen-values, eigen-functions in the Hilbert space $L^2(K_p)$ of the operator T_α, and do we have a complete orthogonal function system in $L^2(K_p)$?

To discuss the eigen-value problem of operator T_α in $L^2(K_p)$, we consider the equation

$$T_\alpha \psi = \lambda \psi, \quad \psi \in L^2(K_p). \tag{6.2.7}$$

By Theorem 6.2.10, any eigen-value λ of T_α is positive, i.e., $\lambda \geqslant 0$.

Let $\lambda = 0$, then (6.2.7) becomes $T_\alpha \psi = 0$, this implies $\psi = 0$, thus $\lambda = 0$ is not an eigen-value.

For $\lambda > 0$, we take a form $(T_\alpha - \lambda)\psi = 0$ of (6.2.7), and take the Fourier transformation

$$0 = (T_\alpha \psi - \lambda \psi)^\wedge(\xi) = (\langle\xi\rangle^\alpha - \lambda)\psi^\wedge(\xi),$$

then it shows that the eigen-values of T_α have forms $\lambda_N = p^{N\alpha}$, $N \in \mathbb{P} = \{0, 1, 2, \cdots\}$.

Suppose that there exists a complete orthogonal system $\{\psi_N(x)\}$ consisted of eigen-functions of operator T_α in $L^2(K_p)$, then the Fourier transformation of function $\psi_N(x)$ is

$$(\psi_N)^\wedge(\xi) = \begin{cases} \Phi_{\{|\xi|=p^{-N}\}}(\xi)\rho_N(\xi), & N > 0, \\ \Phi_{\{|\xi|\leqslant 1\}}(\xi)\rho_0(\xi), & N = 0, \end{cases}$$

where

$$\Phi_{\{|\xi|=p^{-N}\}}(\xi) = \begin{cases} 1, |\xi| = p^{-N}, \\ 0, |\xi| \neq p^{-N}, \end{cases}$$

$$\int_{\{|\xi|=p^{-N}\}} |\rho_N(\xi)|^2 d\xi = 1, \quad N \in \mathbb{N} = \{1, 2, \cdots\},$$

and

$$\Phi_{\{|\xi|\leqslant 1\}}(\xi) = \begin{cases} 1, & |\xi| \leqslant 1, \\ 0, & |\xi| > 1, \end{cases} \quad \int_{\{|\xi|\leqslant 1\}} |\rho_0(\xi)|^2 d\xi = 1.$$

By induction, it follows:

Theorem 6.2.11 *Let $\alpha \in \mathbb{R}$. Then the set of eigen-values $\{\lambda_N\}_{N=0}^{+\infty}$ of operator T_α is*

$$\{\lambda_N\}_{N=0}^{+\infty} = \begin{cases} \{1, p^\alpha, p^{2\alpha}, \cdots\}, & \alpha > 0, \\ \{1\}, & \alpha = 0, \\ \{\cdots, p^{2\alpha}, p^\alpha, 1\}, & \alpha < 0. \end{cases} \quad (6.2.8)$$

To construct a complete orthogonal system by the eigen-functions of operator T_α in $L^2(K_p)$, we prove two lemmas.

Lemma 6.2.4 *Let $\psi(x) = \chi_{p^{-1}}(x)\Phi_{B^0}(x)$. Then $\psi(x)$ is an eigenfunction of T_α, i.e.,*

$$T_\alpha \psi(x) = p^\alpha \psi(x), \quad \alpha \in \mathbb{R}.$$

Proof. We have

$$\psi^\wedge(\xi) = \int_{K_p} \chi_{p^{-1}}(x)\Phi_{B^0}(x)\overline{\chi_\xi}(x) dx = \int_{B^0} \chi((p^{-1}-\xi)x) dx = \Phi_{p^{-1}+\Gamma^0}(\xi),$$

and
$$\langle\xi\rangle^\alpha \psi^\wedge(\xi) = \langle\xi\rangle^\alpha \Phi_{p^{-1}+\Gamma^0}(\xi) = p^\alpha \Phi_{p^{-1}+\Gamma^0}(\xi), \quad \xi \in \Gamma_p.$$

Thus,
$$T_\alpha \psi(x) = (\langle\cdot\rangle^\alpha \psi^\wedge(\cdot))^\vee (x) = p^\alpha \int_{p^{-1}+\Gamma^0} \chi_\xi(x) d\xi$$
$$= p^\alpha \int_{\Gamma^0} \chi_{p^{-1}}(x) \chi_\xi(x) d\xi = p^\alpha \chi_{p^{-1}}(x) \int_{\Gamma^0} \chi_\xi(x) d\xi$$
$$= p^\alpha \chi_{p^{-1}}(x) \Phi_{B^0}(x) = p^\alpha \psi(x).$$

Lemma 6.2.5 Let $\psi(x) = \chi_{p^{-1}}(x) \Phi_{B^0}(x)$, $a, b \in K_p$, $a \neq 0$. Then
$$T_\alpha \psi(ax+b) = \begin{cases} p^\alpha |a|^\alpha \psi(ax+b), & |a| > p^{-1}, \\ \psi(ax+b), & |a| \leqslant p^{-1}. \end{cases} \quad (6.2.9)$$

Proof. The Fourier transformation of $\psi(ax+b)$ is
$$(\psi(ax+b))^\wedge(\xi) = |a|^{-1} \chi_\xi(a^{-1}b) \psi^\wedge(a^{-1}\xi)$$
$$= |a|^{-1} \chi_\xi(a^{-1}b) \Phi_{a(p^{-1}+\Gamma^0)}(\xi),$$

so
$$T_\alpha \psi(ax+b) = (\langle\cdot\rangle^\alpha (\psi(ax+b))^\wedge(\cdot))^\vee(x)$$
$$= \int_{\Gamma_p} \langle\xi\rangle^\alpha |a|^{-1} \chi_{a^{-1}b}(\xi) \Phi_{a(p^{-1}+\Gamma^0)}(\xi) \chi_x(\xi) d\xi$$
$$= |a|^{-1} \int_{a(p^{-1}+\Gamma^0)} \langle\xi\rangle^\alpha \chi_{x+a^{-1}b}(\xi) d\xi.$$

If $|a| \leqslant p^{-1}$, then $a(p^{-1}+\Gamma^0) \subset \Gamma^0$, thus
$$T_\alpha \psi(ax+b) = |a|^{-1} \int_{a(p^{-1}+\Gamma^0)} \langle\xi\rangle^\alpha \chi_{x+a^{-1}b}(\xi) d\xi$$
$$= \int_{\Gamma^0} \chi_{x+a^{-1}b}(a(p^{-1}+\xi)) d\xi$$
$$= \int_{\Gamma^0} \chi(p^{-1}(ax+b)) \chi(\xi(ax+b)) d\xi$$
$$= \chi(p^{-1}(ax+b)) \Phi_{B^0}(ax+b) = \psi(ax+b).$$

If $|a| > p^{-1}$, then $\forall \xi \in a(p^{-1}+\Gamma^0)$ implies $|\xi| = p|a|$, thus
$$T_\alpha \psi(ax+b) = \int_{a(p^{-1}+\Gamma^0)} |\xi|^\alpha |a|^{-1} \chi_{x+a^{-1}b}(\xi) d\xi$$

$$= p^\alpha |a|^\alpha \int_{a(p^{-1}+\Gamma^0)} |a|^{-1} \chi_{x+a^{-1}b}(\xi) d\xi = p^\alpha |a|^\alpha \psi(ax+b).$$

Lemma 6.2.6 Let $\alpha \in \mathbb{R}$. Then the set of eigen-functions of T_α,

$$\{\psi_{N,j,I} : N \in \mathbb{Z}, j = 1, \cdots, p-1, I = z_I + B^0\} \quad (6.2.10)$$

is a complete orthogonal base in $L^2(K_p)$, where

$$\psi_{N,j,I}(x) = p^{\frac{-N}{2}} \chi_j \left(p^{N-1}x\right) \Phi_{B^0} \left(p^N x - z_I\right),$$
$$N \in \mathbb{Z}, \ j = 1, \cdots, p-1, \ I = z_I + B^0, \quad (6.2.11)$$

and

$$T_\alpha \psi_{1-N,j,I}(x) = \begin{cases} p^{N\alpha} \psi_{1-N,j,I}(x), & N > 0, \\ \psi_{1-N,j,I}(x), & N \leqslant 0. \end{cases} \quad (6.2.12)$$

Proof. **First step.** Prove (6.2.12).

$$\psi_{N,j,I}(x) = p^{\frac{-N}{2}} \chi_j \left(p^{N-1}x\right) \Phi_{B^0} \left(p^N x - z_I\right)$$
$$= p^{\frac{-N}{2}} \chi_j \left(p^{-1}\left(p^N x\right)\right) \Phi_{B^0} \left(p^N jx - jz_I\right)$$
$$= p^{\frac{-N}{2}} \chi_j \left(p^{-1}z_I\right) \chi_j \left(p^{-1}\left(p^N x - z_I\right)\right) \Phi_{B^0} \left(p^N jx - jz_I\right)$$
$$= p^{\frac{-N}{2}} \chi_j \left(p^{-1}z_I\right) \psi \left(p^N jx - jz_I\right).$$

If $N < 1$, then $|p^N| > p^{-1}$, thus

$$T_\alpha \psi_{N,j,I}(x) = p^{\frac{-N}{2}} \chi_j \left(p^{-1}z_I\right) T_\alpha \psi \left(p^N jx - jz_I\right)$$
$$= p^{\frac{-N}{2}} \chi_j \left(p^{-1}zI\right) p^\alpha \left|p^N\right|^\alpha \psi \left(p^N jx - jz_I\right)$$
$$= p^{(1-N)\alpha} \psi_{N,j,I}(x).$$

If $N \geqslant 1$, then $|p^N| \leqslant p^{-1}$, thus

$$T_\alpha \psi_{N,j,I}(x) = p^{\frac{-N}{2}} \chi_j \left(p^{-1}z_I\right) T_\alpha \psi \left(p^N jx - jz_I\right)$$
$$= p^{\frac{-N}{2}} \chi_j \left(p^{-1}z_I\right) \psi \left(p^N jx - jz_I\right) = \psi_{N,j,I}(x).$$

Changing N into $1-N$, we get (6.2.12).

Second step. Prove the orthogonal property of $\{\psi_{N,j,I}\}$.

Consider the inner product $(\psi_{N,j,I}, \psi_{N',j',I'})$ in $L^2(K_p)$, we have

$$(\psi_{N,j,I}, \psi_{N',j',I'}) = \int_{p^{-N}I \cap p^{-N'}I'} p^{\frac{-N}{2}} \chi_j \left(p^{N-1}x\right) p^{\frac{-N'}{2}} \overline{\chi}_{j'} \left(p^{N'-1}x\right) dx$$
$$= \delta_{NN'} \int_{p^{-N}(I \cap I')} p^{-N} \chi_j \left(p^{N-1}x\right) \overline{\chi}_{j'} \left(p^{N'-1}x\right) dx$$

$$= \delta_{NN'}\delta_{II'} \int_{p^{-N}I} p^{-N}\chi_{j-j'}\left(p^{N-1}x\right)dx = \delta_{NN'}\delta_{II'}\delta_{jj'}.$$

Moreover, $\forall \psi_{N,j,I}$ implies $\int_{K_p} \psi_{N,j,I}(x)dx = 0$. So that the second step is complete.

Third step. Prove the completeness of $\{\psi_{N,j,I}\}$.

Consider the Fourier coefficient $(\Phi_{B^0}, \psi_{N,j,I})$ of Φ_{B^0}, we have

$$(\Phi_{B^0}, \psi_{N,j,I}) = p^{\frac{-N}{2}} \int_{B^0 \cap p^{-N}I} \overline{\chi}_j\left(p^{N-1}x\right)dx.$$

If $N \leqslant 0$, then $(\Phi_{B^0}, \psi_{N,j,I}) = 0$. If $N \geqslant 0$, then $(\Phi_{B^0}, \psi_{N,j,I}) = p^{\frac{-N}{2}}\delta_{I,B^0}$, thus

$$\sum_{N,J,I} |(\Phi_{B^0}, \psi_{N,j,I})|^2 = (p-1)\sum_{N=1}^{+\infty} p^{-N} = 1 = \|\Phi_{B^0}\|_{L^2(K_p)}^2.$$

Hence, Parserval equality holds for Fourier coefficients of Φ_{B^0}.

The proof is complete.

6.2.2 Further problems on fractal analysis over local fields

To establish the framework of the fractal PDE on a local field K_p, we need more fundamental theory including pseudo-differential operator theory.

What cornerstone theory is needed? We propose one of them, for example, the "Weyl problem".

In 1912, H. Weyl proved the famous "Weyl asymptotic formula"[33],[106]: if the domain $\Omega \subset \mathbb{R}^n$ has a boundary Γ with enough smoothness, then the eigen-values of Dirichlet problem

$$\begin{cases} \sum_{j=1}^{n} \dfrac{\partial^2 u}{\partial x^j} + \lambda u = 0, \quad (x_1, \cdots, x_n) \in \Omega, \\ u|_\Gamma = 0 \end{cases}$$

have the asymptotic formula $N(\lambda) \approx \varphi(\lambda)$, $\lambda \to +\infty$, where $N(\lambda) = \#\{q \geqslant 1 : \lambda_q \leqslant \lambda\}$ is said to be a **counting function**, and the eigenvalues are denoted by $0 < \lambda_1 \leqslant \lambda_2 \leqslant \cdots \leqslant \lambda_q \leqslant \cdots$; the function $\varphi(\lambda) = \dfrac{1}{(2\pi)^n} B_n |\Omega|_n \lambda^{\frac{n}{2}}$ is said to be the **Weyl term**, and B_n is the volume of unit ball in $\mathbb{R}^n$; $|\Omega|_n$ is the Lebesgue measure of Ω.

The work of Weyl is a cornerstone in the study of eigen-value problem of fractal PDE, lots of mathematicians in the world pay their great attention to his work, and more excellent works on the topic appear, especially, the studies of Weyl conjecture under various conditions, such as, $\Gamma = \partial \Omega$ is a fractal with Hausdorff dimension $\dim_H \Gamma = d$, then what about the asymptotic formula of $N(\lambda)$?

In [33], the author considers the Weyl–Berry conjecture, and shows that: for von Koch curve as a boundary with $d = \dfrac{3}{2}$ of a fractal drum problem, the estimate for the counting function is

$$N(\lambda) = \frac{|\Omega|_2}{4\pi}\lambda + O\left(\lambda^{\frac{d}{2}}\right), \quad \lambda \to +\infty,$$

where $|\Omega|_2$ is the Lebesgue measure of Ω with the boundary Γ.

We may consider the **Weyl conjecture** problems for the two kinds of PDE in the section 6.1, what about the Weyl term of two problems in 6.1? And for general fractal PDE, what about the Weyl problem?

The topic of fractal PDE is a relatively new one, see, for example, [21], [72], [102], [109]~[111]. And the topics on Harmonic analysis and fractal analysis over local fields are very interesting and challenging, there are lots of open problems in the area of mathematics theory, and in the area of applications in other scientific study, such as, physics, astronomy, geology, meteorology, biology, medicine science, and so on. It is worth to devote one' s efforts to this area.

Exercises

1. Prove: $(\kappa_\alpha)^\wedge \in \mathbb{E}(K_p)$.
2. Prove Theorem 6.2.3.
3. Prove Theorem 6.2.5.
4. What is the Weyl type conjecture on a local field K_p?
5. What preparation work needs to do for establishing the complete theory about PDE on a local field K_p?
6. What open problems will you consider when you want to study fractal PDE on the product group K^*?

Chapter 7

Applications to Medicine Science

Many scientists pay great attention to applying the mathematics science to medicine science for a long time. Because the structure of the human body is very complex, vital mechanism and functions are depending on lots of factors connected with many scientific areas, thus, scientists in life science hope to work with mathematicians, physicists, chemists, biologists, and experts in computer science, to establish a "system item" to study the problems about the structure of body, and uplift the health of mankind.

Mathematics science plays an important role in the natural science, societal science, information technology and engineering technology, and also in the medicine science. For example, the mathematical model of spread of contagious diseases, problem of density of plasma concentration, growth of population, growth of microbial colony, rate of flow of blood, and so on. Also, in clinical diagnosis, physicians and surgeons use some mathematical models to help their diagnoses and treatments.

In this chapter, we show some applications in the diagnoses and treatments of liver cancers. In Section 7.1 we focus on the derivations of problems, some ideas of solving the considered problems; and in Section 7.2 we show some examples in clinical medicine.

7.1 Determine the malignancy of liver cancers

7.1.1 *Terrible havocs of liver cancer, solving idea*

1. Health and life of Chinese are seriously threatened by liver cancer

The primary hepatic carcinoma (simply, liver cancer) is one of very common malignant tumors in the clinical diagnosis. Recently, the morbidity of liver cancer is going up in the world, as the fifth in all malignant tumors. The numbers of patients of liver cancer in China are 55% of that in the world, it is the highest morbidity. Compared with other malignancies, the liver cancer is at the third. The health and life of Chinese are threatened by liver cancer seriously.

The liver cancer is concealed, metastasis and recurrence are fast, the mortality is higher. A patient can only live 3 to 6 months without effective medical treatment, so called "the king of carcinoma". The mortality of liver cancer in China is 45% of that in the world. It is a terrible information for the Chinese people.

2. Developments of medical imaging and surgery operations solve parts of diagnosis of liver cancer

Since the developments of medical imaging and surgery operations, such that the diagnoses and treatments of liver cancer make great strides forward. For example, B ultrasound technique, multi-slice spiral CT, magnetic resonance imaging. As modern medical imaging science, plays important role in the assessment of characteristics, localization and clinical staging of various kinds of tumors. The applications of new techniques are successful not only for the description of morphology of tumor lesion, but also for analysis of biological characteristics of focuses.

In fact, the marginal shape of tumor lesion can show the growing pattern and degree of invasion. The shape of boundary of a tumor of liver cancer is related closely with the alpha-fetoprotein (AFP), proliferating cell nuclear antigen (PCNA), and other serum and molecular markers. Their relationships show the biology action of the malignancy of tumors in a certain extent, partially. On the other hand, the invasion and metastasis characteristics of liver cancer are connected with the abundant blood vessels in tumors. The generations of blood vessels in tumors in liver cancer are determined and effected by various growth factors, including P53 antibody and vascular endothelial growth factor (VEGF), basic fibroblast growth factor (bFGF). Experiment study shows that: the expression of these factors are connected with the representations of medical imaging of tumors, they have intrinsic relationships. Thus, when doctors try to diagnose cases of liver cancer, they must combine those medical imaging results films and hematologic parameters to analyze the biology characteristics, metastasis and recurrence characteristics, carefully, then they can make the accuracy decision and proper treatments.

3. Clinical medicine science calls for math to build a fractal model of liver cancer

To determine the quantification of shape of a tumor is a difficult problem for medicine science. On one hand, it is because a malignancy tumor is in invasive growth, the boundary of a focus of tumor is so irregular that the classical tools of mathematics fail to describe the patterns. On the other hand, the arrangements of blood vessels in tumors are disordered, lots of abnormal grafting appear in blood vessels, vascular diameters are irregular with many branches, vascular endothelial cells are arranged discontinuously. Recently, the 3-dimensional rebuilt techniques by scan of CT imaging provide certain new methods for describing boundaries of tumors and shapes of blood vessels in tumors. A 3-dimensional picture for focus of tumor of patients is rebuilt by computer to deal with a series continuous 2-dimensional pictures by boundary detection and segmentation. The 3D reconstruction technique enables us to display the complex 3-dimensional structures of living beings exactly, so that doctors not only can observe and operate by rotating and slitting the 3-dimensional pictures, but also can measure the sides of tumors.

However, the measure of boundaries of tumors and shapes of blood vessels used by doctors is by visual test, which could not give exact descriptions for irregular margins of tumors and complex blood vessels quantitatively based upon the Euclidean geometry.

In 1982, the "fractal geometry" was born, B. B. Mandelbrot introduced the concept of fractal firstly in his book *The Fractal Geometry of Nature*[38]. His idea is: there exists a lot of complex objects in the universe, such as, mountains, rivers, clouds, trees, stocks and futures, as well as distributions of blood vessels, boundaries of tumors, ···, they are nonlinear, and not differentiable, even are not continuous. In the point of view of classical calculus, these graphs or functions are irregular, so that it is difficult or unable to be treated by the classical mathematical methods. However, they have certain inherence of themselves, and the fractal geometry is a suitable mathematics science branch to reveal or deal with problems for these complex objects in nature. These complex objects are called fractals by Mandelbort. Then many new concepts, theory and methods appeared, for example, the fractal dimensions, such as the Hausdorff dimension, Box dimension, Packing dimension, Fourier dimension, ···, can be used to describe and deal with the complexities of fractal objects. To establish certain mathematical models, patterns, and to combine with the study of clinicine medical science, mankind then can dominate and triumph over the terrible tumors.

4. Combine mathematics science, medical imaging, molecular biology and clinical medicine science, will bring new concept, method, and benefit to humakind

The applications of fractal analysis to clinical medicine science is a very new and hot topic in the world. For instance, Masters[39] studies retinal blood vessels by fractal analysis, he computes the fractal dimension of retinal blood vessels, it is 1.7. Oczeretko et al[41] study a new pattern of angiogenesis of lung cancer by the area-circumference method and Box dimension method, respectively, and have got the conclusion: since the fractal dimension is greater than the topological dimension, thus the tumor blood vessels are fractal objects. Guidolin et al[24] measure the variations of the blood vessel net before and after the effects of angiostatic activity of docetaxel, then they have a motivation: fractal method can be used to determine curative effect of anti-angiogenesis. Sabo et al[64] discover by studying of renal cell carcinoma: the fractal dimensions of blood vessel not only describe complexity of blood vessel, but also provide some prognosis information about tumors.

We also have some effective study results in China, for example, by the Box dimension of boundaries of liver cancer, combining with the clinical data, we conclude that: there are close relationship between Box dimension of boundary of liver tumor and ability of invasion and metastasis of liver cancer. This shows that: the shape of tumor boundary and the system of blood vessels have certain very complex space multi-structures, thus a coalition of fractal theory and method, as well as 3-dimensional imaging technique will open a new stage of studying liver cancer. We will use certain quite new mathematical concepts and techniques, such as, the theory of harmonic analysis and fractal analysis on local fields, combined with medical imaging, molecule biology, clinical medicine science, then it is sure that a new area will be formed in the diagnoses and treatments of liver cancer[23],[53],[61],[68].

7.1.2 The main methods in studying of liver cancers

The main idea and process in studying the problems of liver cancers are: combining the research approaches in medicine science and that in mathematics science closely, cooperating each other mutually, then we may obtain a mathematical model of generation, development, diagnosis, treatment of a liver tumor, and thus may guide the clinical medicine science.

1. Construct a terrace of 3D reconstruction technique for studying liver cancer with blood vessels

To construct a platform of 3-dimensional reconstruction technique is very important, since all data obtained from imaging pictures, B ultrasound, Computed Tomography, magnetic resonance imaging, are in 2-dimensional aspects, they are not enough to determine the degree of malignance of tumors. However, the imaging pictures of tumor lesion blood vessels reconstructed by 3-dimensional reconstruction technique may help doctors to make definite diagnosis and to design a treatment protocol, directly and exactly, thus the data of 3-dimensional reconstruction technique will play very important auxiliary role.

Based on 3-dimensional reconstruction technique, an origin imaging of focus of tumors with vessels can be generated. Recently, 3-dimensional reconstruction technique is developing, scientists and engineers do their best to optimize the techniques and apply to clinical medicine science. But the 3D reconstruction technique used in rebuilding of shapes of tumor lesion with blood vessels during tumor invasion, metastasis and recurrence are difficult, and new imaging techniques are expected.

2. Various clinical data of patients to be used to establish mathematical model

The materials of B ultrasound technique, multi-slice spiral CT, magnetic resonance imaging of patients are used to rebuild the focus of tumor with blood vessels by using 3-dimensional reconstruction technique; at the same time, by serum tests, molecule biology tests, a lot of data and related materials are obtained, then preparations for establishing mathematical models are complete. Combining all materials, specially, observing the effects of fractal dimension of tumor boundary to the shapes of focus of tumor with blood vessels, discussing mathematical mutual dependence of them, then we may determine an accurate mathematical model.

3. Correct treatment protocol is determined by virtue of mathematical model

Doctors can make a definite diagnosis by the established mathematical model, by experiences, and by 3-dimensional reconstruction image, as well as by the medical biochemical tests, then seek out the causes of illness, determine correct treatment protocol: or surgical operation, or interventional treatment, or molecule-targeted therapy, or chemotherapy, and so on.

4. Steps

(1) **Animal experiments**

The animal test is a traditional research method in medicine science. A 3-dimensional reconstruction of focus of tumor lesion with blood vessels of liver cancer of animals will supply some information and indication.

Take rabbit hepatocellular carcinoma VX2 cell line, then resuscitate, centrifuge, purify as cell suspension. Moreover, take tumor cell suspension 0.1ml (number of cells is 105), inject into livers of New Zealand white rabbits. Observe tumor formation in the liver of rabbits ten days later, do spiral CT scanning for rabbits on day 10, 15, 20 and 25, respectively, then by using 3-dimensional imaging technique to rebuild 3-dimensional pictures of focuses of liver tumors with blood vessels. Moreover, transform pictures to a suitable sides of gray scale, increase contrast ratios, and so on. Then, transform 2-dimensional numerical matrix to 3-dimensional numerical matrix, and then by virtue of visual technique, 3-dimensional numerical matrix is transformed to a 3-dimensional picture.

(2) **Establish fractal mathematical model of focus of liver cancer with blood vessels**

Combining data obtained in step (1), by using modern mathematical tools, such as, local fields, p-type calculus, fractals, random fractals, fractal dimensions, pseudo-differential operators, partial differential equations on fractals, fractal dynamics, $\cdots$, we establish a fractal mathematical model of growth, invasion, metastasis of focus of liver cancer with blood vessels. These idea and tools are quite new, thus they maybe attract interests of scientists, doctors and mathematicians.

By virtue of certain principle of fractal analysis and by using various mathematical tools, we design a software for obtained mathematical model, and deal with the 3-dimensional numerical pictures of focus of liver cancer by the software. Moreover, by using multi-iterated of software, we evaluate the values at different time points of each picture. Then, by using the statistics method, we may compare the obtained data to observe and discuss variations of fractal dimensions in processes of growth, invasion, metastasis of focuses of liver cancer with blood vessels.

(3) **Clinical research**

① Relativity between materials of medical imaging, pathology with fractal mathematical model.

Choose complete materials of medical imaging, pathology and chemical examinations in 5 years, and rebuild the 3-dimensional graphs for CT and arteriography. Then, compare the materials of clinical data with those obtained from fractal mathematical models, so that the relativity between parameters of fractal mathematics of focus of liver cancer with blood vessels, types of tumor, relapse and metastasis of tumor, as well as survival rate of patients are revealed.

On the other hand, by using tissue microarray technique to test molecule targets, such as the AFP in tissues of liver cancer, PCNA, P53 antibody, VEGF, bFGF, $\cdots$, then the relationships between the mathematical parameters and molecule targets can be analyzed.

② Design individual surgery project for a patient by fractal mathematical models of liver cancer.

Choose a suitable patient with liver cancer, do a series test of medical imaging, serology, and molecular biology, then analyze the materials of 3-dimensional reconstruction to obtain corresponding fractal mathematical parameters. By follow-up and data collection, we summarize the internal relations of the fractal mathematical parameters of differential focuses of liver cancers with the curative effects of various treatments for directing the choices of clinical therapy.

In summary, the synthetic study of liver cancer of human by clinical medicine science and mathematics science is to establish a fractal mathematical model firstly by virtue of some modern medical devices, such as the imaging technologies, gene chips, molecular biology tests, and some modern mathematical tools, such as analysis on local fields and fractal analysis. Then, test and complete the accuracy and clearness of obtained mathematical models by comparisons of those chemical examinations obtained by making use of design and evaluate of computer programs. Then, diagnose, choose and determine clinical surgery project by guiding the obtained mathematics models to doctors. Finally, we can obtain science conclusions by follow-up the patients.

Combining the idea and methods of modern mathematical science with tools in abstract harmonic analysis, studying invasion, metastasis of tumors and complexity of blood vessels, are not only provide mathematical base for clinical medicine science, but also reveal and express the universality and applicability of mathematics science; not only solve certain important clinical medical problems, such as the classifications of focus of liver cancer; choices of treatments, survival and recurrence of tumors, but also unite the clinical medicine science with various molecular biology, medics imaging, computer science and mathematical science closely, apply to guide clinicians to choose the appropriate therapy. This is one of the advanced topics in China and the world, and is a nice direction according with individual treatment project in liver cancer up to date.

The multi-intercross of vital subjects in modern sciences, penetration and combination of various scientific branches are realized and developed in our study, fully.

7.2 Examples in clinical medicine

By the imaging of CT or MRI (Magnetic Resonance Imaging) of patient's liver, clinicians diagnose and make sure whether it is liver cancer or not, then determine to do surgery or not. However, there are difficulties for clinicians, since they make diagnosis by observing roughness on surface of tumor by experience, such that deviations of veracity may appeared some times. Thus, clinicians hope certain data of the roughness of surface of tumor to make sure the malignancy of liver cancer, then determine how to do for patients, surgery? chemotherapy? or radiotherapy?

7.2.1 Take data from the materials of liver cancers of patients

We collect data from the materials in the medical imaging of patients, mainly, in B ultrasound, slice spiral CT, MRI, on one hand; and on the other hand, collect data from biology examinations, such as, serology and molecular biology examinations. The following are two examples.

(1) Examination and treatment records of Mr. Ma.

There are 5 preoperative lab slips, 2 B ultrasound reports, 4 CT and 1 operation report, 1 MRI reports, 2 photos of operation, 2 photos of lesions of liver cancer and 1 pathological report, 1 postoperative lab slip.

(2) Examination and treatments records of Mr. Wang.

There are 1 resident admit note (RAN), 1 hepatic function reserve report, 1 B ultrasound report, 1 gastroscopy report, 1 PEC/CT report, 3 biochemical test reports, 5 CT and MRI reports, 10 pictures of operation, 2 photos of lesions of liver cancer, 2 pathological reports, 2 postoperative lab slips and 1 diagnostic record.

7.2.2 Mathematical treatment for data

We study the effect about fractal dimension of boundary of lesion on malignancy of liver cancer in this section[53],[61].

As an example, we take Fig. 7.2.3, it is a 417 × 409 picture of blood vessels in a liver, and then use Matlab.

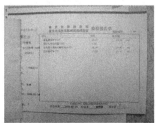

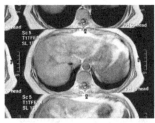
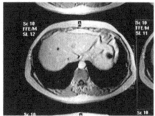
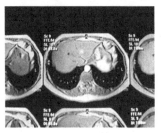
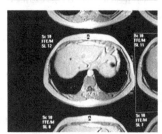

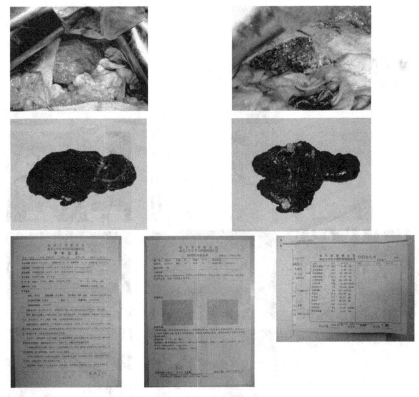

Fig. 7.2.1

Fig. 7.2.4 is the gradation of the color picture 7.2.3, that is
$$\text{gray} = r*0.299 + g*0.587 + b*0.114.$$

1. Gray equalization

We use gray equalization to improve the gray distributions such that the boundary of a picture can be determined.

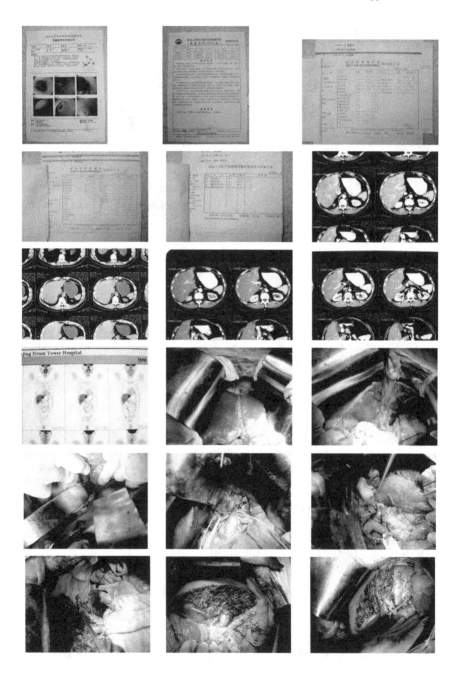

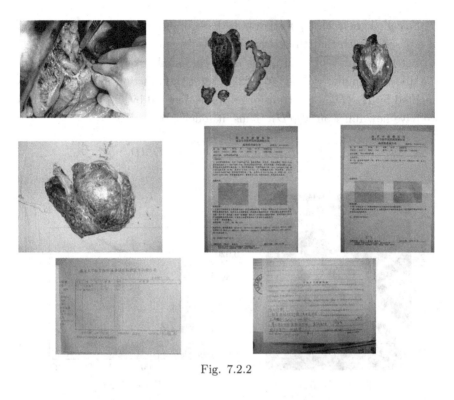

Fig. 7.2.2

Fig. 7.2.3 Fig. 7.2.4

Generally, we deal with a gray equalization problem in Fig. 7.2.5, the contrast ratio of graph is not good.

Fig. 7.2.6 is the relationship between grayscales (0~255) and pixels with the total pixel value 256. There is a pick at grayscale = 50, and the grayscales are not equalized. We will equalize the grayscale to guarantee the gradient of the boundary can be evaluated.

So called grayscale equalization, is that taking a function transformation of grayscale, such that the grayscale tends to equalization.

Let the number of total pixels is N, the number of grayscale is D_m (in the above example, we take it as 256), and the function in the Figures is denoted by $H_A(D_A)$, the transformation is denoted by $D_B = f(D_A)$, then we have new function $H_B(D)$. Thus, it follows that

$$\int_0^{D_A} H_A(D)dD = \int_0^{D_B} H_B(D)dD,$$

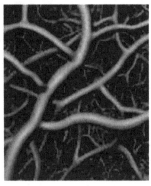

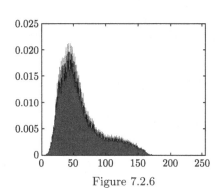

Figure 7.2.5   Figure 7.2.6

since the numbers $D \leqslant D_A$ (before transform) and $D \leqslant D_B$ (after transform) are equal each other. Differentiate both side for D_A, we have

$$H_A(D_A) = H_B(D_B)\frac{dD_B}{dD_A} = H_B(D_B)f'(D_A),$$

hence $H_B(D_B) = \dfrac{H_A(D_A)}{f'(D_A)}$. To guarantee the grayscale equalization, it should hold $H_B(D_B) = \dfrac{D_m}{N}$. Thus, $f'(D_A) = \dfrac{D_m}{N}H_A(D_A)$. Integrate both side about $t \equiv D_A$, then we get

$$f(D) = \frac{D_m}{N}\int_0^D H_A(t)dt,$$

this is the transformation function. Discretize it, and called CDF function, then

$$\text{CDF}(t) = \frac{1}{N}\sum_{i=1}^{t} H_A(t).$$

Use CDF $(t)D_m$ as new grayscale value, instead of t. Thus, the new Figures 7.2.7 and 7.2.8 are obtained, instead of the old two. It is clear that the contrast ratio of graph is increased.

Fig. 7.2.7

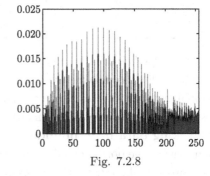

Fig. 7.2.8

2. Boundary test

(1) Doing Gaussian filtering

It is necessary to filter noise before we deal with boundary test since some figures have lots of noises, specially at the particular parts. We use the Gaussian filtering method, for simply, Gauss method, as follows.

The two dimension Gauss function is

$$G(x,y) = \exp\left\{-\frac{x^2+y^2}{2\sigma^2}\right\}.$$

It has rotation symmetry property, so that smoothing scales are same at each direction. Moreover, when we take Gauss model as weight function, the weight is smaller at a boundary, then the details of boundary are not obscured. Take $\sigma^2 = 2$, $n = 5$, then as in Table 7.2.1.

Table 7.2.1

$[i,j]$	-2	-1	0	1	2
-2	0.105	0.287	0.135	0.287	0.105
-1	0.287	0.606	0.779	0.606	0.287
0	0.135	0.779	1	0.779	0.135
1	0.287	0.606	0.779	0.606	0.287
2	0.105	0.287	0.135	0.287	0.105

Doing integerization and normalization, then we have

Table 7.2.2

$[i,j]$	-2	-1	0	1	2
-2	1	2	3	2	1
-1	2	4	6	4	2
0	3	6	7	6	3
1	2	4	6	4	2
2	1	2	3	2	1

The ratio coefficient is $1/100$.

This is the filter equal effective matrix with $n = 5$. Take the convolution of original figure with this matrix, then we have complete the filtering process. And then we may do boundary test.

(2) **Taking data from boundary**

Boundary is the border of a domain in which grayscales have varied very quickly, also it is the intersection line of object and backdrop in a figure.

The case of variation of grayscale in a figure can be described by its gradient, so that we can take differentiation of local figure to test boundary.

We start to construct a boundary test operator at a neighborhood of pixel in original figure.

Suppose that a given continuous function $f(x, y)$, its directional derivative takes local maximal value at the normal line of boundary. Thus, at the θ direction, we have

$$\frac{\partial f}{\partial r} = \frac{\partial f}{\partial x}\frac{\partial x}{\partial r} + \frac{\partial f}{\partial y}\frac{\partial y}{\partial r} = f_x \cos\theta + f_y \sin\theta.$$

Then, $\dfrac{\partial f}{\partial r}$ takes maximal value if $\dfrac{\partial}{\partial \theta}\left(\dfrac{\partial f}{\partial r}\right) = 0$. That is,

$$-f_x \sin\theta_g + f_y \cos\theta_g = 0,$$

with $\theta_g = \arctan\dfrac{f_y}{f_x}$; Hence, the maximal value of gradient is

$$g = \max\left(\frac{\partial f}{\partial r}\right) = \sqrt{f_x^2 + f_y^2}.$$

Use $g = |f_x| + |f_y|$ instead of $g = \sqrt{f_x^2 + f_y^2}$, approximately.

About algorithm, there are many ways: Sobel boundary test operator, Prewitt boundary test operator, Roberts boundary test operator, Marr boundary test operator, Canny boundary operator, $\cdots$.

We take Canny operator since its accuracy is quite good.

The main steps of Canny boundary test operation are:

Step1. Smoothing a figure by Gaussian filtering operation.

Step2. Computing value and direction of the gradient by finite difference of one order partial derivatives.

One may take any one order operator to compute the gradients of two orthogonal directions on the figure. We take an equal effective matrix

$$H = \begin{bmatrix} 1 & 0 & -1 \\ 1 & 0 & -1 \\ 1 & 0 & -1 \end{bmatrix}.$$

Step3. Inhibiting from the non-maximal values of gradient.

To determine the boundary, we have to keep the points at which gradient values take local maximal, and inhibit those of non-maximal. The method is as in Fig. 7.2.9. For a point, let its local gradients be G_x, G_y, and by G_x/G_y to compare their values in its 8 neighborhoods, then inhibit those non-maximal.

Step4. Testing and linking border points by bi-thresh value algorithm.

By the original figure of global grayscale we compute the high-threshold and low-threshold, sometimes, we take

low-threshold=high-threshold/2.

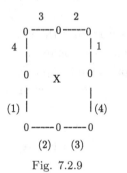

Fig. 7.2.9

Those points which values less then low-threshold are not boundary points, and which values bigger then high-threshold are boundary points; if its values are between the low and high-threshold, then they are boundary points, if and only if there are boundary points in its 8 neighborhoods.

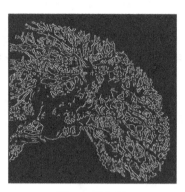

Figure 7.2.10

(3) Thinning algorithms

After boundary test, an obtained boundary is always rough, we thin the figure by thinning algorithms to increase the effect of boundary.

We use Zhang-Suen skeleton thinning algorithm, its speed is fast, and the effect is nice.

In the first sub-substitution, if the middle point $I(i,j)$ in 3×3 two value matrix satisfy the following conditions, then it can be deleted:

① There is just one connected component.
② The one number of neighborhood of middle point is between 2 and 6.
③ At least one in $I(i,j+1)$, $I(i-1,j)$, $I(i,j-1)$ is 0.
④ At least one in $I(i-1,j)$, $I(i+1,j)$, $I(i,j-1)$ is 0.

In the second sub-substitution, if the middle point $I(i,j)$ in 3×3 two value matrix satisfy the following conditions, then it can be deleted:

① There is just one connected component.
② The one number of neighborhood of middle point is between 2 and 6.
③ At least one in $I(i-1,j)$, $I(i,j+1)$, $I(i+1,j)$ is 0.
④ At least one in $I(i,j+1)$, $I(i+1,j)$, $I(i,j-1)$ is 0.

For Fig. 7.2.1, we have the result Figure in Fig. 7.2.10.

7.2.3 Compute fractal dimensions

1. Evaluate by definitions

After equalizing grayscale and obtaining boundary F of figure, we take a sequence δ_k as the length of net in a net cube, then analyze the relationship

Table 7.2.3

δ_k	2	3	4	5	6	7
$\ln \delta_k$	0.69	1.10	1.39	1.61	1.79	1.95
$\ln N_{\delta_k}$	9.88	9.31	8.87	8.48	8.14	7.84
δ_k	8	9	10	11	12	13
$\ln \delta_k$	2.08	2.20	2.30	2.40	2.48	2.56
$\ln N_{\delta_k}$	7.58	7.35	7.15	6.96	6.79	6.63
δ_k	14	15	16	17	18	19
$\ln \delta_k$	2.64	2.71	2.77	2.83	2.89	2.94
$\ln N_{\delta_k}$	6.47	6.35	6.22	6.09	5.99	5.89
δ_k	20	21	22	23	24	25
$\ln \delta_k$	3.00	3.04	3.09	3.14	3.18	3.22
$\ln N_{\delta_k}$	5.77	5.70	5.59	5.53	5.44	5.35

between the sequence $\log N_{\delta_k}(F)$ and δ_k, so that by virtue of Box dimension to search relationship between obtained data of boundary F of figure and the Box dimension of F.

Take a sequence as $\{\delta_k\} = \{2, 3, 4, \cdots, 24, 25\}$ (unit is pixel), by getdim.m function, we have

However, since the unit of length δ_k of net, it may appear the following problems:

① 24 data $-\log N_{\delta_k}(F)/\log \delta_k$ all are negative, this is not reasonable for the Box dimension.

② When $\delta_k = 1$, the definition can not be evaluated.

The problems appear since the denominator does not have unit clearly. For convenience, we may choose the unit of δ_k is 1 pixel, but it is not a real unit in case, so that we may suppose that 1 unit is $1 = C\delta_k$ with a constant C.

The Box dimension also can be expressed as

$$\dim_B F = \lim_{\delta \to 0} \frac{\log N_\delta(F)}{-\log \delta + C},$$

we try to find a constant C_0, such that the convergence rate of above limit is the fastest when $C = C_0$.

2. Linear fit for data

The maximal value of length of $\{\delta_k\}_{k=2}^{25}$ is very small comparing with pixel 409×417, so we may suppose that the sequence $\{\lambda_k\}_{k=2}^{25}$ is near the value of $\dim_B F$ within the area of error, and $\lambda_k = \dfrac{\log N_{\delta_k}(F)}{-\log \delta_k + C_0}$.

Thus, the sequence $\{\log N_{\delta_k}(F)\}$ and $\{-\log \delta_k\}$ have nice linear relation. In fact, we have the relations described in Fig. 7.2.11 and Fig. 7.2.12.

The fit result by linear recurrence to 24 data is

$$\log N_\delta(F) = -1.6712 \log \delta + 10.7422,$$

that is the approximation dimension is 1.6712.

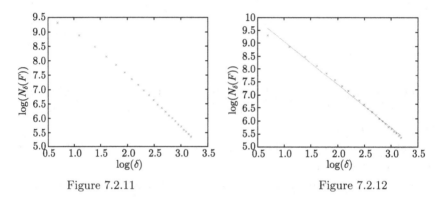

Figure 7.2.11   Figure 7.2.12

3. Hyperbolic fitting extrapolation for data

By definition of Box dimension, the true value of dimension should be computed at infinite subdivided cases. However, since pixels have some restrictions, thus the subdivision has restrictions also. For example, pixel is $400 * 400$, we may divide at most to 1 pixel. But we have found that $\log N_\delta(F)$ and $\log \delta$ have nice linear relation not at the case the length tends to 1 pixel. Since

① As δ tends to 1, there exist noise in figure, and noise can not be filtered clearly, so that at this case the effects of noise may vary strongly;

② As δ tends to 0, the slope is near to true value, and $\lambda_k = \dfrac{\log N_{\delta_k}(F)}{-\log \delta_k + C_0}$, so the smaller δ is, the noisier λ_k is.

Hence we have to give up two final points nearest to 1.

The other fit method — hyperbolic fitting can be used such that the final result may be nice to coincide with the limit process.

Take a δ-net with $\delta_i = i$, $i = 2, 3, \cdots, 24, 25$, we have $\log N_\delta(F)$ and $\log \delta$ in the Fig.7.2.13 and Fig.7.2.14:

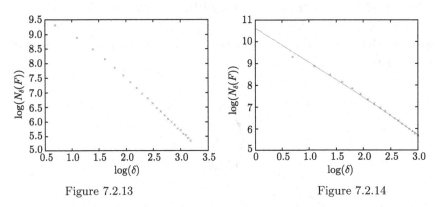

Figure 7.2.13   Figure 7.2.14

By hyperbolic equation, we have

$$\frac{(x - 6.6978)^2}{0.6772^2} - (y - 1.2445)^2 = 3.1473^2,$$

and the fitting degree is $r = 0.0839$. So that Box dimension of F is $\widetilde{\dim}_B F = \dfrac{1}{0.6772} = 1.4768$. The value of $\widetilde{\dim}_B F$ is an approximate one, there is error $\varepsilon = \widetilde{\dim}_B F - \dim_B F$ with the true value $\dim_B F$ of figure F, certainly. When we establish mathematical model of liver cancer, this error has to be considered so that a true model may be constructed.

7.2.4   Induce to obtain mathematical models

After analyzing more than 300 data sets of liver cancer, we conclude that: the malignancy of liver cancer has approximate linear relation with fractal dimension of boundary of lesion of liver cancer. This is an approximate mathematical model of liver cancer. Then the job of us is to examine the accuracy of the mathematical model.

We have two ways to check the accuracy: (i) follow-up research of patients; (ii) using obtained mathematical model to new patients to check whether the model is accurate or not.

The above work is ongoing in our research work.

7.2.5 Other problems in the research of liver cancers

1. Auxiliary partial orthotopic liver transplantation (APOLT)

In living related liver transplantation(LRLT), firstly, take off the bad part from recipient liver, then take a nice part from donor liver, and finally undergo the transplantation to make sure it is beneficial to patient and donor both. See Fig. 7.2.15.

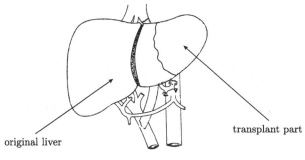

Figure 7.2.15

The problem is: how big volume we have to transplant such that the recipient liver is not harmed, and the transplant part of liver of patient can be lived?

Evaluate the volume which is best for solving the above problem, then provide a consultation to surgeons.

Certainly, this is a very difficult and complex research topic, the volume will be related with malignancy of liver cancer and quality of healthy liver, as well as other factors. Many mathematician and doctors in the world pay their great attentions to this topic.

2. Find genes which control liver cancer

By gene chips supported from clinical doctors and using statistics method, determine the main genes which control liver and liver cancer, then the theory of genetic engineering in medicine science can be established. This is a great engineering related to life science, clinical medicine science, mathematical science and many other scientific areas, and will be developed in the new century.

Bibliography

[1] 齐民友. 线性偏微分算子引论（上）. 科学出版社, 1996.
[2] 沙震, 阮火军. 分形与拟合. 浙江大学出版社, 2005.
[3] 苏维宜. 近代分析引论. 北京大学出版社, 2000.
[4] 文志英等. 分形几何理论与应用. 浙江科学技术出版社, 1998.
[5] 文志英. 分形几何的数学几何基础. 上海科技教育出版社, 2000.
[6] 张禾瑞. 近世代数基础. 人民教育出版社, 2000.
[7] 郑维形, 苏维宜, 任福贤. 沃尔什函数理论与应用. 上海科技出版社, 1983.
[8] Barnsley M. Fractals Everywhere. *Academic Press, INC.*, 1988.
[9] Butzer P L and Nessel R J. Fourier Analysis and Approximation, I. *Academic Press, New York*, 1971.
[10] Butzer P L and Wagner H J. Walsh–Fourier series and the concept of a derivative. *Applicable Anal.*, 1973, 3: 29–46.
[11] Butzer P L and Wagner H J. On dyadic analysis based on the point-wise dyadic derivative. *Anal. Math.*, 1975, 1: 171–196.
[12] Butzer P L and Wagner H J. An extension of the dyadic calculus with fractional order derivatives: further theory and applications. *Comp. & Math. with Appls.*, 1986, 12A(8): 921–943.
[13] Falconer K J. The Geometry of Fractal Sets. *Cambridge Univ. Press*, 1985.
[14] Falconer K J. Fractal Geometry: Mathematical Foundations and Applications. *Chichester, Wiley*, 1990.
[15] Falconer K J. Techniques in Fractal Geometry. *Chichester, Wiley*, 1998.
[16] Gát G. On the two dimensional point-wise dyadic calculus. *J. Approx. Theory*, 92: 191–215, 1998.
[17] Gát G. Convergence and divergence of Féjer means of Fourier series on one and two dimensional Walsh and Vilenkin groups. *Facta Universitatis, Series: Electronics and Energetics, University of Nis*, 2008, 21(3): 291–307.
[18] Gibbs J E and Millard M J. Walsh functions as solution of a logical differential equations. *NPL DES Rept.* 1, 1969.
[19] Gibbs J E and Millard M J. Some methods of solution of linear ordinary

logical differential equations. *NPL DES Rept.* 2, 1969.
[20] Gibbs J E and Ireland B. Some generalizations of the logical derivatives. *NPL DES Rept.* 8, 1971.
[21] Gibbs J E and Stankovic R S. Why IWGD-89? A look at the bibliography of Gibbs derivatives. *Theory and Applications of Gibbs Derivatives, Proc. of the First International Workshop on Gibbs Derivatives*, Butzer P L and Stankovic R S (Eds), Beograd, xi–xxiv, 1990.
[22] Gibbs J E. Fourier analysis in a space of characteristic functions of subsets of the rational numbers. *Walsh and Dyadic Analysis, Proceedings of the workshop dedicated to the memory of James Edmund Gibbs*, 2007: 3–14. Held in Oct. 18–19, 2007, Niš, Serbia.
[23] Gu Q S. Wave equations with fractal boundaries. *Undergraduate Thesis of Nanjing University, Department of Mathematics*, 2010.
[24] Guidolin D, Vacca A, Mussdorfer G G et al. A new image analysis method based on topological and fractal parameters to evaluate the angiostatic activity of docetaxel by using the Matrigel assay in vitro. *Microvasc Res.*, 2004, 67(2): 117–124.
[25] He Z L and Mustard D. Convergence properties of a class of Walsh–Fourier integral operators. *Proc. of the First International Workshop on Gibbs Derivatives*, Kupari–Dubrovnik Yugoslavia, 145–156, 1990.
[26] He Z L. Notes on approximations of Walsh functions (Chinese). *J. of Nanjing Univ.*, 1981, 4: 409–418.
[27] He Z L. An approximation theorem on p-adic Walsh–Féjer operators with some corollaries (Chinese). *J. of Nanjing Univ.*, 1982, 3: 585–597.
[28] He Z L. The derivatives and integrals of fractional order in Walsh–Fourier analysis with applications to approximation theory. *J. of Approx. Theory (U.S.A.)*, 1983, 39: 361–373.
[29] He Z L. A class of approximation operators and best approximation over $L^p(G)$, $1 < p < \infty$, $G = R, T, Z$. *Chinese J. of Contemporary Math.*, 1988, 9(2): 215–224.
[30] Hewitt E and Ross K A. Abstract Harmonic Analysis, I. *Berlin: Springer-Verlag*, 1979.
[31] Jiang H K. The kernels of de la Vallée–Poussin type on p-adic fields. *Approx. Theory & Appl.*, 1990, 6(1): 65–79.
[32] Jiang H K. The derivatives and integrals of functional order on a-adic groups. *Chin. Ann. of Math., Ser.B*, 1993, 14(4): 515–526.
[33] Lapidus M L. Vibrations of fractal drum, the Riemann hypothesis, waves in fractal media and the Weyl–Berry conjecture. *Trans. Amer. Math. Soc.*, 1991, 325: 465–529.
[34] Li Y and Su W Y. Random a-adic groups and random net fractals. *Chaos, Soliton & Fractals*, 2008, 37: 807–816.
[35] Ma L T and Su W Y. Two dimensional wave equations with fractal bound-

aries. *Acta Mathematica Sinica, Eng. Series*, 2013, 29(12): 2321–2342.
[36] Mandelbrot B B. How long is the coast of Britain? Statistical self-similarity and fractional dimension. *Science*, 1967, 155: 636–638.
[37] Mandelbrot B B. Fractals: Form, Chance and Dimension. *San Francisco*: W.H. Freeman & Co., 1977.
[38] Mandelbrot B B. The Fractal Geometry of Nature. *San Francisco*: W.H. Freeman & Co., 1982.
[39] Masters B R. Fractal analysis of the vascular tree in the human retina. *Annu Rev Biomed Eng.*, 2004, 6(4): 27–52.
[40] Mattila P. Geometry of Sets and Measures in Euclidean Spaces: Fractals and Rectifiability. *Cambridge: Cambridge Univ. Press*, 1995.
[41] Oczeretko E, Juczewska M and Kasacka I. Fractal geometric analysis of lung cancer angiogenic patterns. *Folia Histochem Cytobiol*, 2001, 39(2): 75–76.
[42] Onneweer C W and Su W Y. Homogeneous Besov spaces on locally compact Vilenkin groups. *Studia Mathematica*, 1989, T.XCIII: 17–39.
[43] Onneweer C W. Differentiability for Rademacher series on group. *Atca Sci. Math.*, 1977, 39: 121–128.
[44] Onneweer C W. Fractional differentiation on the group of integers of a p-adic or p-series field. *Anal. Math.*, 1977, 3: 119–130.
[45] Onneweer C W. Differentiation of fractional order on p-groups, approximation properties. *Anal. Math.*, 1978, 4: 297–302.
[46] Onneweer C W. Differentiation on p-aidc or p-series field. *Linear Spaces and Approximation, (Proc. Conf. Math. Res. Inst., Oberwolfach, 1977) Internal. Ser. Numer. Math.*, Birkhauser, Basel, 1978, 40: 187–198.
[47] Onneweer C W. Fractional derivatives and Lipschitz spaces on local fields. *Trans. Amer. Math. Soc.*, 1980, 258: 155–165.
[48] Onneweer C W. The Fourier transform of Herz Spaces on certain groups. *Monatsh. f. Math.*, 1984, 97: 297–310.
[49] Onneweer C W. Multipliers on weighted L_p-space over certain totally disconnected groups. *Trans. Amer. Math. Soc.*, 1985, 288: 347–362.
[50] Pál J and Simon P. On the generalized Butzer–Wagner type a.e. differentiability of integral function. *Annales Univ. Scient.*, Budapest Eötvös, Sect. Math., 1977, 20: 157-165.
[51] Pál J. On the connection between the concept of a derivative defined on the dyadic field and the Walsh–Fourier transform. *Annales Univ. Scient.*, Budapest Eötvös, Sect. Math., 1975, 18: 49–54.
[52] Pál J. On a concept of a derivative among functions defined on the dyadic field. *SIAM J. Math. Anal.*, 1977, 8(3): 347–391.
[53] Peng Y. Fractal dimension calculation of two dimension images with computer programming. *Undergraduate Thesis of Nanjing University, Department of Mathematics*, 2010.
[54] Qiu H and Su W Y. Weierstrass-like functions on local fields and their$p$-adic

derivatives, *Chaos, Soliton & Fractals*, 2006, 28: 958–965.
[55] Qiu H and Su W Y. Measures and dimensions of fractal sets in local fields, *Progress in Natural Sci.*, 2006, 16(12): 1260–1268.
[56] Qiu H and Su W Y. 3-adic Cantor function on local fields and its p-adic derivative. *Chaos, Soliton & Fractals*, 2007, 33: 1625–1634.
[57] Qiu H and Su W Y. The connection between the orders of p-adic calculus and the dimensions of Weierstrass type function in local fields. *Fractals*, 2007, 15(3): 279–287.
[58] Qiu H, Su W Y and Li Y. On the Hausdorff dimension of certain Riesz product in local fields, *Analysis in Theory and Appl.*, 2007, 23(2): 147–161.
[59] Qiu H and Su W Y. Distributional dimension of fractal sets in local fields, *Acta. Math. Sinaca (Eng.)*, 2008, 24(1): 147–158.
[60] Qiu H and Su W Y. Pseudo-differential operators on p-adic fields (Chinese). *Science in China*, 2011, 41(1): 1–15.
[61] Qiu Y D, Su W Y, Chen J and Qiu H. The malignant case of a cancer, determined by fractal dimensions of boundaries(Chinese). to appear.
[62] Ren F X, Su W Y and Zheng W X. The generalized logical derivatives and its applications (Chinese). *J. of Nanjing Univ.*, 1978, 3: 1–8.
[63] Ruan H J, Su W Y and Yao K. Box dimension and fractional integral of linear fractal interpolation functions, *J. of Approx. Theory*, 2009, 161, 187–197.
[64] Sabo E, Boltenko A, Sova Y, et al. Microscopic analysis and significance of vascular architectural complexity in renal cell carcinoma. *Clin Cancer Res.*, 2001, 7(3): 533–537.
[65] Sadosky C. Interpolation of Operators and Singular Integrals. *Springer-Verlag*, 1979.
[66] Schipp F, Wade W R, Simon P, Pál J. Walsh Series — An Introduction to the Dyadic Harmonic Analysis. *Adam Hilger*, Bristol, New York, London, 1990.
[67] Schipp F, Wade W R. Fast Fourier transforms on binary fields. *Approx. Theory and Appls.*, 1998, 14(1): 91–100.
[68] Shen K M. Study on two dimensional wave equations with fractal boundaries on local fields (Chinese). *Undergraduate Thesis of Nanjing University, Department of Mathematics*, 2010.
[69] Stankovic R S and Astola J. Gibbs derivatives. *Walsh and Dyadic Analysis, Proceedings of the workshop dedicated to the memory of James Edmund Gibbs*, 2007: 153–170. Held in Oct. 18–19, 2007, Niš, Serbia.
[70] Stankovic R S and Astola J. Remarks on the development and recent results in the theory of Gibbs derivatives. *Facta Universitatis, Series: Electronics and Energetics, University of Nis*, 2008, 21(3): 349–364. Proceedings of discrete analysis and related areas, Workshop on Discrete Analysis and Applications, held in Sep. 27–29, 2008, Thessaloniki, Greece.
[71] Su W Y and Chen G X. Lipschitz classes on local fields. *Science in China*,

Ser.A, Math., 2007, 50(7): 1005–1014.
[72] Su W Y and Qiu H. p-adic calculus and its applications to fractal analysis and medical science. Facta Uinversitatis, Series: Electronics and Energetics, University of Nis, 2008, 21(3): 339–348. Proceedings of discrete analysis and related areas, Workshop on Discrete Analysis and Applications, held in Sep. 27–29, 2008, Thessaloniki, Greece.
[73] Su W Y and Xu Q. Function spaces on local fields. Science in China, Ser.A, Math., 2006, 49(1): 66–74.
[74] Su W Y and Zheng W X. On the theory of approximation operators over local fields. Approx. Theory V (Texas U.S.A.), 1986: 579–582.
[75] Su W Y. An introduction to Walsh functions (Chinese). Proceedings in Digit Techniques and Applications, Wuxi, China, 1976: 298–328.
[76] Su W Y. On an extremum problem for n-variable Walsh transformations (Chinese). J. of Nanjing Univ., 1980, 2: 6–14.
[77] Su W Y. The kernels of Abel-Poisson type on Walsh system. Chin. Ann. of Math., Ser.B, 1981, 2: 81–92.
[78] Su W Y. A proof of the Jackson type theorem in Walsh system (Chinese). Proc. of the Conference on Approximation Theory, Huangshan, China, 1982: 181–188.
[79] Su W Y. The derivatives and integrals on local fields. Nanjing University Biquarterly, 1985, 1: 32–40.
[80] Su W Y. The kernels of product type on local fields (I), Approx. Theory & its Appl., 1985, 1(2): 93–109.
[81] Su W Y. The kernels of product type on local fields (II), Approx. Theory & its Appl., 1986, 2(2): 95–111.
[82] Su W Y. The approximate identity kernels of product type for the Walsh system. J. of Approx. Theory, 1986, 47(4): 284–301.
[83] Su W Y. Kernels of Poisson type on local fields. Science in China, Ser.A, 1988, 31(6): 641–653.
[84] Su W Y. Pseudo-differential operators in Besov spaces over local fields. Approx. Theory & Appl., 1988, 4(2): 119–129.
[85] Su W Y. Approximation theory and harmonic analysis on locally compact groups. Approximation, Optimization and Computing: Theory and Appl., Elsevier Science Publishers B.V. North-Holand, 1990: 181–184.
[86] Su W Y. Para-product operators over locally compact Vilenkin groups. A Friendly Collection of Mathematical Papers I, Proc. in Celebration of 70's Birthday of Professor Shu Lizhi, 1990: 1–5.
[87] Su W Y. Fractal and harmonic analysis over locally compact groups (Chinese). Proc. of non-linear problems in Science and Techniques, Nanjing, Jiangsu, China, 1991: 17–20.
[88] Su W Y. Psuedo-differential operators and derivatives on locally compact Vilenkin groups, Science in China, Ser.A, 1992, 35(7): 826–836.

[89] Su W Y. Gibbs derivative and its applications to approximation theory and fractals. *Approx. Theory VII, Austin U.S.A.*, 1992: 61–63.

[90] Su W Y. Walsh analysis in the last 25 years (Chinese). *Porc. of the Fifth International Workshop on Spectral Techniques, Univ. of Aeronautics and Astronautics, Beijing, China*, 1994: 117–127.

[91] Su W Y. Operators-derivatives-spaces-differential equations on locally compact Vilenkin groups. *Harmonic Analysis in China, Edited by Yang C C et al., Kluwer Academic Publishers, Hong Kong*, 1995: 240–255.

[92] Su W Y. Gibbs derivatives and their applications. *Numer. Funct. Anal. and Optimiz.*, 1995, 16(5&6): 805-824.

[93] Su W Y. Para-product operators and para-linearization on locally compact Vilenkin groups. *Science in China, Ser. A*, 1995: 38(11), 1304–1312.

[94] Su W Y. Gibbs derivatives and differential equations on Vilenkin groups. *Recent Developments in Abstract Harmonic Analysis with Applications in Signal Processing, Edited by Stankovic R S. et al., Nauka Belgrade, Yugoslavia*, 1996: 79-94.

[95] Su W Y. Gibbs-Butzer differential operators on locally compact Vilenkin groups. *Science in China, Ser.A*, 1996, 39(7): 718–727.

[96] Su W Y. The boundedness of certain operators on Hölder and Sobolev spaces. *Approx. Theory & Appl.*, 1997, 13(1): 18–32.

[97] Su W Y. Calculus on fractals based upon local fields. *Approx. Theory & Appl.*, 2000, 16(1): 92-100.

[98] Su W Y. Gibbs derivatives — the development over 40 years in China. *Walsh and Dyadic Analysis, Proceedings of the workshop dedicated to the memory of James Edmund Gibbs*, 2007: 15–30. Held in Oct. 18–19, 2007, Niš, Serbia.

[99] Su W Y. Two dimensional wave equations with fractal boundary. *Applicable Analysis*, 2011, 90(3–4), 533–543. *Proc. of dedicated to the 80's Birth of Porf. Dr. P.L.Butzer*, Held in Mar. 21–24, 2009, Lindau, Germany.

[100] Taibleson M H. Fourier Analysis on Local Fields. *Princeton Univ. Press*, 1975.

[101] Triebel H. Theory of Function Spaces. *Basel: Birkauser Verlag*, 1983.

[102] Triebel H. Fractals and Spectra. *Basel: Birkauser Verlag*, 1997.

[103] Van der Waerden B L. Algebra. *Springer- Verlag*, 1955.

[104] Vladimirov V S, Volovich I V and Zelenoc E I. p-adic Analysis and Mathematical Physics. *World Scientific Singapore*, 1994.

[105] Wang Z X. Chains of function spaces over Euclidian spaces and local fields. *Approx. Theory & its Appl.*, 2009, 25(2): 92–100.

[106] Weyl H. Das asymptotische verteilungsgesetz der eigenwerte linearer partieller differentialgleichungen, *Math. Ann.*, 1912, 71: 441–479.

[107] Wu B Y and Su W Y. The type of convolution operators on Vilenkin groups. *J. of Nanjing University*, 1999, 35(4): 393–400.

[108] Wu B and Su W Y. Eigen-frequencies of non-isotropic fractal drums. *Chaos,*

Soliton & Fractals, 2009, 42(5): 3210–3218.
[109] Xu N and Su W Y. On eigenvalues of spherical fractal drums. *Science in China, Ser.A, Eng.*, 2003, 46(1): 39–47.
[110] Xu Y and Su W Y. Modification and generalization of fractal percolation. *Progress in Natural Sci.*, 1997, 7(2): 148–154.
[111] Yao K, Su W Y and Zhou S P. On the fractal calculus functions of a fractal function. *Appl. Math. J. Chinese Univ., Ser.B*, 2002, 17(4): 377–381.
[112] Zheng S J and Liu J M. Representation thoerems on local fields. *J. of Nanjing Univ. Biquart.*, 1993, 29(4): 533–540.
[113] Zheng S J and Liu J M. A Note on Riesz means over the ring of p-adic integers. *J. of Nanjing Univ. Biquart.*, 1996, 13(1): 58–63.
[114] Zheng S J and Zheng W X. Almost everywhere convergence of sequences of multiplier operators on local fields. *Science in China*, 1997, 40(1): 10–21.
[115] Zheng S J. On Riesz type kernels over local fields. *Approx. Theory & its Appl.*, 1995, 11(4): 24–34.
[116] Zheng S J. Riesz type kernels over the ring of integers of a local fields. *J. Math. Anal. and Appl.*, 1997, 208: 528–552.
[117] Zheng W X, Su W Y and Jiang H K. A note for the concept of derivatives on local fields. *Approx. Theory & its Appl.*, 1990, 6(3), 48–58.
[118] Zheng W X, Su W Y and Ren F X. Walsh Analysis (Chinese). *Proceedings of the Conference on Approximation Theory, Hangzhou, China*, 1978: 42-50.
[119] Zheng W X and Su W Y. The logical derivatives and integrals (Chinese). *J. Math. Res. & Exposition*, 1981, 1: 79–90.
[120] Zheng W X and Su W Y. The best approximation on Walsh system (Chinese). *J. of Nanjing University*, 1982, 2: 254–262.
[121] Zheng W X and Su W Y. Walsh analysis and approximation operators (Chinese). *Advances in Mathematics*, 1983, 12(2): 81–93.
[122] Zheng W X and Su W Y. The logical derivatives and integrals II. *J. Math. Res. & Exposition*, 1987, 2: 217–224.
[123] Zheng W X and Zheng S J. Remarks on self-similar fractal sets. *J. of Nanjing Univ. Biquart.*, 1999, 6(1): 1–7.
[124] Zheng W X. Generalized Walsh transform and on extreme problem (Chinese). *Acta Math., Sinica*, 1979, 22(3): 362–374.
[125] Zheng W X. The approximation identity kernels on Walsh system (Chinese). *Chin. Ann. of Math., A*, 1983, 2: 177–184.
[126] Zheng W X. A class of approximation identity kernels. *Approx. Theory & Appl.*, 1984, 1(1): 65–76.
[127] Zheng W X. A note to Hilbert transforms on local fields. *J. of Nanjing Univ. Biquart.*, 1984, 2: 124–131.
[128] Zheng,W X. Derivatives and approximation theorems on local fields. *Rocky Mountain J. of Math.*, 1985, 15(4): 803–817.
[129] Zheng W X. Further on a class of approximation identity operators on local

fields. *Scientia Sinica, Ser. A*, 1987, 30(9): 641–653.
[130] Zheng W X. On a class approximation operators over local fields. *Proc. of Constructive Theory of Functions, Bulgaria*, 1988: 498–505.
[131] Zheng W X. Remarks on the kernel of product type. *A Friendly Collection of Mathematical Papers I, Proc. in Celebration of 70's Birthday of Professor Shu Lizhi*, 1990: 43–46.
[132] Zheng W X. On p-adic Cantor functions. *Lecture Notes in Math.*, 1990, 1(491): 219–226.
[133] Zheng W X. Expansion of self-similar functions. *Proc. Asian Math. Confer.'90*, 1991: 564–569.
[134] Zheng W X. Self-similar functions on local fields. *Chin. Ann. Math. A*, 1993, 14(1): 93–98.
[135] Zheng W X. Approximation operators and self-similarity over p-adic field. *J. Math. Study*, 1994, 27(1): 9–13.
[136] Zheng W X. On self-similarity of functions. *Harmonic Analysis in China, Edited by Yang C C et al., Kluwer Academic Publishers, Hong Kong*, 1995: 256–265.
[137] Zheng W X. p-adic analysis and its applications to fractals. *Recent Developments in Abstract Harmonic Analysis with Applications in Signal Processing, Edited by Stankovic R. et al., Nauka Belgrade, Yugoslavia*, 1996: 95–108.
[138] Zheng W X. On generalized Koch curve. *Approx. Theory & its Appl.*, 1999, 15(4): 6–14.
[139] Zhou G C and Su W Y. Elementary aspects of $B_{pq}^s(K_n)$ and $F_{pq}^s(K_n)$ spaces, *Approx. Theory & its Appl.*, 1992, 8(2): 11–28.
[140] Zhou G C and Su W Y. Local Hardy spacees on local fields, *J. Math. Res.& Exposition*, 1994, 12(2): 245–248.
[141] Zhou G C. Abel–Poisson type kernels on dyadic fields. *Approx. Theory & its Appl.*, 1991, 7(4): 68–75.
[142] Zhu Y P and Zheng W X. Multiplier in weighted Hardy spaces over locally compact Vilenkin groups. *Chin. Bull. Science*, 1998: 2041–2045.
[143] Zhu Y P and Zheng W X. Besov spaces and Herz spaces on local fields. *Science in China, Ser. A*, 1998, 41(10): 1051–1060.
[144] Zhu Y P and Zheng W X. Cesaro summability of two dimensional Walsh–Fourier series on the ring of integers in a p-series field. *Northeast Math. J.*, 1998, 14(3): 317–324.
[145] Zhu Y P and Zheng W X. Weighted Hardy spaces on homogeneous groups. *Approx. Theory & its Appl.*, 1999, 15(2): 15–21.
[146] Zhu Y P and Zheng W X. Maximal functions and Fourier transforms on local fields. *J.of Nanjing Univ.*, 2000, 36(3): 317–322.
[147] Zhu Y P and Zheng W X. BMO and singular integers over certain disconnected groups. *J. Math. Res. & Exposition*, 2000, 36(3): 317–322.

[148] Zhu Y P and Zheng W X. Weighted local Hardy spaces over locally compact Vilenkin groups. *Acta Math. Scient.*, 2000, 20(Add.): 614–624.

Index

$(c, 1)$-mean, 111
2-dimension affine transformation(similar transformation), 172
3-adic Cantor type set, 213
3-adic von Koch type curve, 260
$C(K_p), C_0(K_p), C_C(K_p)$, 41
$M_n(E)$, 190
$N_n(E)$, 190
$N_n^*(E)$, 190
k-cut function, 62
k-cutout operator, 51
m-order difference, 154
m^{th}-order Lipschitz class, 155
m^{th}-order continuous modulus, 154
p-adic number field, 6
p-adic von Koch type curve, 255
p-series field, 6
p-type derivative of distribution, 92
p-type derivative operator, 267
p-type integral of distribution, 92
p-type integral operator, 267
s-dimension packing measure, 196
s-dimensional Hausdorff measure, 184
s-dimensional Hausdorff net measure, 186
$s - (r, s)$ type operator, 59
3-adic Cantor type function, 213

Abel–Poisson kernel, 126
Abel–Poisson type kernel, 120
Abelian group, 1
affine transformation, 172
algebraic element, 3
algebraic extension field, 3
analytic transformation, 173
annihilator, 29
approximation identity kernel, 105
attractor of IFS, 177

B-type space, 129
backward k-iterate, 176
ball in K_p, 14
base for neighborhood system of K_p, 10
basic character, 27
Besov space, 135
Bessel potential space, 135
best approximation, 153, 156, 157
BMO space, 135
bmo space, 135
Borel probability measure, 239
bounded variation function, 115
Box dimension, 190

Cantor type set, 164
Cantor type set, 16, 164, 181
Cauchy–Poisson kernel, 128

315

character, 21
character group, 22
character group of group, 22
characteristic function, 16
characteristic number, 2
circle group, 22
closed ball net, 186
closed set net, 186
collage of IFS, 179
compact set net, 186
compatible, 5
condensation mapping, 178
condensation set, 178
congruence class, 3
continuous modulus, 153, 154
contraction factor, 174
contraction mapping, 174
convolution of distributions, 78
convolution operator, 50
coset, 11
covering of set, 183

de la Vallée–Poussin kernel, 127
degree of extension field, 4
diameter of the set, 183
dilation of distribution, 74
dilation of set, 164
dilation operator, 44
Dini convergence theorem, 110
Dini–Lip class, 161
Dirac distribution δ, 70
Dirac distribution δ, 72
Dirac distribution δ_{t_0}, 76
Dirichlet kernel, 105, 107, 126
discrete group, 23
distance, 163, 164
distribution dimension, 201
distribution space $\mathbb{S}^*(K_p)$, 71
distribution space on $\mathbb{S}(K_p)$, 71
dyadic von Koch type curve, 244

eigen-equation, 94, 97, 101

eigen-function, 94, 97, 101
eigen-value, 97, 101
equivalent theorem, 97, 98, 101
extension field, 3

F-type space, 129
Fejér integral operator, 126
Fejér kernel, 126, 127
field, 2
finite extension field, 4
finite field, 2
fixed point, 175
forward k-iterate, 176
foundational function, 86
foundational function space, 86
Fourier dimension, 210
Fourier coefficient, 107
Fourier series, 107
Fourier transformation of
    L^1-function, 42
Fourier transformation of
    L^2-function, 62
Fourier transformation of
    L^r-function, 68
fractal, 214
fractal space, 214
fractal space on local field, 214
fractional ideal, 10

Galois field, 3
Gauss–Weierstrass kernel, 127
generalized fractal space, 163
generator, 11
Gibbs derivative, 88
graph of a fractal function, 225
group, 1

H–L maximal operator, 58
Hölder type space, 136
Haar integral, 8
Haar measurable function, 7
Haar measure, 7

Hardy–Littlewood maximal operator, 58
Hausdorff dimension, 185
Hausdorff distance, 163
Hausdorff measure, 183
higher order logical derivatives, 95
homogeneous B-type space, 135
homogeneous F-type space, 135
homogeneous unit decomposition, 131
hyperbolic IFS, 178

identity transformation, 173
image measure, 240
image set of a transformation, 171
index pair, 48
infinite extension field, 4
intervals in K_p, 15
inverse Fourier transformation, 56, 57
invertible mapping, 171
iterated function system (IFS), 177

Jackson kernel, 127
Jackson–de la Vallée–Poussin kernel, 128

kernel, 105

L–P decomposition of a distribution, 131
Lebesgue type space, 141
linear fractional transformation, 172
linear transformation, 172
Lipschitz class, 97, 101, 118, 123
Littlewood–Paley decomposition of a distribution, 131
local field, 6
locally compact field, 6
locally compact group, 7
locally constant function, 268
locally homogeneous Hardy space, 135
locally non-homogeneous Hardy space, 135
logical derivative, 94, 95
lower box dimension, 192

Möbius transformation, 172
medicine science, 283
mod p operations, 11, 12
modular function, 8
multiplication formula, 63
multiplication with function, 73

net, 186
non-Archimedean valued field, 7
non-Archimedean valued norm, 6
non-increasing norm, 53, 68
norm preserved property, 63
null sequence, 49

open ball net, 186
open set net, 186
order, 2
order of non-zero element, 2
order structure in a local field, 15
orthogonal transformation, 173
orthonormal complete system, 105

packing dimension, 197
packing measure, 196
Parseval formula, 63, 70, 109
Parseval formula of distribution, 75
partial sum of Fourier series, 107
Picard kernel, 128
point-wise p-type derivative, 88
point-wise p-type integral, 89
point-wise logical derivative, 95
point-wise logical integral, 99
Poisson type kernel, 125
polynomial transformation, 171
Pontryagin dual theorem, 23
pre-Packing dimension, 196
prime field, 4
prime ideal in K_p, 10

pseudo-differential operator, 85, 266, 267
pseudo-differential operators on distribution space, 87, 267

radial approximation identity kernel, 125
ramification, 27
ramified degree, 27
reflection of a distribution, 74
reflection operator, 44
reflection transformation, 173
regular distribution, 72
Riemann–Lebesgue lemma, 54, 109
ring of integers in K_p, 9
Rogosinski kernel, 126
rotation transformation, 173

second Weierstrass type function, 233
semi-group, 173
singular distribution, 72
singular integral operator, 123
smooth modulus, 154
Sobolev type space, 141
space $B_{rt}^{s,\Theta}(K_p)$, 201
strong p-type derivative, 88
strong logical derivative, 95
strong logical integral, 97
support of a Borel measure, 200
support set of distribution, 76
symbol, 82

symbol class, 82

test function, 46
test function class, 46
the dyadic compact Abelian group, 95
total variation, 115
totally disconnected field, 6
transformation group, 173
transformation set, 173
translation, 172
translation of distribution, 73
translation operator, 44
two norms, 130
typical means kernel, 121

ultra-metric inequality, 7
ultra-metric space, 9
unit decomposition, 150
unit prime group, 10
upper box dimension, 192

w-$(1,1)$ type operator, 59
W-class, 161
Walsh function system, 256
Weierstrass kernel, 127
Weierstrass type function, 226
Weyl conjecture, 282
Wiener covering lemma, 59

zero distribution, 76
Zygmund class, 161

Printed in the United States
By Bookmasters